FUNDAMENTALS OF PIPE FLOW

OTHER BOOKS BY THE AUTHOR

Handbook of Generalized Gas Dynamics
Plenum Press, 1966 (with W. G. Steltz)

Handbook of Specific Losses in Flow Systems
Plenum Press, 1966 (with N. A. Carlucci)

Manual on the Use of Thermocouples in
Temperature Measurement
STP470, STP470A, & STP470B, American Society for
Testing and Materials, 1970, 1974, & 1980 (Editor)

Journey Away From God
Fleming H. Revell Co., 1972

Fundamentals of Temperature, Pressure, and
Flow Measurements
John Wiley and Sons, 1969 and 1977

FUNDAMENTALS OF PIPE FLOW

Robert P. Benedict
Fellow Mechanical Engineer
Westinghouse Electric Corporation
Steam Turbine Division
Adjunct Professor of Mechanical Engineering
Drexel University
Evening College
Philadelphia, Pennsylvania

A WILEY-INTERSCIENCE PUBLICATION
JOHN WILEY & SONS
New York • Chichester • Brisbane • Toronto • Singapore

The quotations used as chapter headnotes are from *The Body Has a Head* by Gustav Eckstein © 1969, and are reprinted by permission of Harper and Row, Publishers, Inc.

Copyright © 1980 by John Wiley & Sons, Inc.

All rights reserved. Published simultaneously in Canada.

Reproduction or translation of any part of this work beyond that permitted by Sections 107 or 108 of the 1976 United States Copyright Act without the permission of the copyright owner is unlawful. Requests for permission or further information should be addressed to the Permissions Department, John Wiley & Sons, Inc.

Library of Congress Cataloging in Publication Data

Benedict, Robert P
  Fundamentals of pipe flow.

  "A Wiley-Interscience publication."
  Includes bibliographical references and index.
  1. Pipe—Fluid dynamics. I. Title.

TJ935.B46   621.8'67   79-23924
ISBN 0-471-03375-8

Printed in the United States of America

10 9 8 7 6

*To My Three Sons
Mark, John, James*

PREFACE

The study of fluid flow in pipes may well be traced back to Hero of Alexandria, the Greek historian and scientist who, in his *Dioptra* in the second century B.C., gave the earliest known expression of the relationship between cross-sectional area, velocity, volume, and time.* Two Roman engineers, Vitruvious and Frontinus, of the first century B.C. and the second century A.D., respectively, wrote valuable treatises on Roman methods for distributing water through pipes and aqueducts. About 1500 it was that genius of all trades, Leonardo da Vinci, who first clearly enunciated the continuity principle. Pascal, Newton, Leibniz —all made their contributions to fluid flow, but it was not until about 1750, with the advent of the gifted friends Leonhard Euler and Daniel Bernoulli, that we find hydraulics advanced to the state of a science. The real fluid effects of viscosity were dealt with experimentally by Hagen, Poiseuille, and Weisbach, to name only a few in the nineteenth century, while Navier and Stokes formalized these same effects mathematically in the same period to found the science of hydrodynamics. It remained for Prandtl and his school at Gottingen to bridge the gap between hydraulics and hydrodynamics with his important concepts of the boundary layer and the friction factor.

In this book an attempt is made to summarize the most important results that are available in the field of fluid flow in pipes. In Part I the basic conservation principles of mass, momentum, and energy are presented in Chapter 1. The flow of ideal liquids is given in Chapter 2, while ideal gas flows are treated in Chapter 3. In Part II the deviations from an ideal fluid as described by the boundary layer are considered in Chapter 4, velocity profiles in Chapter 5, and the friction factor in Chapter 6. In Part III the flow of real liquids and gases and that of two-phase fluids are treated in Chapters 7, 8, and 9, respectively. Then

---

History of Hydraulics, by H. Rouse and S. Ince, Dover Publications, New York, 1963, p. 21.

the specific losses in various piping components are given in equation, curve, and tabular form in Chapter 10, while Chapter 11 introduces the idea of flow network analysis, wherein several piping components are combined in various series and parallel arrangements. Part IV is concerned with the measurement of the thermodynamic quantities of temperature, pressure, and flow rate in Chapters 12, 13, and 14, respectively. Finally, Chapter 15 gives a brief account of some of the specialized measurements made in pipes.

The book represents experience that I gained in teaching fluid mechanics, gas dynamics, and instrumentation at Drexel University, Evening College, as well as industrial experience gained at the Westinghouse Electric Corporation, Steam Turbine Division, in the Development Engineering Department.

The book is written primarily as a reference work for practicing engineers (mechanical, chemical, aeronautical, and civil) and scientists whose main field of endeavor is not necessarily pipe flow. It is also believed that the book will serve as a text for courses in fluid mechanics and gas dynamics, at both the graduate and the undergraduate levels. There are numerous worked-out examples, and the book is self-contained in the sense that it provides all the tables and curves necessary for the application of the principles presented herein.

I would like to close with a general acknowledgment to the many contributors to the field of fluid flow in general and to pipe flow in particular down through the years. These acknowledgments are covered in part by the many references to the original works given at the end of each chapter, where, incidentally, I have attempted to shield the reader from the overabundance of papers that now flood our journals by selecting only those I have used and am familiar with. There is not space to single out my friends and coworkers in this field, but I must thank my wife, Ruth, for her patience and skill in preparing the manuscript, and for encouraging me all along the way.

My final thought is to send my best wishes to you the student, you the teacher, and you the researcher who use this work.

ROBERT P. BENEDICT

Holly Hill, Pennsylvania
February 1980

CONTENTS

PART I
FLOW OF IDEAL FLUIDS IN PIPES

1 The Conservation Equations for Ideal Pipe Flow   3

    1.1   General Remarks, 3

         1.1.1   Continuity, 4
         1.1.2   Momentum, 4
         1.1.3   Energy, 5

    1.2   General Conservation Principle, 5

    1.3   Conservation of Mass, 6

         1.3.1   Local Rate of Increase of Mass within Volume, 7
         1.3.2   Net Rate of Efflux of Mass across Surface of Volume, 7
         1.3.3   Strength of Sources of Change of Mass, 7
         1.3.4   Summation, 7
         1.3.5   One-Dimensional Steady Flow, 8

    1.4   Conservation of Momentum, 9

         1.4.1   Local Rate of Increase of Momentum within Volume, 9
         1.4.2   Net Rate of Efflux of Momentum across Surface of Volume, 9
         1.4.3   Strength of Sources of Change of Momentum, 10
         1.4.4   Summation, 11
         1.4.5   One-Dimensional Steady Flow, 12

x Contents

- **1.5 Conservation of Energy, 13**
  - 1.5.1 Local Rate of Increase of Energy within Volume, 13
  - 1.5.2 Net Rate of Efflux of Fluid Energy across Surface of Volume, 14
  - 1.5.3 Strength of Sources of Change of Fluid Energy, 14
  - 1.5.4 Summation, 15
  - 1.5.5 One-Dimensional Steady Flow, 15
- **1.6 Relation Between Momentum and Energy, 16**

References, 17

Nomenclature, 18

2 Solutions to Ideal-Incompressible Pipe Flow   20

- **2.1 General Remarks, 20**
- **2.2 The One-Dimensional Bernoulli Energy Equation, 20**
- **2.3 The One-Dimensional Bernoulli Work Equation, 24**
- **2.4 The One-Dimensional Bernoulli Pressure Equation, 28**
- **2.5 Two- and Three-Dimensional Solutions, 29**
  - 2.5.1 Characteristics of Vector Fields, 30
    Continuity and the Stream Function, 30
    Circulation and the Potential Function, 33
    The Laplace Equation, 34
  - 2.5.2 Two-Dimensional Electric-Fluid Flows, 35
    Electric Current Field, 35
    Fluid Flow Field, 36
    Conducting Sheet Analog, 36
  - 2.5.3 The Graphical Flow Net, 43
  - 2.5.4 Numerical Solutions, 47
    Relaxation Methods, 47
    Finite Element Methods, 52

2.6 Several Idealized Solutions, 53

    2.6.1 Potential Solutions to an ASME Nozzle, 53
    2.6.2 Momentum Solutions to Various Orifices, 56
    2.6.3 Ideal Pressure Recovery across an Abrupt Enlargement, 62

References, 63

Nomenclature, 64

3 Solutions to Ideal-Compressible Pipe Flow   66

3.1 General Remarks, 66

3.2 Thermodynamic Concepts, 66

    3.2.1 Zero-th Law, 68
    3.2.2 First Law, 69
    3.2.3 General Energy, 75
    3.2.4 Second Law, 77
    3.2.5 The Perfect Gas, 81
    3.2.6 Thermodynamic Processes, 86
    3.2.7 Stagnation States, 89

3.3 The Mach Number, 91

3.4 The One-Dimensional Bernoulli Energy Equation, 92

3.5 One-Dimensional Compressible Pressure Equations, 96

3.6 Comparisons between Compressible and Incompressible Treatments, 98

    3.6.1 Flow Numbers and Expansion Factors, 99
    3.6.2 Hydraulic-Gas Analogy, 107

3.7 Idealized Solutions, 117

    3.7.1 Several Nozzle Solutions, 117
    3.7.2 Generalized Contraction Coefficient, 120

References, 128

Nomenclature, 129

PART II
REAL FLUID CONCEPTS

4 The Boundary Layer   133

  4.1   General Remarks, 133

  4.2   The Concept, 133

       4.2.1   Ideal Relations, 134
       4.2.2   Exact Relations, 135
       4.2.3   Empirical Relations, 136
       4.2.4   Basic Boundary Layer Concept, 137
       4.2.5   Boundary Layer Relations, 138
       4.2.6   The Blasius Solution, 139

  4.3   Boundary Layer Regimes, 140

       4.3.1   Laminar, 140
       4.3.2   Transition, 141
       4.3.3   Turbulent, 142
       4.3.4   Separation, 143

  4.4   Boundary Layer Parameters, 144

       4.4.1   Boundary Layer Thickness (δ), 145
       4.4.2   Displacement Thickness (δ^*), 146
       4.4.3   Momentum Thickness (θ), 148
       4.4.4   Energy Thickness (δ^{**}), 150
       4.4.5   Comparisons, 152
       4.4.6   Shape Factors, 154

  4.5   Simplified Applications, 155

       4.5.1   General Continuity Considerations, 155
       4.5.2   General Momentum Considerations, 156
       4.5.3   General Flat Plate Considerations, 156
       4.5.4   Laminar Boundary Layer on a Flat Plate, 158
       4.5.5   Turbulent Boundary Layer on a Flat Plate, 159

  4.6   More Complex Solution Methods, 160

       4.6.1   Momentum Integral Equation, 160
       4.6.2   Energy Integral Equation, 163

4.6.3 Walz Stepping Equations, 164
4.6.4 Application of Walz Solution to a Nozzle, 167

References, 173

Nomenclature, 175

5 Velocity Distributions 178

5.1 General Remarks, 178

5.2 Laminar Flow, 181

5.3 Transition, 185

5.4 Turbulent Flow in Smooth Pipes, 186

    5.4.1 Power Law, 187
    5.4.2 Smooth Law of the Wall, 190
        Logarithmic Layer, 193
        Laminar Sublayer, 196
        Buffer Zone, 198
        Summary of Law of the Wall, 199

    5.4.3 Further Discussion of Law of the Wall, 201
        Nikuradse Shift, 201
        Average Velocity for Turbulent Flow, 201
        Comparison of Power Law and Logarithmic Profile, 203

    5.4.4 Other Forms of Law of the Wall, 206
        Reichardt's Expression, 206
        Deissler's Equations, 206

    5.4.5 Pressure Gradient Effect on Law of Wall, 210

5.5 Turbulent Flow in Rough Pipes, 210

    5.5.1 Wall Roughness, 211
    5.5.2 Rough Law of the Wall, 211
    5.5.3 Pipe Factor, 213

5.6 Universal Law of the Wall, 215

5.7 Kinetic Energy Coefficient, 217

    5.7.1 Parabolic Profile of Laminar Flow, 218
    5.7.2 1/7 Power Law Profile of Turbulent Flow, 219
    5.7.3 Logarithmic Profile of Turbulent Flow, 219

xiv Contents

- 5.8 Momentum Correction Factor, 221
  - 5.8.1 Parabolic Profile of Laminar Flow, 222
  - 5.8.2 1/7 Power Law Profile of Turbulent Flow, 222
  - 5.8.3 Logarithmic Profile of Turbulent Flow, 223

  References, 224

  Nomenclature, 225

6 The Friction Factor 228

- 6.1 General Remarks, 228
- 6.2 Dimensional Analysis, 229
- 6.3 Laminar Flow, 230
- 6.4 Critical Zone, 231
- 6.5 Turbulent Flow in Smooth Pipes, 232
- 6.6 Turbulent Flow in Fully Rough Pipes, 236
- 6.7 Transition between Smooth and Rough Pipes, 237
- 6.8 Reasons for Differences in the Transition Region, 242
- 6.9 Engineering Charts for Determining Friction Factor, 243
  - 6.9.1 Moody Plot, 243
  - 6.9.2 Rouse Plot, 247
- 6.10 The Skin Friction Coefficient, 247
  - 6.10.1 Clauser Plot, 249
- 6.11 Other Expressions for the Turbulent Friction Factor, 249
  - 6.11.1 For Smooth Pipes, 250
  - 6.11.2 For Transition, 250
  - 6.11.3 For Fully Rough Pipes, 250
  - 6.11.4 For Skin Friction Coefficient, 253
- 6.12 Effect of Additives on Fluid Friction, 254
- 6.13 Friction in Developing Boundary Layers, 255

  References, 257

  Nomenclature, 259

PART III
FLOW OF REAL FLUIDS IN PIPES

7 Flow of Real Liquids in Pipes   263

    7.1   General Remarks, 263

    7.2   The One-Dimensional Energy Equation with Losses, 263

    7.3   The Head Loss Equation, 265

    7.4   The Loss Coefficient, 265

         7.4.1   Definition, 265
         7.4.2   Combining Loss Coefficients, 266
         7.4.3   Deriving a Loss Coefficient, 267

    7.5   Solution of Real Liquid Flow in Pipes, 270

         7.5.1   Solution Steps, 270
         7.5.2   Numerical Examples, 272

    7.6   Generalized Constant Density Pipe Flow, 280

         7.6.1   Generalized Flow Function, 280
         7.6.2   Generalized Flow Map, 285
         7.6.3   Generalized Examples, 288

    References, 290

    Nomenclature, 290

8 Flow of Real Gases in Pipes   292

    8.1   General Remarks, 292

    8.2   The One-Dimensional Energy Equation with Losses, 292

    8.3   The Head Loss Equation, 293

    8.4   Compressible Loss Coefficients, 293

         8.4.1   Adiabatic Loss Coefficient, 294
         8.4.2   Isothermal Loss Coefficient, 298
         8.4.3   Comparison of Compressible and Incompressible Loss Coefficients, 301

8.4.4 Combining Compressible Loss Coefficients, 304
8.4.5 Deriving a Loss Coefficient, 308

8.5 Solution of Real Flow of Gas in Pipes, 311

8.5.1 Solution Steps, 311
8.5.2 Numerical Examples, 311

8.6 Generalized Compressible Pipe Flow, 314

8.6.1 Generalized Flow Function, 314
8.6.2 Compressible Processes, 318
8.6.3 Generalized Flow Maps, 323
8.6.4 Generalized Examples, 329

References, 340

Nomenclature, 341

9 Flow of Liquid-Vapor Mixtures in Pipes   343

9.1 General Remarks, 343

9.2 Liquid Flow Situation, 344

9.3 Flashing Flow Situation, 345

9.3.1 Maximum Flashing Flow, 346
9.3.2 Pipe Critical Pressure, 347
9.3.3 Pseudo-Isentropic Exponent, 350

9.4 The Control Valve, 353

9.4.1 Valve Critical Pressure, 353

9.5 Determination of Pressure Downstream of Valve, 354

9.5.1 Choked Flow, 354
9.5.2 Subsonic Flow, 355

9.6 Pipe Sizing for Flashing Flow, 359

9.7 Flow Rate Determination for Flashing Flow, 359

References, 361

Nomenclature, 362

10 Loss Characteristics of Piping Components   363

10.1 General Remarks, 363

10.2 Viscous Pipes, 363

10.3 Elbows, 364
10.4 Tees, 372
10.5 Reducers, 376
10.6 Diffusers, 377
10.7 Abrupt Enlargements, 380
10.8 Abrupt Contractions, 383
10.9 Inlets and Exits, 390
10.10 Differential Pressure Type Fluid Meters, 393
10.11 Valves, 407
10.12 Screens, 413
References, 416
Nomenclature, 419

11 Piping Networks 421

11.1 General Remarks, 421
11.2 The Series Network, 422
11.3 The Parallel Network, 424
11.4 The Branching Network, 431
11.5 Forward and Backward Solutions, 434
References, 437
Nomenclature, 438

PART IV
THERMODYNAMIC MEASUREMENTS IN PIPES

12 Temperature Measurement in Pipes 443

12.1 General Remarks, 443
12.2 Temperature Measurement in Moving Fluids, 444

xviii Contents

    12.3   Heat Transfer Effects on Temperature Measurement, 451

    12.4   Temperature Measurement Systems, 459

    12.5   The Usual Engineering Approximations, 463

    References, 464

    Nomenclature, 464

13 Pressure Measurement in Pipes   466

    13.1   General Remarks, 466

    13.2   Pressure Measurement in Moving Fluids, 466

    13.3   Pressure Measuring Systems, 477

    References, 481

    Nomenclature, 482

14 Flow Measurement in Pipes   483

    14.1   General Remarks, 483

    14.2   The Basic Equations, 483

    14.3   The Discharge Coefficient, 485

    14.4   The Expansion Factor, 494

    References, 498

    Nomenclature, 499

15 Special Measurements in Pipes   500

    15.1   General Remarks, 500

    15.2   Skin Friction, 500

        15.2.1   The Preston Tube, 500
        15.2.2   Balance Measurements for Obtaining τ_0, 505
        15.2.3   Inference from Velocity Measurements, 505

15.3 **Boundary Layer Measurements, 506**

    15.3.1 Hot Wire, 507
    15.3.2 Pitot Tube, 509

15.4 **Laser-Doppler Velocimeter, 512**

    15.4.1 The Laser, 513
    15.4.2 The LDV, 513
    15.4.3 LDV Arrangements, 516
    15.4.4 LDV Results, 517

15.5 **Nonintrusive Flow Measurement, 518**

References, 519

Nomenclature, 521

Name Index, 523

Subject Index, 527

FUNDAMENTALS OF
PIPE FLOW

I

FLOW OF IDEAL FLUIDS IN PIPES

Here we discuss the hydrodynamics of Daniel Bernoulli and Leonhard Euler for incompressible fluid flow and the latter-day treatment of compressible fluid flow, both of these in the complete absence of losses, irreversibilities, entropy increases, and so on. In other words, the ideal potential flow of mathematics is considered here. Comparisons are drawn between constant density and compressible fluids in the form of flow numbers and expansion factors. Such ideal approaches often can be used to approximate solutions to real fluid flow problems with certain very important exceptions, that is, ideal solutions fail in very narrow regions near solid bodies and fail whenever losses in piping components dominate the flow. Such real flow situations are treated in Parts II and III.

1
The Conservation Equations for Ideal Pipe Flow

...from the pumps go the pipelines—go the day-and-night deliveries....—Gustav Eckstein

1.1 GENERAL REMARKS

In approaching the subject of pipe flow, it is convenient to treat first a hypothetical ideal fluid. Ideal, in this case, simply means that the fluid is assumed to have no viscosity. Such a fluid is termed *inviscid*. In the flow of an ideal fluid, there will be no frictional effects either between the various layers of the fluid or between the fluid and the pipe walls. But, although the ideal moving fluid will support no tangential (shear) stresses, it will support normal (pressure) stresses and will be subject to any field forces such as gravity.

The momentum equations, which result from application of Newton's second law, will be greatly simplified in the inviscid case but will yet yield solutions to pipe flow problems that are often adequate for engineering purposes. Later, of course, we will remove the inviscid restriction, but for now, on with ideal pipe flow.

Several basic conservation concepts are evoked in solving pipe flow problems. These physical concepts are embodied in several differential equations which will be developed in this opening chapter. First, we will state each conservation equation in its final form, so that we may know where we are headed, and so that we can easily display some of the significant characteristics of the equations before becoming bogged down in mathematical details.

4 The Conservation Equations for Ideal Pipe Flow

1.1.1 Continuity

The continuity equation, whose concept was first announced [1] for *incompressible* flow by Leonardo da Vinci (1452–1519) in 1502 by the words

> ...a river in each part of its length in an equal time gives passage to an equal quantity of water whatever the width, the depth, the slope, the roughness, the tortuosity,...,

describes the *conservation of mass* in a system through which a fluid flows. In its final form, for three-dimensional, unsteady, *compressible* flow we have

$$\frac{\partial \rho}{\partial t} + \frac{\partial}{\partial x}(\rho u_1) + \frac{\partial}{\partial y}(\rho u_2) + \frac{\partial}{\partial z}(\rho u_3) = 0, \qquad (1.1)$$

where ρ is the fluid density, u_1, u_2, u_3 are the velocity components in the three mutually perpendicular directions x, y, z, and t is the time.

1.1.2 Momentum

The momentum equation, which was first given [2] in its *inviscid* form by Leonhard Euler (1707–1783) in 1750, interprets Newton's second law as applied to fluid flow, and hence describes the *conservation of momentum* of a system through which a fluid flows. In its final form, for unsteady flow, in the x direction, for example, we have

$$\frac{du_1}{dt} = -\frac{g_c}{\rho}\frac{\partial p}{\partial x} - g\frac{\partial Z}{\partial x}, \qquad (1.2)$$

where p is the compressive stress (i.e., pressure), and Z is the vertical height above some arbitrary datum. Similar expressions can be written for the y and z directions. Note that g is the local value for the gravitational acceleration, while g_c is the gravitational constant, which serves as the proportionality constant in Newton's second law in the pound mass-pound force system, that is,

$$F = \frac{1}{g_c} Ma, \qquad (1.3)$$

where F is the unbalanced force (lbf) and M is the mass (lbm).

1.1.3 Energy

The energy equation, which was first given [3] in its *incompressible, inviscid* form by Daniel Bernoulli (1700–1782) in 1738, describes the *conservation of mechanical energy* in a system through which a fluid flows. In its final form, for three-dimensional, unsteady, *diabatic*, compressible flow we have

$$\frac{\partial}{\partial t}\left(\rho E + \frac{\rho V^2}{2g_c} + \frac{\rho g Z}{g_c}\right) dV + \frac{\partial}{\partial x}\left[\left(E + \frac{V^2}{2g_c} + \frac{gZ}{g_c}\right)(\rho u_1)\right] dV$$
$$+ \frac{\partial}{\partial y}\left[\left(E + \frac{V^2}{2g_c} + \frac{gZ}{g_c}\right)(\rho u_2)\right] dV$$
$$+ \frac{\partial}{\partial z}\left[\left(E + \frac{V^2}{2g_c} + \frac{gZ}{g_c}\right)(\rho u_3)\right] dV$$
$$= -\left[\frac{\partial}{\partial x}(pu_1) + \frac{\partial}{\partial y}(pu_2) + \frac{\partial}{\partial z}(pu_3)\right] dV + \delta\dot{W} + \delta\dot{Q}, \qquad (1.4)$$

where $\dot{Q}$ is the rate of heat transfer across the system boundaries, $\dot{W}$ is the rate of work transfer across the system boundaries, E is the internal energy of the fluid per unit mass, V is the resultant velocity at any point in the flow field, and dV represents a differential control volume as discussed in the next section.

We now consider separately a general conservation formulation, and then go on to develop each of the three basic conservation equations, noting their lineage as we go.

1.2 GENERAL CONSERVATION PRINCIPLE

Consider a differential control volume, dV, fixed in space and extent, and containing at a particular instant the fluid mass, dM. The various fluid and thermodynamic properties (e.g., ρ, u, p) are specified at the *center* of this element of volume, where they are functions of time only (see Figure 1.1). However, all of these properties may also vary with position if they are considered at some position away from the center—at the surface of the control volume, for example. Thus, in general, each arbitrary property (P) is a function of both the time (t) and space (x,y,z) coordinates. At the center of the element (where the space coordinates are fixed), we have

$$\left(\frac{dP}{dt}\right)_{x,y,z} = \frac{\partial P}{\partial t}. \qquad (1.5)$$

6 The Conservation Equations for Ideal Pipe Flow

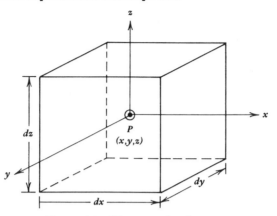

Figure 1.1 The control volume.

At the surfaces of the element (for a given time, and neglecting all second-order effects), each arbitrary property is expressed as

$$P \pm \frac{\partial P}{\partial x_k} \frac{dx_k}{2}, \tag{1.6}$$

where the minus sign is used at the inlet surfaces and the plus sign at the outlet. The Cartesian tensor index, k, can be either 1, 2, or 3, according to the specific coordinate being considered.

For any conserved quantity (N), in any flow field, we may write [4]

Local rate of increase of N within volume + net rate of efflux of N across surface of volume = strength of sources of change of N.

(1.7)

The relation given by (1.7) embodies the general conservation principle and states in effect that the net rate of accumulation of N within the volume plus the net amount of N convected out of the volume in the same time must equal the total rate of change in N caused by the sources. The latter may occur by the spontaneous generation of N within the element, by the action of the resultant of all forces on the fluid within the element, or by changes in N caused by external energy transfer to or from the fluid within the volume element.

1.3 CONSERVATION OF MASS

A basic premise of pipe flow is that mass is one of the conserved quantities. Thus we proceed to write a mass balance for the volume element $d\mathbf{V}$ in accordance with (1.7).

1.3.1 Local Rate of Increase of Mass within Volume

Since a fluid is deformable, both the mass and the density of the fluid enclosed by the control volume may, in general, change with time. The rate of accumulation of mass may be expressed in terms of the rate of change of density as

$$\frac{\partial}{\partial t}(dM) = \frac{\partial}{\partial t}(\rho d\mathbf{V}) = \frac{\partial \rho}{\partial t} d\mathbf{V}. \tag{1.8}$$

1.3.2 Net Rate of Efflux of Mass across Surface of Volume

The mass per unit time convected across any particular surface may be expressed as the product of the appropriate density, velocity, and area terms. Thus the net difference between the mass per unit time leaving the volume element and that entering it is

$$\left[\left(\rho + \frac{\partial \rho}{\partial x_i}\frac{dx_i}{2}\right)\left(u_i + \frac{\partial u_i}{\partial x_i}\frac{dx_i}{2}\right)dA_j\right]_{\text{out}} - \left[\left(\rho - \frac{\partial \rho}{\partial x_i}\frac{dx_i}{2}\right)\left(u_i - \frac{\partial u_i}{\partial x_i}\frac{dx_i}{2}\right)dA_j\right]_{\text{in}}$$
$$= \frac{\partial}{\partial x_i}(\rho u_i) d\mathbf{V}, \tag{1.9}$$

where the index i indicates a summation operation, and the subscript j indicates perpendicularity to direction of the term under consideration. When expanded, the final form of (1.9) is to be interpreted as

$$\left[\frac{\partial}{\partial x}(\rho u_1) + \frac{\partial}{\partial y}(\rho u_2) + \frac{\partial}{\partial z}(\rho u_3)\right]d\mathbf{V}.$$

1.3.3 Strength of Sources of Change of Mass

We consider, in pipe flow, a flow field having no points where mass is spontaneously introduced or removed, that is, a flow with no sources or sinks.

1.3.4 Summation

Applying (1.7), we have via (1.8) and (1.9)

$$\frac{\partial \rho}{\partial t} + \frac{\partial}{\partial x_i}(\rho u_i) = 0, \tag{1.10}$$

which mathematically expresses the principle of conservation of mass and is usually called the *continuity equation*. Equation 1.1 is the expanded form of (1.10).

In the steady state there can be no accumulation of mass within the element (i.e., $\partial \rho/\partial t = 0$), and the conservation of mass is achieved by a balance of the convected quantities alone. The continuity equation then reduces to

$$\frac{\partial}{\partial x_i}(\rho u_i) = 0. \qquad (1.11)$$

If incompressible flow is considered (i.e., $\rho = C$), (1.11) reduces to

$$\frac{\partial u_i}{\partial x_i} = 0. \qquad (1.12)$$

1.3.5 One-Dimensional Steady Flow

If one-dimensional, steady flow is considered, we have from (1.9)

$$\frac{\partial}{\partial x}(\rho u_1 \, dx \, dy \, dz) = 0 \qquad (1.13)$$

or, more simply,

$$\dot{m} = \rho V A, \qquad (1.14)$$

Table 1.1 Summary of Various Forms of the Steady Flow Continuity Equation

Situation	Continuity Equation	
	Compressible	Incompressible
General	$\dfrac{\partial}{\partial x_i}(\rho u_i) = 0$	$\dfrac{\partial u_i}{\partial x_i} = 0$
Three-dimensional	$\dfrac{\partial}{\partial x}(\rho u_1) + \dfrac{\partial}{\partial y}(\rho u_2) + \dfrac{\partial}{\partial z}(\rho u_3) = 0$	$\dfrac{\partial u_1}{\partial x} + \dfrac{\partial u_2}{\partial y} + \dfrac{\partial u_3}{\partial z} = 0$
One-dimensional	$\dfrac{\partial}{\partial x}(\rho u_1) = 0$	$\dfrac{\partial u_1}{\partial x} = 0$
	$\dot{m} = \rho A V$	$Q = AV$
	(mass flow rate)	(volumetric flow rate)

where, for a given piping system, $\dot{m}$ is a constant called the mass flow rate, and V is the volumetric average velocity.

The various forms of the continuity equation for steady flow are summarized in Table 1.1.

1.4 CONSERVATION OF MOMENTUM

Another premise which is basic to pipe flow is that the total momentum of any system may change only under the action of forces. We can write a momentum-force balance on the fluid within the control volume according to (1.7).

1.4.1 Local Rate of Increase of Momentum within Volume

The momentum of the fluid enclosed by the volume element may change with time since mass and velocity are both functions of time. The rate of accumulation of momentum may be expressed in terms of (1.3) as

$$\frac{1}{g_c}\frac{\partial}{\partial t}(dM\,u_k) = \frac{1}{g_c}\frac{\partial}{\partial t}(\rho\,dV\,u_k) = \frac{1}{g_c}\frac{\partial}{\partial t}(\rho u_k)\,dV \qquad (1.15)$$

where the index k can be either 1, 2, or 3, and indicates the various components of momentum under consideration.

1.4.2 Net Rate of Efflux of Momentum across Surface of Volume

The momentum per unit time convected across any particular surface may be expressed as the product of the appropriate mass per unit time and velocity terms. Thus the net difference between the momentum per unit time leaving the volume element and that entering it is

$$\frac{1}{g_c}\left[\left(\rho + \frac{\partial \rho}{\partial x_i}\frac{dx_i}{2}\right)\left(u_i + \frac{\partial u_i}{\partial x_i}\frac{dx_i}{2}\right)dA_j\left(u_k + \frac{\partial u_k}{\partial x_i}\frac{dx_i}{2}\right)\right]_{\text{out}}$$

$$- \frac{1}{g_c}\left[\left(\rho - \frac{\partial \rho}{\partial x_i}\frac{dx_i}{2}\right)\left(u_i - \frac{\partial u_i}{\partial x_i}\frac{dx_i}{2}\right)dA_j\left(u_k - \frac{\partial u_k}{\partial x_i}\frac{dx_i}{2}\right)\right]_{\text{in}}$$

$$= \frac{1}{g_c}\frac{\partial}{\partial x_i}(\rho u_i u_k)\,dV, \qquad (1.16)$$

where the index i indicates a summation of terms at a particular k, which consistently indicates the component under consideration. The expanded

10 The Conservation Equations for Ideal Pipe Flow

form of the right-hand side of (1.16) is

$$\frac{1}{g_c}\left[\frac{\partial}{\partial x}(\rho u_1 u_1) + \frac{\partial}{\partial y}(\rho u_2 u_1) + \frac{\partial}{\partial z}(\rho u_3 u_1)\right]dV \quad \text{in the } k=1 \text{ direction,}$$

$$\frac{1}{g_c}\left[\frac{\partial}{\partial x}(\rho u_1 u_2) + \frac{\partial}{\partial y}(\rho u_2 u_2) + \frac{\partial}{\partial z}(\rho u_3 u_2)\right]dV \quad \text{in the } k=2 \text{ direction,}$$

$$\frac{1}{g_c}\left[\frac{\partial}{\partial x}(\rho u_1 u_3) + \frac{\partial}{\partial y}(\rho u_2 u_3) + \frac{\partial}{\partial z}(\rho u_3 u_3)\right]dV \quad \text{in the } k=3 \text{ direction.}$$

1.4.3 Strength of Sources of Change of Momentum

In the absence of viscous shearing stresses, as in the ideal pipe flow under discussion here, we will consider only forces arising from internal normal stresses (i.e., pressures), and those arising because the fluid within the element is in an external conserved force field such as the earth's gravitational field.

The net difference between the *pressure forces* acting on the inlet and the outlet surfaces of the volume element is

$$\left[\left(p - \frac{\partial p}{\partial x_k}\frac{dx_k}{2}\right)dA_j\right]_{\substack{\text{inlet}\\\text{surfaces}}} - \left[\left(p + \frac{\partial p}{\partial x_k}\frac{dx_k}{2}\right)dA_j\right]_{\substack{\text{outlet}\\\text{surfaces}}} = -\frac{\partial p}{\partial x_k}dV \quad (1.17)$$

where the minus sign indicates that the net compressive force is in the direction of the flow, and the index k indicates the various components of interest. Note in (1.17) that pressure, being a tensor rather than a vector, is independent of direction.

Restricting the field forces to those due to gravity alone, we note that each fluid particle has potential energy as a function of its position in the force field. The potential energy (P.E.) can be expressed as the weight of the fluid in the volume element times the vertical distance (Z) above some arbitrary datum, that is,

$$\text{P.E.} = \left(\frac{1}{g_c}dMg\right)Z = \frac{g}{g_c}Z(\rho dV). \quad (1.18)$$

The *gravitational field force*, that is, the rate of change of P.E. with distance, is given in terms of (1.18) as

$$-g\frac{\partial Z}{\partial x_k}\left(\frac{\rho dV}{g_c}\right), \quad (1.19)$$

where the minus sign is introduced because the force acts vertically downward whereas the positive direction of Z is vertically upward.

1.4.4 Summation

Applying (1.7), we have in terms of (1.15), (1.16), (1.17), and (1.19)

$$\frac{\partial}{\partial t}(\rho u_k) + \frac{\partial}{\partial x_i}(\rho u_i u_k) = -g_c \frac{\partial p}{\partial x_k} - \rho g \frac{\partial Z}{\partial x_k}. \qquad (1.20)$$

Expanding the left side of (1.20) and using the continuity expression of (1.10) yields

$$\frac{\partial u_k}{\partial t} + u_i \frac{\partial u_k}{\partial x_i} = -\frac{g_c}{\rho} \frac{\partial p}{\partial x_k} - g \frac{\partial Z}{\partial x_k}, \qquad (1.21)$$

which mathematically expresses the principle of conservation of momentum and is usually called the *Euler equation of motion*. Equation 1.2 is the expanded form of (1.21) *in the x direction* (i.e., for $k=1$ and $i=1$). This is clarified by the following brief development.

Since $u_1 = f(x,y,z,t)$, it follows that

$$du_1 = \frac{\partial u_1}{\partial x} dx + \frac{\partial u_1}{\partial y} dy + \frac{\partial u_1}{\partial z} dz + \frac{\partial u_1}{\partial t} dt,$$

and dividing through by dt gives

$$\frac{du_1}{dt} = u_1 \frac{\partial u_1}{\partial x} + u_2 \frac{\partial u_1}{\partial y} + u_3 \frac{\partial u_1}{\partial z} + \frac{\partial u_1}{\partial t},$$

where $u_1 = dx/dt$, $u_2 = dy/dt$, and $u_3 = dz/dt$. But, expanding the left side of (1.21), for $k=1$ we obtain

$$\frac{\partial u_1}{\partial t} + u_1 \frac{\partial u_1}{\partial x} + u_2 \frac{\partial u_1}{\partial y} + u_3 \frac{\partial u_1}{\partial z},$$

which obviously equals du_1/dt. Equation 1.2 thus follows directly from (1.21).

12 The Conservation Equations for Ideal Pipe Flow

1.4.5 One-Dimensional Steady Flow

For steady, one-dimensional, horizontal, *constant diameter* pipe flow, we have, respectively,

$$\frac{du_k}{dt} = 0,$$

$$k = i = 1,$$

$$\frac{\partial Z}{\partial x} = 0,$$

and

$$\frac{\partial u_1}{\partial x} = 0.$$

Thus (1.21) becomes

$$0 = -\frac{g_c}{\rho}\frac{\partial p}{\partial x}. \tag{1.22}$$

It *further* follows for a zero viscosity, constant density fluid that $\partial p/\partial x = 0$, which simply states that under these ideal conditions there can be no pressure gradient in the pipe.

For steady, one-dimensional, horizontal, *variable area* pipe flow, (1.21) can be written as

$$0 + u_1 \frac{\partial u_1}{\partial x}(\rho d\mathbf{V}) = -g_c \frac{\partial p}{\partial x} d\mathbf{V} - 0, \tag{1.23}$$

which indicates that a pressure gradient and a velocity gradient are to be expected under these conditions.

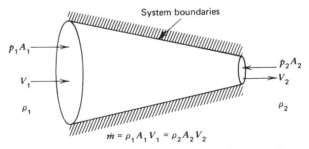

Figure 1.2 Notation for a one-dimensional force balance.

Table 1.2   Summary of Various Forms of the Momentum Equation for the Steady Flow of an Inviscid Fluid

Situation	Momentum Equation
General	$u_i \dfrac{\partial u_k}{\partial x_i} = -\dfrac{g_c}{\rho}\dfrac{\partial p}{\partial x_k} - g\dfrac{\partial Z}{\partial x_k}$
Three-dimensional x direction $k=1$ $i=1,2,3$	$u_1\dfrac{\partial u_1}{\partial x} + u_2\dfrac{\partial u_1}{\partial y} + u_3\dfrac{\partial u_1}{\partial z} = -\dfrac{g_c}{\rho}\dfrac{\partial p}{\partial x} - g\dfrac{\partial Z}{\partial x}$
One-dimensional x direction $k=i=1$	$u_1\dfrac{\partial u_1}{\partial x} = -\dfrac{g_c}{\rho}\dfrac{\partial p}{\partial x} - g\dfrac{\partial Z}{\partial x}$
One-dimensional Horizontal	$u_1\dfrac{\partial u_1}{\partial x}(\rho d\mathbf{V}) = -g_c \dfrac{\partial p}{\partial x}d\mathbf{V}$
	$g_c(p_1 A_1 - p_2 A_2) = (\rho A V)(V_2 - V_1)$

In the more familiar form of an elementary force balance, that is, $g_c F = \dot{m}\Delta V$, (1.23) becomes

$$g_c(p_1 A_1 - p_2 A_2) = (\rho A V)(V_2 - V_1), \qquad (1.24)$$

where the symbols may be clarified by reference to Figure 1.2.

The various forms of the momentum equation for the steady flow of an inviscid fluid are summarized in Table 1.2.

1.5   CONSERVATION OF ENERGY

A third basic premise involved in pipe flow is that the total energy of any isolated system must be conserved. For the more general case, where external energy transfers are involved, we can write an energy balance on the fluid in the volume element of Figure 1.1 according to the relation of (1.7).

1.5.1   Local Rate of Increase of Energy within Volume

The total energy of the fluid enclosed by the volume is made up of the internal energy of the fluid mass and the kinetic and potential energies of

14 The Conservation Equations for Ideal Pipe Flow

the fluid continuum. In general, each of these can change with time. Thus the rate of accumulation of fluid energy can be expressed as

$$\frac{\partial}{\partial t}\left(\rho E + \frac{\rho V^2}{2g_c} + \frac{\rho g Z}{g_c}\right) dV \qquad (1.25)$$

where E is the internal energy per unit mass, V is the directed resultant velocity of the fluid continuum (i.e., $V = iu_1 + ju_2 + ku_3$, where u_1, u_2, u_3 are the x,y,z components of V, and i,j,k in this relation are unit vectors), and Z is the vertical distance above some arbitrary horizontal datum where the potential energy is assumed to be zero.

1.5.2 Net Rate of Efflux of Fluid Energy across Surface of Volume

The fluid energy per unit time convected across any particular surface may be expressed as the product of the appropriate terms for fluid energy per unit mass per unit time. Thus the net difference between values of fluid energy per unit time leaving and entering the volume element is

$$\left[\left(E + \frac{\partial E}{\partial x_i}\frac{dx_i}{2}\right) + \frac{1}{2g_c}\left(V^2 + \frac{\partial V^2}{\partial x_i}\frac{dx_i}{2}\right) + \frac{g}{g_c}\left(Z + \frac{\partial Z}{\partial x_i}\frac{dx_i}{2}\right)\right]$$
$$\times \left[\left(\rho + \frac{\partial \rho}{\partial x_i}\frac{dx_i}{2}\right)\left(u_i + \frac{\partial u_i}{\partial x_i}\frac{dx_i}{2}\right) dA_j\right]_{\text{out}}$$
$$- \left[\left(E - \frac{\partial E}{\partial x_i}\frac{dx_i}{2}\right) + \frac{1}{2g_c}\left(V^2 - \frac{\partial V^2}{\partial x_i}\frac{dx_i}{2}\right) + \frac{g}{g_c}\left(Z - \frac{\partial Z}{\partial x_i}\frac{dx_i}{2}\right)\right]$$
$$\times \left[\left(\rho - \frac{\partial \rho}{\partial x_i}\frac{dx_i}{2}\right)\left(u_i - \frac{\partial u_i}{\partial x_i}\frac{dx_i}{2}\right) dA_j\right]_{\text{in}}$$
$$= \frac{\partial}{\partial x_i}\left[\left(E + \frac{V^2}{2g_c} + \frac{g}{g_c}Z\right)(\rho u_i)\right] dV. \qquad (1.26)$$

1.5.3 Strength of Sources of Change of Fluid Energy

We here consider the net rate of internal pressure-volume work done on the fluid, the net rate of transfer of work out of the fluid by boundary deformation (i.e., by external paddle wheel type work), and the net rate of heat transfer into the fluid from external sources.

Conservation of Energy

The net difference between the pressure-volume work done on the fluid at the inlet surfaces and that done at the outlet surfaces in a unit of time is

$$\left[\left(p - \frac{\partial p}{\partial x_i}\frac{dx_i}{2}\right)\left(u_i - \frac{\partial u_i}{\partial x_i}\frac{dx_i}{2}\right)dA_j\right]_{\text{inlet surfaces}}$$

$$-\left[\left(p + \frac{\partial p}{\partial x_i}\frac{dx_i}{2}\right)\left(u_i + \frac{\partial u_i}{\partial x_i}\frac{dx_i}{2}\right)dA_j\right]_{\text{outlet surfaces}}$$

$$= -\frac{\partial}{\partial x_i}(pu_i)\,dV, \qquad (1.27)$$

where the minus sign indicates that pressure-volume work is done *by* the fluid.

We represent the net rate of transfer of external work by boundary deformation by the term $\delta \dot{W}$, while the net rate of heat transfer from external sources is simply $\delta \dot{Q}$.

1.5.4 Summation

Applying (1.7), we have via (1.25), (1.26), and (1.27)

$$\frac{\partial}{\partial t}\left(\rho E + \frac{\rho V^2}{2g_c} + \frac{\rho g Z}{g_c}\right)dV + \frac{\partial}{\partial x_i}\left[\left(E + \frac{V^2}{2g_c} + \frac{gZ}{g_c}\right)(\rho u_i)\right]dV$$

$$= -\frac{\partial}{\partial x_i}(pu_i)\,dV + \delta \dot{W} + \delta \dot{Q}, \qquad (1.28)$$

which expresses the principle of conservation of energy and is usually called the *general* energy equation*. Equation 1.4 is the expanded form of (1.28).

1.5.5 One-Dimensional Steady Flow

In the steady state there can be no accumulation of fluid energy within the element, that is, $\partial/\partial t(\rho E + \rho V^2/2g_c + \rho g Z/g_c) = 0$. Then, by separating

---

*The question naturally arises as to why (1.28) may be called a *general* energy equation when only the normal stresses were considered. Actually, shearing stresses set up at the same time both negative work terms done on the fluid and positive heat terms added to the fluid. Thus we have in no way assumed an inviscid fluid flow in deriving the energy equation. We have simply omitted two counterbalancing terms.

16 The Conservation Equations for Ideal Pipe Flow

Table 1.3 Summary of Various Forms of the Steady Flow Energy Equation

Situation	Energy Equation
General	$\dfrac{\partial}{\partial x_i}[(E + \dfrac{V^2}{2g_c} + \dfrac{g}{g_c}Z)(\rho u_i)]d\mathbf{V} = -\dfrac{\partial}{\partial x_i}(pu_i)d\mathbf{V} + \delta \dot{W} + \delta \dot{Q}$
Three-dimensional	$\dfrac{\partial}{\partial x}[(E + \dfrac{V^2}{2g_c} + \dfrac{g}{g_c}Z)(\rho u_1)]d\mathbf{V} + \dfrac{\partial}{\partial y}[(E + \dfrac{V^2}{2g_c} + \dfrac{g}{g_c}Z)(\rho u_2)]d\mathbf{V}$
	$+ \dfrac{\partial}{\partial z}[(E + \dfrac{V^2}{2g_c} + \dfrac{g}{g_c}Z)(\rho u_3)]d\mathbf{V}$
	$= -[\dfrac{\partial}{\partial x}(pu_1) + \dfrac{\partial}{\partial y}(pu_2) + \dfrac{\partial}{\partial z}(pu_3)]d\mathbf{V} + \delta \dot{W} + \delta \dot{Q}$
One-dimensional	$\delta Q + \delta W = dE + p\,dv + v\,dp + \dfrac{V\,dV}{g_c} + \dfrac{g}{g_c}dZ$

internal and external effects, and by making use of continuity as given by (1.11), the general energy equation of (1.28) becomes in the steady state

$$(\delta \dot{Q} + \delta \dot{W})_{\text{external}} = \left[(\rho u_i\, dA_j)\dfrac{\partial}{\partial x_i}\left(E + \dfrac{p}{\rho} + \dfrac{V^2}{2g_c} + \dfrac{g}{g_c}Z\right)dx_i\right]_{\text{internal}}. \quad (1.29)$$

Equation 1.29 can be expressed in the familiar thermodynamic form of energy per unit mass by dividing through by the mass flow rate, $\dot{m}$ of (1.14). Thus

$$\delta Q + \delta W = dE + p\,dv + v\,dp + \dfrac{V\,dV}{g_c} + \dfrac{g}{g_c}dZ, \quad (1.30)$$

where each term has the conventional units of foot-pounds force per pound mass, and $v = 1/\rho$ and represents the specific volume of the fluid.

The various forms of the energy equation for steady flow are summarized in Table 1.3.

1.6 RELATION BETWEEN MOMENTUM AND ENERGY

It is well known that the Euler momentum equation of (1.21) can be written in the form of an energy equation. For steady flow we may obtain the Bernoulli equation as follows. Equation 1.21 is first rearranged slightly and then integrated along a streamline* to yield

$$\int_s \left(\frac{du_k}{dt} + \frac{g_c}{\rho} \frac{\partial p}{\partial x_k} + g \frac{\partial Z}{\partial x_k} \right) dx_k = \text{constant}. \quad (1.31)$$

At any point along this streamline there is a resultant velocity V and its components u_1, u_2, and u_3. Thus we may expand, regroup, and evaluate (1.31) along a streamline in the steady state, noting that $dx/u_1 = dy/u_2 = dz/u_3 = dt$, and that g acts only in the negative z direction, to obtain

$$\int_s \left[(u_1 du_1 + u_2 du_2 + u_3 du_3) + \frac{g_c}{\rho} \left(\frac{\partial p}{\partial x} dx + \frac{\partial p}{\partial y} dy + \frac{\partial p}{\partial z} dz \right) + g \, dZ \right] = \text{constant}$$

or, finally,

$$\int_s \left(V \, dV + \frac{g_c}{\rho} dp + g \, dZ \right) = \text{constant}. \quad (1.32)$$

On an energy per unit mass basis, (1.32) becomes

$$\frac{dp}{\rho} + \frac{V \, dV}{g_c} + \frac{g}{g_c} dZ = 0, \quad (1.33)$$

which is the Bernoulli equation.

But (1.33) coincides with the general energy equation of (1.30) when the fluid is inviscid and there is not external work. This inviscid restriction becomes clear upon consideration of the first law of thermodynamics (to be discussed in Chapter 3), namely,

$$\delta Q + \delta F_f = dE + p \, dv, \quad (1.34)$$

where F_f, the frictional energy loss per unit mass, is necessarily zero in the inviscid case.

REFERENCES

1. H. Rouse and S. Ince, *History of Hydraulics*, Dover Publications, 1963, p. 49.
2. C. Truesdell, *Essays in the History of Mechanics*, Springer-Verlag, 1968, p. 114.

---

*If conditions of irrotationality (i.e., $\partial u_k / \partial x_i = \partial u_i / \partial x_k$) are assumed or satisfied, we obtain the same Bernoulli equation without the restriction of having to follow a particular streamline.

3. D. Bernoulli, *Hydrodynamica*, Strasbourg, 1738. Translated from the Latin by T. Carmody and H. Kobus at the Iowa Institute of Hydraulic Research, Dover Publications, 1968.
4. R. P. Benedict, "Analog simulation," *Electro-Technol.*, Science and Engineering Series No. 60, December 1963, p. 73.

NOMENCLATURE

Roman

- a acceleration
- A area
- E internal energy per pound mass
- f function
- F force
- F_f frictional energy loss per pound mass
- g local gravity
- g_c gravitational constant
- M mass
- N conserved quantity
- p pressure
- P arbitrary property
- P.E.   potential energy
- Q heat per pound mass, volumetric flow rate
- t time
- u velocity component
- v specific volume
- V resultant velocity
- $\mathbf{V}$ volume
- W work per pound mass
- x, y, z Cartesian coordinates
- Z vertical height

Greek

- ρ fluid density

Mathematical Symbols

- d total derivative
- ∂ partial derivative

Subscripts

- 1, 2, 3   in x, y, z directions
- i, k Cartesian tensor indices, generally 1 to 3 (i signifies a summation process, while k signifies the particular component under consideration)
- j geometric subscript signifying perpendicularity to component or term under consideration

2
Solutions to Ideal-Incompressible Pipe Flow

> ...pipelines are coming, pipelines are going. So much of us is pipes...—Gustav Eckstein

2.1 GENERAL REMARKS

In this chapter we apply the conservation equations of Chapter 1 to the flow in pipes of a hypothetical ideal-incompressible fluid. Our purpose is to gain some insight into the basic behavior of incompressible pipe flow. Thus we examine here solutions to inviscid, constant density liquid flow in pipes and through various piping components.

We have already said that under such conditions all frictional effects are to be overlooked. Hence the equations will be simplified, and we can see the main aspects of pipe flow solutions more clearly without burdening ourselves with the small perturbations on the main flow such as are introduced by boundary layer growth, eddy formation, turbulence, and so on. In other words, we apply the equations for the conservation of mass, of momentum, and of energy to pipe flow problems involving an idealized fluid to obtain *approximate* solutions.

2.2 THE ONE-DIMENSIONAL BERNOULLI ENERGY EQUATION

The streamline equation (1.33) also can be applied to larger and larger stream tubes, up to and including complete pipe cross-sectional areas, whenever the streamlines are without curvature. This means quite simply that a one-dimensional analysis will apply very well across any straight section of pipe in which flows an inviscid fluid. Conversely, even for the ideal fluid, the one-dimensional analysis will break down whenever there is an *increase* in the pipe diameter (such as is encountered in a sudden enlargement in the pipe diameter, or in the gradually enlarging diameter of a diffusing section); whenever there is a *decrease* in the pipe diameter (as encountered in the sudden reduction in diameter of an orifice section, or in the gradually reducing diameter of a nozzle section); whenever there is a valve or elbow in the piping; and so on (see Figure 2.1).

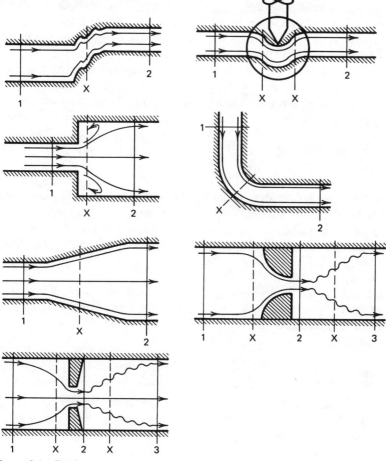

Figure 2.1 Regions in various piping sections where one-dimensional analysis can and cannot be applied. Numbered sections indicate acceptable regions; X, sections, where one-dimensional analysis does not apply.

22 Solutions to Ideal-Incompressible Pipe Flow

But for any straight pipe section we can write the Bernoulli equation, in the ideal incompressible case, as

$$\frac{p}{\rho} + \frac{V^2}{2g_c} + \frac{g}{g_c} Z = \text{constant}, \qquad (2.1)$$

where both pressure and elevation are to be specified at the pipe centerline.

We saw in Chapter 1 that (2.1) follows from *either* momentum or energy considerations, and that it is *independent* of heat transfer effects in the inviscid case. Here we note further that the sum of the three terms of (2.1), namely, the pressure head (p/ρ), the velocity head $(V^2/2g_c)$, and the elevation head (gZ/g_c), represents the *total energy* of the fluid at any point in the pipe. It is important to observe that all terms in (2.1) are on a foot-pounds force per pound mass basis, and that, whenever $g_{local} = g_{standard}$, the pound force numerically equals the pound mass, and hence each term in (2.1) can be looked upon as so many *feet of head*.

In the absence of elevation change (i.e., in a horizontal pipe section), it is quite clear that the Bernoulli principle of (2.1) requires that as velocity *increases* (as is caused by a decrease in flow area, according to the continuity principle of (1.14), where $A_1 V_1 = A_2 V_2$), the pressure must *decrease*.

Finally, we make the very important *practical* observation that the streamlines *need not* be straight and parallel *throughout* the piping system. In other words, there can be piping bends, abrupt area changes, and so on, as long as the streamlines are straight and parallel at the inlet and outlet stations across which (2.1) is applied. This principle is illustrated by Figure 2.2 and Example 2.1.

Example 2.1. Referring to Figure 2.2, estimate the pressure heads at A and B. Water, at a density of 62.4 lbm/ft^3, is flowing at a volumetric flow rate of 0.47856 ft^3/sec. Assume that local gravity equals 32.174 ft/sec^2 (i.e., numerically equals the standard gravity value).

Solution:

1. The ideal Bernoulli energy equation (2.1) can be written between any two points in the system. However, we encounter a discrepancy at once if we apply (2.1) between reservoir surfaces 1 and 2. This is explained by the fact that in *real* flow the potential energy available from the elevation difference between 1 and 2 is just used up by the flow losses between 1 and 2, and such losses have been neglected in this chapter.

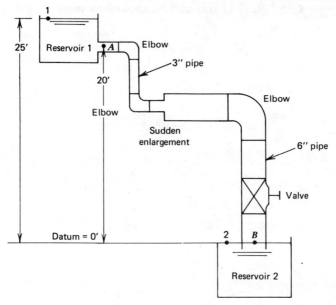

Figure 2.2 General piping system such as can be analyzed by the Bernoulli equation.

2. Between 1 and A, however, (2.1) can be applied with some success:

$$\frac{p_1}{\rho} + \frac{V_1^2}{2g_c} + Z_1 = \frac{p_A}{\rho} + \frac{V_A^2}{2g_c} + Z_A.$$

3. But p_1 is at atmospheric pressure, that is, at 0 psig; also, $V_1 \simeq 0$ compared with the pipe velocities at A and B, and

$$\frac{V_A^2}{2g_c} = \frac{V_3^2}{2g_c} = \frac{(Q/A_3)^2}{2g_c} = \left[\frac{0.47856}{(\pi/4)(\frac{1}{4})^2}\right]^2 \times \frac{1}{2g_c} = 1.47706 \text{ ft-lbf/lbm}.$$

Hence we have

$$0 + 0 + 25 = \frac{p_A}{\rho} + 1.47706 + 20$$

or

$$\left(\frac{p_A}{\rho}\right)_{\text{ideal}} = 25 - 20 - 1.47706 = 3.523 \text{ ft-lbf/lbm},$$

which is the *ideal* pressure head at A.

24 Solutions to Ideal-Incompressible Pipe Flow

4. Between A and B, (2.1) also can be applied to yield

$$\frac{p_A}{\rho} + \frac{V_A^2}{2g_c} + Z_A = \frac{p_B}{\rho} + \frac{V_B^2}{2g_c} + Z_B.$$

But the total energy at A is clearly 25 ft-lbf/lbm, so we have

$$\frac{p_B}{\rho} = 25 - \frac{V_B^2}{2g_c} - Z_B,$$

where

$$\frac{V_B^2}{2g_c} = \frac{V_6^2}{2g_c} = \frac{(Q/A_6)^2}{2g_c} = \left[\frac{0.47856}{(\pi/4)(\frac{1}{2})^2}\right]^2 \times \frac{1}{2g_c} = 0.09232.$$

Then

$$\left(\frac{p_B}{\rho}\right)_{\text{ideal}} = 25 - 0.09232 - 0 = 24.908 \text{ ft-lbf/lbm},$$

which is the *ideal* pressure head at B.

Although this example has served to illustrate the application of the Bernoulli energy equation to a piping problem, the results are far from realistic. In Part III this same example will be done again as Example 7.3, with appropriate losses assumed or calculated for the various piping elements, to show how unreliable are these ideal answers whenever losses are as important as they are in this example.

2.3 THE ONE-DIMENSIONAL BERNOULLI WORK EQUATION

Although (2.1) is independent of heat transfer effects in the ideal case, of course it must be modified to account for work addition according to the relation

$$(W_{\text{on}})_{1,2} = \frac{p_2 - p_1}{\rho} + \frac{V_2^2 - V_1^2}{2g_c} + \frac{g}{g_c}(Z_2 - Z_1), \quad (2.2)$$

where $(W_{\text{on}})_{1,2}$ signifies the net amount of work transferred across the pipe walls *into* the fluid (i.e., work done *on* the fluid) between the arbitrary axial

The One-Dimensional Bernoulli Work Equation

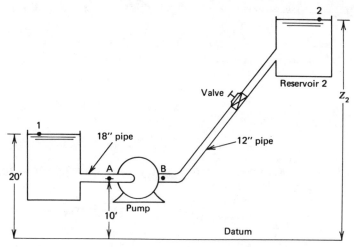

Figure 2.3 General piping system with work addition.

stations, 1 and 2. Note that this work addition completely accounts for the difference in total energy between inlet and exit, that is, (2.2) can also be written as

$$E_{in} + (W_{on})_{1,2} = E_{out}, \qquad (2.3)$$

where E here represents the total energy of the fluid at an axial station, and as before each term in (2.3) represents energy per pound mass.

Example 2.2. Referring to Figure 2.3, estimate the elevation of water that can be maintained in reservoir 2 by a pump delivering 70 horsepower to the system at a volumetric flow rate of 8 ft³/sec. Also estimate the pressures at A and B on either side of the pump. Assume that local gravity equals the standard gravity value of 32.174 ft/sec², and that the density of water is 62.4 lbm/ft³.

Solution:

1. The ideal Bernoulli work equation (2.2) can be written between any two points in a system where the streamlines are straight and parallel. It is convenient to write the energy equation between points where some of the terms drop out, as at the reservoir surfaces 1 and 2, where (2.2) becomes

$$\frac{p_1 - p_2}{\rho} + \frac{V_1^2 - V_2^2}{2g_c} + \frac{g}{g_c}(Z_1 - Z_2) + (W_{on})_{1,2} = 0.$$

26 Solutions to Ideal-Incompressible Pipe Flow

2. But $p_1 = p_2 =$ atmospheric (or $p_1 = p_2 = 0$ psig); hence the first term drops out.
3. But the velocity at the surface of any reservoir or large tank is negligible compared to the velocity in a pipe; hence the second term above drops out.
4. Since local $g =$ standard g, $g/g_c = 1$.
5. There remains

$$(Z_2)_{ideal} = Z_1 + (W_{on})_{1,2}$$

where the work term is obtained from the given horsepower as follows:

$$\text{Power} = \frac{\text{work(ft-lbf)}}{\text{time(sec)}} = \frac{\text{ft-lbf}}{\text{lbm}} \times \frac{\text{lbm}}{\text{sec}} = W_{on} \times \dot{m},$$

where

$$\dot{m} = \rho A V = \rho Q,$$

and

$$\text{Horsepower} = W_{on} \times \dot{m} \times \left(\frac{1}{550} \frac{\text{sec hp}}{\text{ft-lbf}} \right).$$

Therefore

$$(W_{on})_{1,2} = \frac{\text{hp} \times 550}{\rho Q}.$$

6. Thus

$$(Z_2)_{ideal} = 20 + \frac{70 \times 550}{62.4 \times 8} = 20 + 77.12 = 97.12 \text{ ft.}$$

7. The pressure at A (in Figure 2.3) can be estimated by applying the ideal Bernoulli equation (2.1) between 1 and A as follows:

$$\frac{p_1}{\rho} + \frac{V_1^2}{2g_c} + Z_1 = \frac{p_A}{\rho} + \frac{V_A^2}{2g_c} + Z_A$$

$$0 + 0 + 20 = \frac{p_A}{\rho} + \frac{V_{18}^2}{2g_c} + 10.$$

But from continuity

$$V_A = V_{18} = \frac{Q}{A_{18}} = \frac{8}{\pi(1.5)^2/4},$$

and

$$\frac{V_{18}^2}{2g_c} = \left(\frac{32}{\pi \times 2.25}\right)^2 \times \frac{1}{2g_c} = 0.3185 \text{ ft-lbf/lbm.}$$

Hence

$$\left(\frac{p_A}{\rho}\right)_{ideal} = 20 - 10 - 0.3185 = 9.6815 \text{ ft-lbf/lbm}$$

or

$$(p_A)_{ideal} = \frac{9.6815 \times 62.4}{144} = 4.195 \text{ psig.}$$

8. The pressure at B can be estimated by applying the ideal Bernoulli work equation (2.3) between A and B as follows:

$$\left(\frac{p_A}{\rho} + \frac{V_A^2}{2g_c} + Z_A\right) + (W_{on})_{A,B} = \left(\frac{p_B}{\rho} + \frac{V_B^2}{2g_c} + Z_B\right)$$

or

$$(9.6815 + 0.3185 + 10) + 77.12 = \left(\frac{p_B}{\rho} + 1.6124 + 10\right),$$

where

$$\frac{V_B^2}{2g_c} = \frac{V_{12}^2}{2g_c} = \left(\frac{32}{\pi}\right)^2 \times \frac{1}{2g_c} = 1.6124.$$

Then

$$\left(\frac{p_B}{\rho}\right)_{ideal} = 9.6815 + 0.3185 + 77.12 - 1.6124 = 85.5076 \text{ ft-lbf/lbm,}$$

or

$$(p_B)_{ideal} = \frac{85.5076 \times 62.4}{144} = 37.0533 \text{ psig.}$$

28 Solutions to Ideal-Incompressible Pipe Flow

2.4 THE ONE-DIMENSIONAL BERNOULLI PRESSURE EQUATION

The Bernoulli equation (2.1) can be written in terms of pressure, rather than head, simply by multiplying each term of (2.1) by the fluid density. Thus

$$p + \frac{\rho V^2}{2g_c} + \frac{\rho g Z}{g_c} = \text{constant}. \tag{2.4}$$

In (2.4), p is the *static* pressure of the fluid, $\rho V^2/2g_c$ is the *dynamic* pressure of the fluid, and $\rho g Z/g_c$ is the *potential* pressure of the fluid, all pertaining to and evaluated at the centerline of the pipe at a given axial station of interest.

It is convenient and conventional to introduce the concept of *total* pressure at this point in our discussion. The total pressure, p_t, as the name implies, is quite simply defined as the sum of the static and dynamic pressures. Thus

$$p_t = p + \frac{\rho V^2}{2g_c}. \tag{2.5}$$

Example 2.3. In Figure 2.4, a Pitot tube is located in the center of a 12 in. pipe in which flows water at a density of 62.4 lbm/ft³. Estimate the velocity of the fluid from ideal relations, if the Pitot pressure difference is 5 psi.

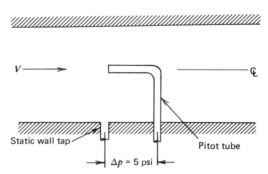

Figure 2.4 Ideal velocity determination by a Pitot tube.

Solution:

1. By the Bernoulli pressure equation (2.5), we have

$$p_t - p = \frac{\rho V^2}{2g_c}$$

or

$$V_{ideal} = \sqrt{\frac{2g_c(p_t - p)}{\rho}}$$

2. Thus

$$V_{ideal} = \sqrt{2 \times \frac{32.174 \text{ ft-lbm}}{\text{sec}^2 \text{ lbf}} \times \frac{5 \text{lbf}}{\text{in}^2} \times \frac{144 \text{ in}^2}{\text{ft}^2} \times \frac{1 \text{ ft}^3}{62.4 \text{ lbm}}}$$

or

$$V_{ideal} = 27.25 \text{ ft/sec}.$$

3. This ideal velocity is much greater than the volumetric average velocity in the pipe in the case of real viscous flow. Hence this example will be done again in Part II as Example 5.1, after our discussion of velocity distributions and the pipe factor in Chapter 5.

2.5 TWO- AND THREE-DIMENSIONAL SOLUTIONS

We have just seen that in general a pipe flow problem can be solved by one-dimensional conservation relations, as long as the axial stations across which the equations are applied show reasonably straight and parallel streamlines. However, when we must consider flow details within the piping, as with flow patterns through a nozzle, an orifice, an abrupt enlargement, and so on, then we must resort to a two-dimensional or even a three-dimensional solution with axial symmetry.

In such two- and three-dimensional problems, where streamlines are curved and not parallel, we must set up and solve a flow network.

Briefly, many problems in engineering involve continuous *potential* fields such as can be described by the Laplace equation [1]. The Laplace equation concerns fields, free of sources or sinks, in the steady state

30 Solutions to Ideal-Incompressible Pipe Flow

condition and is given in terms of the field potential, φ, by

$$\nabla^2 \varphi = \frac{\partial^2 \varphi}{\partial x^2} + \frac{\partial^2 \varphi}{\partial y^2} + \frac{\partial^2 \varphi}{\partial z^2} = 0. \tag{2.6}$$

Such fields, where vector quantities such as velocity are important, can be studied by the use of rigorous mathematical techniques. However, the boundary conditions necessary to solve the differential equations are very difficult to evaluate or manipulate. Thus none but the most elementary field problems can be solved by purely analytical means. Instead, numerical methods (based on relaxation and iteration techniques, to be discussed) are usually used for solving potential field problems. However, in using these approximation methods, two serious difficulties are encountered. First, even with the use of a digital computer, obtaining a solution is time consuming because either method requires that boundary values be written. Furthermore, the complete field must be considered since rough values of the function must be assumed over the whole field and programmed to obtain a solution, despite the fact that interest may bear on only a small portion of the field.

Continuous potential fields also can be studied by the use of electrical analog techniques because of the direct correspondence between the differential equations that characterize the voltage and current fields and those that describe the physical system being studied. By using an electric field analog, a completely relaxed solution can be obtained at the throw of a switch, and only the portions of the field that are of interest need be examined in detail. Boundary conditions can be taken into account through the use of properly constructed, geometrically similar models.

2.5.1 Characteristics of Vector Fields

To arrive at the basic principles required to set up the potential field analogy, certain characteristics of vector fields must be taken into account. A vector field exists when, at every point within a region, some physical characteristic has a definite magnitude and direction. Such fields are encountered in the study of electricity, fluid mechanics, or heat transfer (to mention only a few).

Continuity and the Stream Function. Although the strength of a field may vary from point to point, its value at any one point is related to that at every other point, that is, there is a certain symmetry which must be considered. Examples include electrostatic lines of force, electric current field lines, heat conduction flux lines, and the streamlines of fluid flow. In the steady state, continuity may be defined intuitively by stating that the amount of something entering a given region must equal the amount leaving the same region. This applies whether the "something" is an electric current, a quantity of fluid, or a flow of heat.

A more general approach to continuity, and one that adds a useful function to the field notation, involves the use of the line integral. Consider a two-dimensional region in which there are no sources or sinks. If an arbitrary potential gradient vector in the region at point (x,y) is denoted by V, it also can be expressed in terms of its components, $V\sin\theta$ and $V\cos\theta$. The angle θ is defined with respect to an arbitrary curve in the region going through the point (x,y). The vector $V\sin\theta$ is the vector component perpendicular to the curve, and $V\cos\theta$ is the vector component tangent to the curve. The line integral formed as the limit of the sum of the products of the vector $V\sin\theta$, a characteristic function λ of the field, and small intervals of distance along a continuous curve in the region represents the time rate of transfer of charge (i.e., flux) across the curve. Thus

$$\Delta\psi = \int_c \lambda V \sin\theta \, ds. \tag{2.7}$$

Equation 2.7 defines ψ, the stream function. By expressing $V\sin\theta$ in terms of its x and y components, u_1 and u_2 (see Figure 2.5), ψ can be expressed in the following more useful form:

$$\psi_B - \psi_A = \int_A^B (-\lambda u_2 \, dx + \lambda u_1 \, dy). \tag{2.8}$$

Using Green's theorem, we can also express the line integral as a double integral:

$$\Delta\psi = \int\int \left[\frac{\partial}{\partial x}(\lambda u_1) + \frac{\partial}{\partial y}(\lambda u_2) \right] dx \, dy. \tag{2.9}$$

However, the integral represented by $\Delta\psi$ is independent of the path of integration in the absence of sources and sinks, that is, the net flux crossing any closed path in the region is zero. Thus

$$d\psi = \frac{\partial}{\partial x}(\lambda u_1) + \frac{\partial}{\partial y}(\lambda u_2) = 0, \tag{2.10}$$

which is in the form of (1.11), the continuity equation of fluids. Mathematically, the stream function is an exact differential function of x and y

32 Solutions to Ideal-Incompressible Pipe Flow

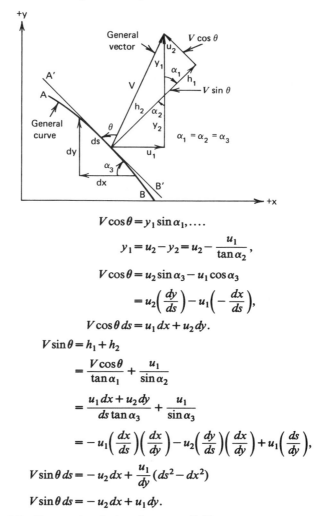

$$V\cos\theta = y_1 \sin\alpha_1, \ldots$$

$$y_1 = u_2 - y_2 = u_2 - \frac{u_1}{\tan\alpha_2},$$

$$V\cos\theta = u_2 \sin\alpha_3 - u_1 \cos\alpha_3$$

$$= u_2\left(\frac{dy}{ds}\right) - u_1\left(-\frac{dx}{ds}\right),$$

$$V\cos\theta\, ds = u_1\, dx + u_2\, dy.$$

$$V\sin\theta = h_1 + h_2$$

$$= \frac{V\cos\theta}{\tan\alpha_1} + \frac{u_1}{\sin\alpha_2}$$

$$= \frac{u_1\, dx + u_2\, dy}{ds\tan\alpha_3} + \frac{u_1}{\sin\alpha_3}$$

$$= -u_1\left(\frac{dx}{ds}\right)\left(\frac{dx}{dy}\right) - u_2\left(\frac{dy}{ds}\right)\left(\frac{dx}{dy}\right) + u_1\left(\frac{ds}{dy}\right),$$

$$V\sin\theta\, ds = -u_2\, dx + \frac{u_1}{dy}(ds^2 - dx^2)$$

$$V\sin\theta\, ds = -u_2\, dx + u_1\, dy.$$

Figure 2.5 Geometric relations in vector fields.

only and can be written as

$$d\psi = \frac{\partial\psi}{\partial x}\, dx + \frac{\partial\psi}{\partial y}\, dy. \tag{2.11}$$

By combining (2.8) and (2.11), we obtain

$$u_1 = \frac{1}{\lambda}\frac{\partial\psi}{\partial y}$$

and

$$u_2 = -\frac{1}{\lambda}\frac{\partial \psi}{\partial x}. \tag{2.12}$$

Note that the slope of a constant ψ line (i.e., at $d\psi=0$), in terms of u_1 and u_2, can be expressed as $(dy/dx)_{\psi=c} = u_2/u_1$. As will be seen shortly, this indicates that streamlines are everywhere tangent to the potential gradient vectors. Thus ψ lines indicate *direction* of flow, since there is no flow across them. Such lines can be mapped out in any vector field and represent lines of vector action or of vector flux.

Circulation and the Potential Function. We consider now the vector component tangent to the arbitrary curve in the field. The line integral, formed as the limit of the sum of the product of the vector $V\cos\theta$ and small intervals of distance along a continuous curve in the region, represents the circulation of the vector about the curve. Thus

$$\Delta\varphi = \int_c V\cos\theta \, dS. \tag{2.13}$$

Equation 2.13 defines φ, the potential function. By expressing $V\cos\theta$ in terms of u_1 and u_2, (2.13) can be expressed in the following line integral form:

$$\varphi_B - \varphi_A = \int_A^B (u_1 \, dx + u_2 \, dy). \tag{2.14}$$

Using Green's theorem again, we can express the line integral as a double integral:

$$\Delta\varphi = \iint \left(\frac{\partial u_2}{\partial x} - \frac{\partial u_1}{\partial y}\right) dx\, dy. \tag{2.15}$$

If this line integral is zero around every closed path in the field, the field is called *conservative* (i.e., irrotational). Thus

$$d\varphi = \frac{\partial u_2}{\partial x} - \frac{\partial u_1}{\partial y} = 0. \tag{2.16}$$

Mathematically, the potential function is an exact differential function of x and y only and can be written as

$$d\varphi = \frac{\partial \varphi}{\partial x} dx + \frac{\partial \varphi}{\partial y} dy. \tag{2.17}$$

34 Solutions to Ideal-Incompressible Pipe Flow

Combining (2.14) and (2.17), we have

$$u_1 = \frac{\partial \varphi}{\partial x}$$

and (2.18)

$$u_2 = \frac{\partial \varphi}{\partial y}.$$

Thus φ is defined so that the *derivative* of φ in any direction gives the *velocity* in that direction. Hence it is often called the *velocity potential*. Furthermore, since $\partial^2\varphi/\partial x\, \partial y = \partial u_1/\partial y = \partial u_2/\partial x$, it follows that $\partial u_2/\partial x - \partial u_1/\partial y = 0$, which means physically that the fluid element has no rotation in space. Thus the assumption of a velocity potential as defined above *requires* that the flow be irrotational, in agreement with (2.16).

Note that the slope of a constant φ line in terms of u_1 and u_2 can be expressed as $(dy/dx)_{\varphi = c} = -u_1/u_2$. This indicates that potential lines are everywhere perpendicular to the potential gradient vectors. It follows that φ and ψ lines are everywhere *orthogonal trajectories* of each other. This means quite simply that ψ and φ lines intersect at right angles. Each constant potential line has a definite value indicating the relative potential of the field compared to an arbitrary reference potential. Thus the relative spacing of equal increments of equipotentials indicates the magnitude of the electric field intensity ($E_x = \partial V/\partial x$), of the temperature gradient ($H_x = \partial T/\partial x$), or of the fluid velocity ($u = \partial \varphi/\partial x$).

The Laplace Equation. Combining (2.12) and (2.18), we have

$$u_1 = \frac{\partial \varphi}{\partial x} = \frac{1}{\lambda}\frac{\partial \psi}{\partial y}$$

and (2.19)

$$u_2 = \frac{\partial \varphi}{\partial y} = -\frac{1}{\lambda}\frac{\partial \psi}{\partial x},$$

which are identical in form to the Cauchy-Riemann equations. Therefore we are assured mathematically that φ and ψ are related by an analytical function (i.e., a function that has a unique derivative everywhere in a given region except at isolated points, called singular points) of a complex variable. Thus

$$f(x + iy) = \varphi + i\psi. \qquad (2.20)$$

Specifically, φ and ψ satisfy generalized Laplace equations, which in

Cartesian coordinates are

$$\frac{\partial}{\partial x}\left(\lambda \frac{\partial \varphi}{\partial x}\right) + \frac{\partial}{\partial y}\left(\lambda \frac{\partial \varphi}{\partial y}\right) = 0$$

and (2.21)

$$\frac{\partial}{\partial x}\left(\frac{1}{\lambda} \frac{\partial \psi}{\partial x}\right) + \frac{\partial}{\partial y}\left(\frac{1}{\lambda} \frac{\partial \psi}{\partial y}\right) = 0.$$

Equations 2.21 can be substantiated by substituting (2.18), the velocity components in terms of the potential function, into (2.10), and by substituting (2.12), the velocity components in terms of the stream function, into (2.16).

For the special case where the field characteristic (λ) is a constant (as in a constant density fluid or a constant resistivity electric field), φ and ψ become interchangeable since both then satisfy identical Laplace equations, namely,

$$\frac{\partial^2 \varphi}{\partial x^2} + \frac{\partial^2 \varphi}{\partial y^2} = 0 = \frac{\partial^2 \psi}{\partial x^2} + \frac{\partial^2 \psi}{\partial y^2}, \qquad (2.22)$$

bringing us full circle back to (2.6).

Note that only one pattern of irrotational flow can exist for given fixed boundaries. This pattern is the *potential* solution.

We will next see how the interchangeability, as expressed by (2.22), makes possible two simplified analogies in the two-dimensional electric field analog.

2.5.2 Two-Dimensional Electric-Fluid Flows

To illustrate the synthesis of a valid field analogy, we consider a two-dimensional steady flow in continuous electric and fluid fields that can be described by (2.22). First the basic continuity equations are given. Then the pertinent analogous terms are discussed and tabulated. Analogous terms also are given for the heat transfer field, which is of general interest.

Electric Current Field. The basic differential equation for continuity for the electric current field is a form of Ohm's law:

$$di = -\gamma \frac{\partial V_e}{\partial y} dx + \gamma \frac{\partial V_e}{\partial x} dy. \qquad (2.23)$$

On comparing (2.23) with (2.8), the general continuity equation, we note

36 Solutions to Ideal-Incompressible Pipe Flow

that for the electric field the current (i) is analogous to the stream function, the voltage (V_e) corresponds to the potential function, the field strength ($\partial V_e/\partial S$) corresponds to the potential gradient, and the conductivity (γ) of the electrical conductor corresponds to the characteristic function of the field.

Fluid Flow Field. The basic differential equation for the fluid flow field is

$$d\dot{m} = -\rho u_2 \, dx + \rho u_1 \, dy. \tag{2.24}$$

By comparing this with (2.8), we find that the mass flow rate ($\dot{m}$) is analogous to the stream function, the velocity potential (φ) corresponds to the potential function, the field strength corresponds to the velocity (V), and the fluid density (ρ) corresponds to the characteristic function of the field.

These terms are tabulated for easy reference in Table 2.1. As indicated previously, there are two possible analogies that can be drawn between the electric and fluid fields.

In analogy A, the electric potential (V_e) is proportional to the velocity potential (φ), and the electric current (i) is proportional to the stream function (ψ). Fluid velocities ($u_1 = \partial \varphi/\partial x$) are proportional to the electric potential gradients in the direction of the velocities and, as such, are easily determined. Streamlines are identified with current direction lines and are inaccessible for measuring purposes.

In analogy B, the electric potential is proportional to the stream function, while the electric current is proportional to the velocity potential. Fluid velocities (now $u_1 = 1/\rho \; \partial \psi/\partial y$) are again proportional to the electric potential gradients, but in this case the gradient considered must be in a direction perpendicular to the velocities. In this analogy the velocity potential lines are identified with the inaccessible current direction lines.

By examining a given model using analogs based on analogies A and B, we can obtain a complete map of the orthogonal φ and ψ coordinates. At the same time the results should be completely interchangeable.

Conducting Sheet Analog. In 1845 Gustav Kirchhoff [2] first proposed that the equipotential lines of an irregular two-dimensional electric field could be determined by the use of a geometrically similar sheet of metal and suitably connected electrodes. He simulated the field on a thin disk of copper, and potential balances were indicated by the deflection of a compass needle. This early analog had certain inherent faults, among which were nonuniform resistivity of the copper sheet and excessively high

Table 2.1 Some Analogous Terms and Equations for Several Two-Dimensional Vector Fields (steady flow in homogeneous, continuous, indestructible fields)

General Terms and Equations	Electric Current Field	Fluid Flow Field	Heat Flow Field
Basic differential equation	$i = -\gamma \dfrac{\partial V_e}{\partial x} A$ (Ohm's law)	$\dot{m} = \rho A V$ (continuity)	$q = -k \dfrac{\partial T}{\partial x} A$ (Fourier's law)
ψ, stream function Constant ψ line	i, electric current Field line	$\dot{m}$, mass flow rate Streamline	q, heat flow rate Flux line
φ, potential function Constant φ line	V_e, voltage Equipotential	φ, velocity potential Equipotential	T, temperature Isotherm
V, potential gradient λ, characteristic field function	E, field intensity γ, electric conductivity	V, fluid velocity ρ, fluid density	H, temperature gradient k, thermal conductivity
u_1, x component of V	$E_x = \dfrac{\partial V_e}{\partial x} = -\dfrac{1}{\gamma}\dfrac{\partial i}{\partial y}$	$u_1 = \dfrac{\partial \varphi}{\partial x} = -\dfrac{1}{\rho}\dfrac{\partial \dot{m}}{\partial y}$	$H_x = \dfrac{\partial T}{\partial x} = \dfrac{1}{k}\dfrac{\partial q}{\partial y}$
u_2, y component of V	$E_y = \dfrac{\partial V_e}{\partial y} = -\dfrac{1}{\gamma}\dfrac{di}{\partial x}$	$u_2 = \dfrac{\partial \varphi}{\partial y} = -\dfrac{1}{\rho}\dfrac{\partial \dot{m}}{\partial x}$	$H_y = \dfrac{\partial T}{\partial y} = -\dfrac{1}{k}\dfrac{\partial q}{\partial x}$

Table 2.1 Continued

General Terms and Equations	Electric Current Field	Fluid Flow Field	Heat Flow Field
	General Differential Equations ($\lambda \neq$ constant)		
Continuity in terms of vector components	$\dfrac{\partial}{\partial x}(\gamma E_x) + \dfrac{\partial}{\partial y}(\gamma E_y) = 0$	$\dfrac{\partial}{\partial x}(\rho u_1) + \dfrac{\partial}{\partial y}(\rho u_2) = 0$	$\dfrac{\partial}{\partial x}(kH_x) + \dfrac{\partial}{\partial y}(kH_y) = 0$
Continuity in terms of potential function	$\dfrac{\partial}{\partial x}\left(\gamma \dfrac{\partial V_e}{\partial x}\right) + \dfrac{\partial}{\partial y}\left(\gamma \dfrac{\partial V_e}{\partial y}\right) = 0$	$\dfrac{\partial}{\partial x}\left(\rho \dfrac{\partial \varphi}{\partial x}\right) + \dfrac{\partial}{\partial y}\left(\rho \dfrac{\partial \varphi}{\partial y}\right) = 0$	$\dfrac{\partial}{\partial x}\left(k \dfrac{\partial T}{\partial x}\right) + \dfrac{\partial}{\partial y}\left(k \dfrac{\partial T}{\partial y}\right) = 0$
Continuity in terms of stream function	$\dfrac{\partial}{\partial x}\left(\dfrac{1}{\gamma} \dfrac{\partial i}{\partial x}\right) + \dfrac{\partial}{\partial y}\left(\dfrac{1}{\gamma} \dfrac{\partial i}{\partial y}\right) = 0$	$\dfrac{\partial}{\partial x}\left(\dfrac{1}{\rho} \dfrac{\partial \dot{m}}{\partial x}\right) + \dfrac{\partial}{\partial y}\left(\dfrac{1}{\rho} \dfrac{\partial \dot{m}}{\partial y}\right) = 0$	$\dfrac{\partial}{\partial x}\left(\dfrac{1}{k} \dfrac{\partial q}{\partial x}\right) + \dfrac{\partial}{\partial y}\left(\dfrac{1}{k} \dfrac{\partial q}{\partial y}\right) = 0$
	Special Differential Equations ($\lambda =$ constant)		
Laplace in vectors, u	$\dfrac{\partial E_x}{\partial x} + \dfrac{\partial E_y}{\partial y} = 0$	$\dfrac{\partial u_1}{\partial x} + \dfrac{\partial u_2}{\partial y} = 0$	$\dfrac{\partial H_x}{\partial x} + \dfrac{\partial H_y}{\partial y} = 0$
Laplace in potentials, φ	$\dfrac{\partial^2 V_e}{\partial x^2} + \dfrac{\partial^2 V_e}{\partial y^2} = 0$	$\dfrac{\partial^2 \varphi}{\partial x^2} + \dfrac{\partial^2 \varphi}{\partial y^2} = 0$	$\dfrac{\partial^2 T}{\partial x^2} + \dfrac{\partial^2 T}{\partial y^2} = 0$
Laplace in streams, ψ	$\dfrac{\partial^2 i}{\partial x^2} + \dfrac{\partial^2 i}{\partial y^2} = 0$	$\dfrac{\partial^2 \dot{m}}{\partial x^2} + \dfrac{\partial^2 \dot{m}}{\partial y^2} = 0$	$\dfrac{\partial^2 q}{\partial x^2} + \dfrac{\partial^2 q}{\partial y^2} = 0$

resistivity of the copper. Thus simulation of boundary conditions was difficult, and the balance-indication scheme was insensitive.

Today, Teledeltos paper, originally developed for use in Western Union facsimile transmitting and recording apparatus, is widely employed as the conducting sheet. The paper is of carbon- or graphite-impregnated stock approximately 0.004 in. thick. Its resistivity is on the order of 4000 Ω per square. There are some anisotropic variations in this resistance, since the resistance measured lengthwise of the roll is lower than the crosswise resistance. This variation is on the order of ±3% of the mean value. Changes in humidity likewise produce anomalies in the resistance of the paper.

The conducting sheet is cut so that it is geometrically similar to the field in question. The larger the size, the less important are inaccuracies incurred in cutting the boundaries of the sheet. Equipotential boundaries are usually obtained by painting on the Teledeltos paper with silver paint (1 to 4 Ω per square).

A Wheatstone bridge circuit is used to sense potentials within the electric field set up in the conducting sheet [3–6]. As the battery current is drained, the sensitivity of the measurement decreases. A standard cell may be incorporated to "standardize" the current. This requires that a current-adjusting potentiometer be used to balance out the standard cell voltage.

A potentiometer circuit is used to sense potential gradients in the conducting sheet. Two independent potentials are arranged to oppose each other, making possible a null balance. A combined potential-potential gradient circuit for a conducting sheet analog control box is shown in Figure 2.6.

Example 2.4. Theoretical solutions are available for certain simple geometric shapes. We now indicate how to compare theoretical solutions with

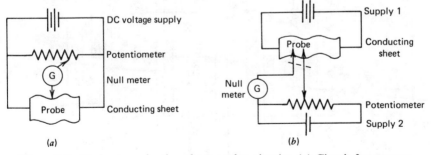

Figure 2.6 Various conducting sheet analog circuits. (*a*) Circuit for measuring potentials. (*b*) Circuit for measuring potential gradients. (*c*) Circuit for measuring both potentials and gradients.

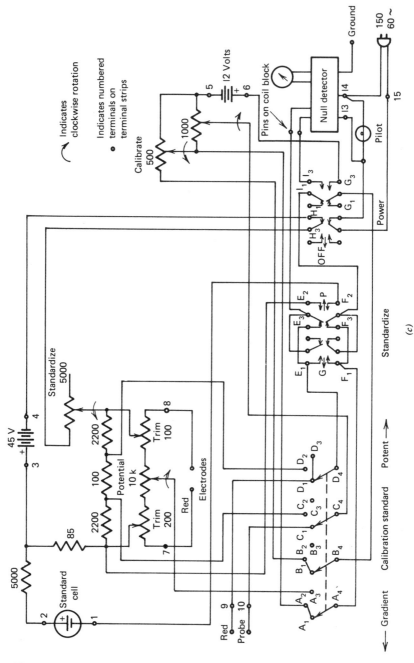

Figure 2.6 Continued

analog results for a circular cylinder [7]. The flow pattern for transverse flow disturbed by a circular cylinder can be described in terms of the velocity potential and stream function as

$$\frac{\varphi}{VR} = \cos\theta\left(\frac{r}{R} + \frac{R}{r}\right)$$

and (2.25)

$$\frac{\psi}{VR} = \sin\theta\left(\frac{r}{R} - \frac{R}{r}\right).$$

In terms of the velocity components, these equations are

$$u_r = \frac{\partial\varphi}{\partial r} = \frac{1}{r}\frac{\partial\psi}{\partial\theta} = V\cos\theta\left[1 - \frac{1}{(r/R)^2}\right]$$

and (2.26)

$$u_\theta = \frac{1}{r}\frac{\partial\varphi}{\partial\theta} = -\frac{\partial\psi}{\partial r} = -V\sin\theta\left[1 + \frac{1}{(r/R)^2}\right].$$

Boundary conditions satisfied by this flow are

$$\left(\frac{\varphi}{VR}\right)_{r=R} = 2\cos\theta$$

and (2.27)

$$\left(\frac{\psi}{VR}\right)_{r=R} = 0.$$

This means that the potential at the surface of the cylinder varies as a cosine wave with an amplitude of 2, and that the stream function has a constant value at the surface of the cylinder. The latter condition requires a cylinder with a nonconducting surface in the electrical analog. Other boundary conditions are

$$\left(\frac{u_\theta}{V}\right)_{r=R} = \frac{1}{VR}\frac{\partial\varphi}{\partial\theta} = -2\sin\theta$$

and (2.28)

$$\left(\frac{u_r}{V}\right)_{r=R} = 0.$$

This means that the velocity distribution around the surface of the cylinder varies as a sine wave with an amplitude of 2, and that the radial velocity

42 Solutions to Ideal-Incompressible Pipe Flow

component is zero at the cylinder surface. A final boundary condition is

$$\left(\frac{u_r}{V}\right)_{\substack{r=\infty \\ \theta=0}} = \left(\frac{u_\theta}{V}\right)_{\substack{r=\infty \\ \theta=\pi/2}} = \left(\frac{u_x}{V}\right)_{x=\infty} = 1. \qquad (2.29)$$

This means that far from the cylinder the various velocity components approach the free stream velocity, V.

Thus, in the analog, two dimensionless quantities are significant: the potential at the surface of the cylinder (φ/VR), and the tangential velocity component at the cylinder surface (u_θ/V). Using analogy A, we can evaluate the potential experimentally as $\varphi = mV_e$, where V_e is the dimensionless electric potential (referred to the total voltage across the analog), and m is a constant equal to Vl, where V is the free stream velocity and l is the length between electrodes. Then

$$\left(\frac{\varphi}{VR}\right)_{r=R} = \frac{1}{R} V_e = 2\cos\theta, \qquad (2.30)$$

and, since fluid velocities are represented by voltage gradients in the analog,

$$\left(\frac{u_\theta}{V}\right)_{r=R} = \frac{(\Delta V_e)_{\text{surface}}}{(\Delta V_e)_{\text{free stream}}} = -2\sin\theta. \qquad (2.31)$$

There are several possible methods for evaluating (2.31). They include numerical differentiation of $(l/R)V_e$, graphical differentiation of $(l/R)V_e$, and electrical differentiation, directly in the analog, by a gradient circuit and a two-electrode probe. Repeated tests have indicated that experimental velocity potentials agree with theoretical potentials within $\pm 1\%$, while experimental velocities (i.e., potential gradients) agree with theoretical velocities within $\pm 5\%$. These figures indicate the degree of confidence we can place in analog results for potentials and potential gradients (see Figure 2.7).

By way of further illustration of the type of results that can be obtained by an electrical analog, we also show the equipotential map of the second lens of an electron gun, as obtained from an electric field analog [8], in Figure 2.8. Similarly, the temperature distribution in a gas turbine rotor, as determined by an analog [9], is shown in Figure 2.9.

When fluid density is a variable, as considered in the next chapter, or when three-dimensional effects are important, an electrolytic analog has

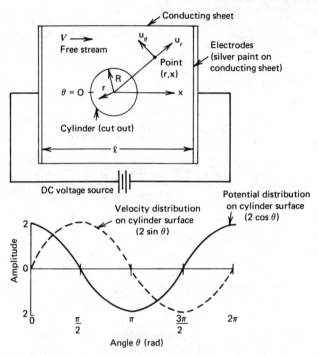

Figure 2.7 Simulation of cylinder in fluid flow field, and results from analogy A.

been devised [10] which uses depth variation to simulate density and/or area variations.

2.5.3 The Graphical Flow Net

Whereas flow problems involving regular boundaries can best be handled by numerical means (to be discussed) or even by analytical means, the Laplace equation for flow problems involving irregular boundaries is usually approximated by analogical methods (just discussed) or by graphical techniques.

The graphical solution is obtained by constructing a flux plot. Briefly, equipotentials (i.e., φ lines) are drawn perpendicular to solid boundaries. Streamlines (i.e., ψ lines) are drawn parallel to the solid boundaries. Thus equipotentials are perpendicular to streamlines, and an orthogonal system of curves is formed. Such a flow net is shown in Figure 2.10.

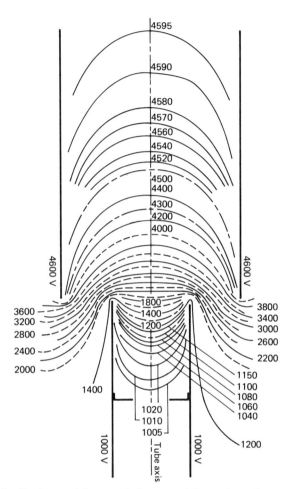

Figure 2.8 Equipotential map of the second lens of an electron gun (after Zworykin [8]).

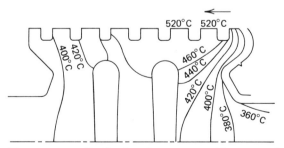

Figure 2.9 Temperature distribution in a gas turbine rotor body (after Baumann [9]).

Two- and Three-Dimensional Solutions   45

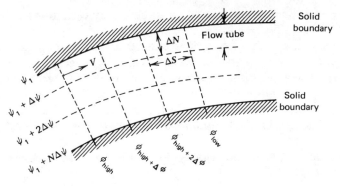

$$\dot{m}_{\text{tube}} = \rho \Delta N \frac{\Delta \varphi}{\Delta S},$$

where $\Delta \varphi / \Delta S = V$.

$$\frac{\dot{m}_{\text{NBE}}}{\rho \Delta \varphi} = \frac{\Delta N}{\Delta S},$$

where squares result if $\Delta N / \Delta S = 1$.

$$\Delta \varphi = \frac{\varphi_{\text{high}} - \varphi_{\text{low}}}{M},$$

M is the number of equipotential intervals.

$$\dot{m}_{\text{total}} = N \dot{m}_{\text{tube}},$$

where N is the number of flow tubes.

$$\dot{m}_{\text{total}} = N \rho \Delta \varphi \quad \text{for a square grid.}$$

$$\dot{m}_{\text{total}} = \left(\frac{N}{M}\right) \rho (\varphi_{\text{high}} - \varphi_{\text{low}}),$$

where N/M is the shape factor of the flux plot since it involves only geometric factors.

Figure 2.10 Various flow definitions applied to a two-dimensional flow net.

Note that our goal in drawing flux plots is to form *squares* such that the change between adjacent equipotential lines and adjacent streamlines is identical, that is, $\Delta \varphi = \Delta \psi$ and hence $\Delta S = \Delta N$. For the incompressible case, when squares are formed, we have by (2.19)

$$V \simeq \frac{\Delta \varphi}{\Delta S} \simeq \frac{\Delta \psi}{\Delta N}, \qquad (2.32)$$

where the approximations are indicated because ΔS and ΔN are finite. Naturally, with such finite spacing, we cannot expect the squares to be perfect. Therefore, by trial and error, involving many erasures, we sketch

46 Solutions to Ideal-Incompressible Pipe Flow

and improve the flow net according to boundary conditions until a net of near-squares is obtained [5].

From the discussion above, it is clear that fluid flow problems can be solved once the velocity potentials are available at every grid point on the flow map. These constitute an approximation to the Laplace equation and thus an approximation to the potential solution. The velocity distribution is then obtained from (2.32), and the pressure distribution from these velocities via the Bernoulli equations of (2.1) and/or (2.5).

Example 2.5. Estimate the velocity distribution in a miter elbow, as shown in Figure 2.11.

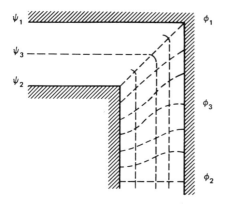

Figure 2.11 Flow net for a miter elbow.

Solution:

1. Make a square flow net within the solid boundaries.
   A. First, construct the vertex φ line as shown.
   B. Then bisect the ψ_1 and ψ_2 space by the constant ψ_3 line.
   C. Take advantage of symmetry, and work only within one section of the elbow.
   D. Bisect the φ_1 and φ_2 space by the constant φ_3 line, observing perpendicularity at solid boundaries and across each ψ line.
   E. Improve the plot to get near-squares by adding φ and ψ lines as shown (erasures may be required).
2. From the resulting flow net, observe that the potential gradient $(\Delta\varphi/\Delta S)$ is greater on the inner wall than on the outer wall (because, for the same $\Delta\varphi$, ΔS is smaller on the inner wall). Conclude that velocity is greater on the inner wall.
3. The gradient in φ in any direction is the velocity in that direction.

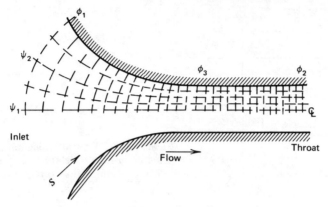

Figure 2.12 Flow net for a flow nozzle.

Example 2.6. Construct a flow net (i.e., a flux plot) for a flow nozzle, as shown in Figure 2.12.

Solution:

1. Divide the flow passage in two by a ψ line of symmetry.
2. Construct equipotentials perpendicular to solid boundaries and to ψ lines.
3. Add ψ and φ lines to improve the plot by getting near-squares.
4. Conclude that, where spacing between equipotentials is large (as at inlet), the velocity is small, whereas at the throat the velocity is large.
5. Thus, from the flow net, we get $u_S = \Delta\varphi/\Delta S$, which is the Laplacian potential distribution.

2.5.4 Numerical Solutions

Relaxation Methods [11, 12]. In relaxation methods, values of φ are first *assumed* at every point of intersection throughout a flow map, which in this case is simply a square grid superimposed over the flow boundaries. The initial assumed value of φ at a given grid point (or node) is improved (or relaxed) in general by setting it equal to the *average* of the values of φ at the four neighboring grid points. Actually, the method is complicated slightly in that special formulas are required to get average values of φ near the boundaries. Of course, it is best to use a high speed digital computer for these reductions, but in principle the numerical relaxation is straightforward.

48 Solutions to Ideal-Incompressible Pipe Flow

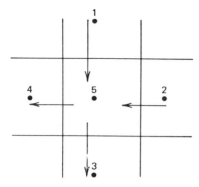

Figure 2.13 Typical node and notation in a continuity relaxation grid (arrows indicate assumed flow direction).

Two approaches are possible in applying relaxation methods to flow problems. In one method we look at the flow map from a *continuity* viewpoint, that is, we apply $\dot{m} = \rho A\, \partial \varphi / \partial x$. In the other method we consider the *energy* viewpoint, which involves the Laplace equation.

Continuity Viewpoint (refer to Figure 2.13)

1. A square grid is superimposed on the flow field.
2. The potential (φ) is specified at the nodes, which are the centers of each square.
3. In the steady state the net flow rate ($\dot{m}$) into each square must equal zero, that is, by continuity

$$\dot{m}_{in} = \dot{m}_{out}$$

$$\dot{m}_{1,5} + \dot{m}_{2,5} = \dot{m}_{5,3} + \dot{m}_{5,4}$$

$$\frac{\rho \Delta x (\varphi_1 - \varphi_5)}{\Delta y} + \frac{\rho \Delta y (\varphi_2 - \varphi_5)}{\Delta x} = \frac{\rho \Delta x (\varphi_5 - \varphi_3)}{\Delta y} + \frac{\rho \Delta y (\varphi_5 - \varphi_4)}{\Delta x}.$$

4. When $\Delta y = \Delta x$, and when ρ is the same in all directions,

$$\varphi_1 - \varphi_5 + \varphi_2 - \varphi_5 = \varphi_5 - \varphi_3 + \varphi_5 - \varphi_4$$

or

$$\varphi_5 = \frac{\varphi_1 + \varphi_2 + \varphi_3 + \varphi_4}{4}, \qquad (2.33)$$

which indicates that the new (relaxed) potential at a given node is the *average* of the surrounding potentials.

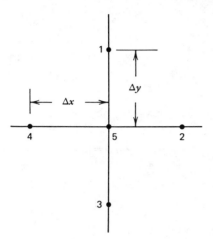

Figure 2.14 Typical node and notation in an energy relaxation grid.

5. Thus, by assuming an initial φ distribution, the final distribution can be determined by relaxing the field, that is, by getting new φ's in terms of surrounding φ's. Of course, this is a highly repetitive process, meaning that many new average values must be obtained until the change in the φ's is negligibly small.

Energy Viewpoint (refer to Figure 2.14)

1. A square grid is once more superimposed on the flow field.
2. The potential (φ) is specified at the nodes.
3. In the steady state, from the Laplace viewpoint, (2.22) applies. This *differential* energy equation is written as the *difference* equation [13, 14]:

$$\frac{\Delta^2 \varphi}{(\Delta x)^2} + \frac{\Delta^2 \varphi}{(\Delta y)^2} = 0. \qquad (2.34)$$

4. In terms of Figure 2.14, we have

$$\frac{\Delta \varphi}{\Delta x} = \frac{\varphi_4 - \varphi_5}{\Delta x}$$

and (2.35)

$$\frac{\Delta}{\Delta x}\left(\frac{\Delta \varphi}{\Delta x}\right) = \frac{(\varphi_4 - \varphi_5) - (\varphi_5 - \varphi_2)}{(\Delta x)^2}.$$

50 Solutions to Ideal-Incompressible Pipe Flow

Similarly,

$$\frac{\Delta}{\Delta y}\left(\frac{\Delta \varphi}{\Delta y}\right) = \frac{(\varphi_3 - \varphi_5) - (\varphi_5 - \varphi_1)}{(\Delta y)^2}. \tag{2.36}$$

5. Inserting (2.35) and (2.36) into (2.34), at $\Delta y = \Delta x$, we obtain

$$(\varphi_4 - \varphi_5) - (\varphi_5 - \varphi_2) + (\varphi_3 - \varphi_5) - (\varphi_5 + \varphi_1) = 0$$

or

$$\varphi_5 = \frac{\varphi_1 + \varphi_2 + \varphi_3 + \varphi_4}{4}, \tag{2.33}$$

which again indicates that the *new* potential at any node is simply the average of the surrounding potentials.

Thus the continuity approach and the energy approach yield consistent results.

Special Boundaries. We have looked only at interior points in the flow field, that is, where four nodes surround the node in question. Special boundary conditions have been treated in the literature (e.g., [15] and [16]), and we need not go into them in any detail here. However, to illustrate the approach briefly, we consider two such special boundaries: the straight three-point boundary, and the sloped two-point boundary.

Straight Three-Point Boundary. By continuity, for $\Delta x = \Delta y$ and for $\rho =$ constant, we have for Figure 2.15:

$$(\varphi_1 - \varphi_5) + (\varphi_2 - \varphi_5) = (\varphi_5 - \varphi_3)$$

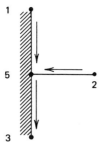

Figure 2.15 Straight three-point boundary with continuity notation.

Figure 2.16 Sloped two-point boundary with continuity notation.

or

$$\varphi_5 = \frac{\varphi_1 + \varphi_2 + \varphi_3}{3}, \qquad (2.37)$$

which once more indicates that φ_{new} is the *average* of the φ's of the surrounding nodes.

Sloped Two-Point Boundary. By continuity, for $\Delta x = \Delta y$ and for $\rho = $ constant, we have for Figure 2.16:

$$\varphi_2 - \varphi_5 = \varphi_5 - \varphi_3$$

or

$$\varphi_5 = \frac{\varphi_2 + \varphi_3}{2}, \qquad (2.38)$$

which again indicates that φ_{new} is the *average* of the φ's of the surrounding nodes.

Any boundary problem can, of course, also be handled by the more general Laplace notation. For example, the sloped two-point boundary problem is handled in the Laplace notation as follows (see Figure 2.17):

$$\frac{\Delta}{\Delta x}\left(\frac{\Delta \varphi}{\Delta x}\right)_N = (-)\left[\frac{\varphi_{N,M} - \varphi_{N+1,M}}{(\Delta x)^2}\right] \qquad (2.39)$$

and

$$\frac{\Delta}{\Delta y}\left(\frac{\Delta \varphi}{\Delta y}\right)_M = +\left[\frac{\varphi_{N,M-1} - \varphi_{N,M}}{(\Delta y)^2}\right]. \qquad (2.40)$$

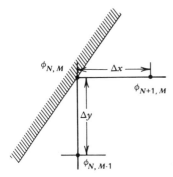

Figure 2.17 Sloped two-point boundary with Laplace notation.

Combining (2.39) and (2.40) according to the Laplace difference equation (2.34), with $\Delta x = \Delta y$, we obtain

$$-\varphi_{N,M} + \varphi_{N+1,M} + \varphi_{N,M-1} - \varphi_{N,M} = 0$$

or

$$\varphi_{N,M} = \frac{\varphi_{N+1,M} + \varphi_{N,M-1}}{2}, \qquad (2.41)$$

once more confirming the continuity results.

General Relaxation Rules. A few general rules can be given for relaxing two-dimensional potential flow fields.

1. Relax the potential value of each node by *averaging* the potentials of the surrounding nodes.
2. Use the *latest* available values of potentials for the averaging to speed the convergence.
3. Continue to relax (update) the potentials throughout the flow field until the values do not change beyond an uncertainty you are willing to accept.

Because relaxation methods are so common and require computer application, in general, they will be discussed no further here.

Finite Element Methods [17]. Another technique fast coming into favor is called the method of finite elements. In this approach the grid need not be square, but makes use of arbitrary elements made up of straight lines that often form triangles. This method will be illustrated in greater detail in the next section when we consider the potential solution to nozzle flow by the finite element technique.

2.6 SEVERAL IDEALIZED SOLUTIONS

By way of illustrating the ideas we have been discussing in this chapter, the ideal solutions to several typical piping element flow problems are given in this section. The potential flow through a nozzle is solved by a finite element numerical technique, the orifice contraction coefficient problem is solved by application of the momentum equation used together with a force defect factor, and the ideal flow through an abrupt enlargement is solved for the pressure recovery by means of the one-dimensional energy equation.

2.6.1 Potential Solutions to an ASME Nozzle

For two-dimensional flow through a nozzle, we could use the conducting sheet analog of Section 2.5.2, the graphical flow net solution of Section 2.5.3, or the numerical relaxation method of Section 2.5.4. However, the flow through a pipe nozzle is inherently three-dimensional with axial symmetry. Hence we must resort to the electrolytic tank analog [10], where depth variations represent changes in flow area with radius, to computerized relaxation techniques [11], or to finite element techniques [17].

Because of complex geometric boundary conditions in the tank bottom of an electrolytic analog, and because of the long time required for solution by the relaxation methods, we have chosen a finite element solution [18] for this illustration.

The boundary conditions for the finite element solution to the potential flow through an ASME nozzle are shown in Figure 2.18. These include the geometric boundaries and arbitrary numerical potential values set at the inlet and exit planes of the nozzle.

The finite element network is shown in part in Figure 2.19. It is at each of these nodes (where the lines intersect) that the Laplace equation is finally solved.

The resulting velocity potential map for the nozzle is shown in Figure 2.20, where we note that the spacing of constant φ lines is directly indicative of the potential velocity. Larger spacing between equipotentials signifies lower velocities; closer spacing of equipotentials, higher velocities.

To obtain nondimensional, and hence generalized, results, the following steps are taken.

54 Solutions to Ideal-Incompressible Pipe Flow

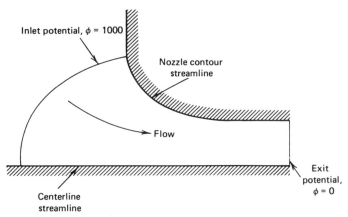

Figure 2.18 Boundary conditions for ASME nozzle analysis by finite element techniques.

1. Between each boundary node, the distance (ΔS) *along the contour* is established.
2. Then, from the potential solution, the quantity $\Delta \varphi / \Delta S$ is developed. This is the ideal fluid velocity along the nozzle contour. Of course, this velocity depends in value on the arbitrary potentials set as boundary conditions.
3. A generalized, nondimensional velocity is next obtained by referring the velocity at every surface node to that of some reference velocity from the same potential solution. Usually, the throat velocity is chosen as the reference value. Hence we obtain the ratio V/V_{throat}, which represents the *required, unique,* potential flow solution for the given geometric boundary conditions.

All of these quantities are given in Table 2.2. This solution for the ASME nozzle with a plenum inlet is valid only for conditions of an inviscid-incompressible fluid in irrotational flow. These numerical results are plotted for easy reference in Figure 2.21.

From the resulting velocity distribution along the contour, we could obtain the surface static pressure distribution via the Bernoulli equation. Conversely, if the pressure distribution were *measured* for the nozzle, we could work backwards and deduce the potential velocity distribution again via the Bernoulli equation, and thus avoid the need for a potential solution.

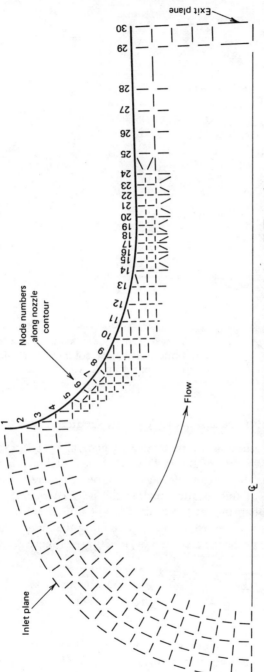

Figure 2.19 Portion of finite element network for ASME nozzle.

56 Solutions to Ideal-Incompressible Pipe Flow

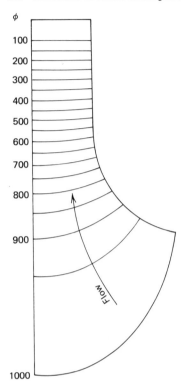

Figure 2.20 Equipotential results for the finite element solution to the Laplace equation for an ASME nozzle.

2.6.2 Momentum Solutions to Various Orifices

Flow through an orifice is characterized in general by a convergence of streamlines in the vicinity of the orifice. At some location downstream of the orifice, however, the streamlines again can be considered to be parallel, and the flow area at this location (called the *vena contracta*) is found in general to be less than that at the geometric opening of the orifice.

Much effort has been exerted toward predicting the amount of contraction of the fluid jet at the vena contracta [19, 20] (see Figure 2.22), where

$$C_c = \frac{A_3}{A_2}, \tag{2.42}$$

and C_c is called the contraction coefficient. For the inviscid-incompressible case, such as we are considering in this chapter, the flow coefficient (K), as

Table 2.2 Potential Solution to ASME Nozzle by Finite Element Technique

Node	Geometry			Potential Results			
	S	ΔS	S/D	φ	$\Delta\varphi$	$\Delta\varphi/\Delta S$	V/V_{throat}
1	0	—	—	1000.	—	—	—
2	0.504	0.504	0.042	988.554	11.446	22.710	0.219
3	1.056	0.552	0.130	974.933	13.621	24.676	0.238
4	1.625	0.569	0.233	957.594	17.339	30.473	0.294
5	2.243	0.618	0.322	935.621	21.973	35.555	0.343
6	2.658	0.415	0.408	917.860	17.761	42.798	0.412
7	3.066	0.408	0.477	896.772	21.088	51.686	0.498
8	3.401	0.335	0.539	877.611	19.161	57.197	0.551
9	3.769	0.368	0.598	855.471	22.190	60.163	0.580
10	4.325	0.556	0.674	817.967	37.504	67.453	0.650
11	4.891	0.566	0.768	774.655	43.312	76.523	0.737
12	5.398	0.507	0.857	728.490	46.165	91.055	0.878
13	5.906	0.508	0.942	680.853	47.637	93.774	0.904
14	6.416	0.510	1.027	630.644	50.209	98.449	0.949
15	6.674	0.257	1.091	604.265	26.379	102.642	0.989
16	6.931	0.258	1.134	577.438	26.827	103.981	1.002
17	7.182	0.251	1.176	550.608	26.830	106.892	1.030
18	7.433	0.251	1.218	523.606	27.002	107.578	1.037
19	7.683	0.250	1.260	496.539	27.067	108.268	1.043
20	7.933	0.250	1.301	469.442	27.097	108.388	1.045
21	8.183	0.250	1.343	442.505	26.937	107.748	1.038
22	8.433	0.250	1.385	416.015	26.490	105.960	1.021
23	8.683	0.250	1.426	389.733	26.282	105.128	1.013
24	8.933	0.250	1.468	363.566	26.167	104.668	1.009
25	9.433	0.500	1.530	311.444	52.122	104.244	1.005
26	9.933	0.500	1.614	259.448	51.996	103.992	1.002
27	10.433	0.500	1.697	207.516	51.932	103.864	1.001
28	10.933	0.500	1.780	155.617	51.899	103.748	1
29	11.933	1.000	1.906	51.868	103.748	103.748	1
30	12.432	0.500	2.030	0	51.868	103.736	1

58 Solutions to Ideal-Incompressible Pipe Flow

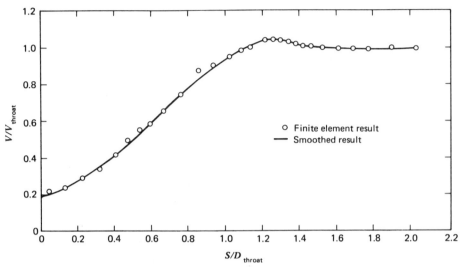

Figure 2.21 Potential solution to ASME nozzle for plenum inlet.

published by the ASME [19], has been related quite simply to the contraction coefficient by

$$C_c = \frac{K}{(1+\beta^4 K)^{1/2}}, \qquad (2.43)$$

where β represents the diameter ratio formed by the geometric orifice diameter (D_2) and the upstream inlet pipe diameter (D_1). Thus

$$\beta = \frac{D_2}{D_1}. \qquad (2.44)$$

Our purpose here is to write a momentum equation for a general orifice in terms of the contraction coefficient of (2.42) and (2.43), to illustrate its application in other than one-dimensional flow.

A momentum balance is written for the element of fluid enclosed between plane 1, upstream of the orifice, and plane 3, at the vena contracta (see Figure 2.22), The force exerted on the fluid stream tube which forms the boundary of the through-flow fluid upstream of the orifice is expressed in two parts: an idealized force based on p_1 (i.e., we first assume p_1 extends over the whole area, $A_1 - A_2$), and then a correction term based on a force defect [21], the latter to account for the nonuniform pressure distribution

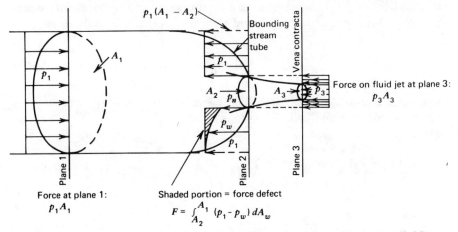

Figure 2.22 Generalized orifice flow (showing notation and forces on fluid).

on the fluid in the vicinity of the orifice. Thus the general expression

$$\Sigma F_x = Ma = \frac{\dot{m}\Delta V}{g_c} \tag{2.45}$$

(in the lbf-lbm system) becomes, in the inviscid-incompressible case,

$$p_1 A_1 - p_1(A_1 - A_2) + \int_{A_2}^{A_1}(p_1 - p_w) dA_w - p_3 A_2 = \frac{\dot{m}(V_3 - V_1)}{g_c}. \tag{2.46}$$

This can be reduced further to

$$(p_1 - p_3)A_2 + F_D = \frac{\dot{m}(V_3 - V_1)}{g_c}, \tag{2.47}$$

where F_D is the force defect as defined by the integral term of (2.46), and represents an additional driving force enhancing the momentum because p_w represents a *decreasing* pressure from the wall to the opening.

To completely solve this momentum equation, as is our goal, the force defect is evaluated in terms of (2.47) as follows. The first term of (2.47) can be written as

$$(p_1 - p_3)A_2 = \left(\frac{\rho V_3^2 - \rho V_1^2}{2g_c}\right) A_2 = \left(\frac{\dot{m}^2}{\rho A_2 g_c}\right)\left(\frac{1}{2C_c^2} - \frac{\beta^4}{2}\right), \tag{2.48}$$

60 Solutions to Ideal-Incompressible Pipe Flow

which follows directly, once we recognize that the total pressure of (2.5) is constant throughout the orifice, and of course when the continuity equation of (1.14) is applied. The second term of (2.47) can be written as

$$F_D = f\left(\frac{\dot{m}^2}{\rho A_2 g_c}\right), \qquad (2.49)$$

where f is a dimensionless force coefficient, and the term in parentheses is based on dimensional considerations. The third term of (2.47) can be written as

$$\frac{\dot{m}(V_3 - V_1)}{g_c} = \frac{\dot{m}}{g_c}\left[\left(\frac{\dot{m}}{\rho A_2}\right)\frac{1}{C_c} - \left(\frac{\dot{m}}{\rho A_2}\right)\beta^2\right] = \left(\frac{\dot{m}^2}{\rho A_2 g_c}\right)\left(\frac{1}{C_c} - \beta^2\right), \qquad (2.50)$$

which follows directly from the continuity expression ($\dot{m} = \rho A_2 V_2 =$ constant).

When these three terms, (2.48), (2.49), and (2.50), are combined according to (2.47), there results the explicit solution for the force coefficient:

$$f = \left(\frac{1}{C_c} - \frac{1}{2C_c^2}\right) - \beta^2\left(1 - \frac{\beta^2}{2}\right) \qquad (2.51)$$

in terms of the contraction coefficient (C_c) of the orifice, as given by (2.43), and the diameter ratio (β), as given by (2.44). Thus the complete momentum equation of (2.47) can be solved.

We consider next several implications of (2.51), as applied to an orifice, a Borda reentrant tube, and a nozzle (see Figure 2.23). For simplicity we

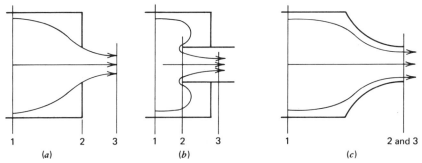

Figure 2.23 Flow through various orifices. (*a*) Orifice. (*b*) Borda tube. (*c*) Nozzle.

Several Idealized Solutions 61

consider flow from a plenum inlet only. A plenum signifies that β approaches zero since the inlet diameter (D_1) is much greater than the orifice diameter (D_2).

Orifice. Here the general momentum equation of (2.47) applies, as does (2.51). Thus

$$f_{\text{orifice}} = \frac{1}{C_c} - \frac{1}{2C_c^2}.$$

Borda Tube. Here the force defect term is missing from (2.47) because the lip of the reentrant tube prevents fluid accelerations as the opening is approached (see Figure 2.23). Thus

$$f_{\text{borda}} = 0,$$

and (2.47) yields

$$(p_1 - p_3)A_2 = \frac{\rho A_3 V_3}{g_c}(V_3 - V_1),$$

which, in the plenum case, where $V_1 \simeq 0$ and $p_1 \simeq p_t$, yields

$$C_{c_{\text{borda}}} = \frac{A_3}{A_2} = \tfrac{1}{2}.$$

Alternatively, when $F_D = 0$, then by (2.49)

$$f_{\text{borda}} = 0,$$

and (2.51) yields, in the plenum case,

$$\frac{1}{C_c} = \frac{1}{2C_c^2}$$

or

$$C_{c_{\text{borda}}} = \tfrac{1}{2}.$$

Nozzle. Here the general momentum equation applies, but there is no contraction of the jet in nozzle flow. Thus

$$C_{c_{\text{nozzle}}} = 1$$

62 Solutions to Ideal-Incompressible Pipe Flow

and

$$f_{\text{nozzle}} = \tfrac{1}{1} - \tfrac{1}{2} = \tfrac{1}{2}.$$

2.6.3 Ideal Pressure Recovery Across an Abrupt Enlargement

An abrupt enlargement in a piping section acts as a diffuser, that is, there is a rise in static pressure across the abrupt enlargement at the expense of a drop in kinetic energy [22].

It is instructive to consider this pressure recovery, both in the ideal case (as in this section) and in the real case (in Part III), when we consider losses across the various piping elements.

In the inviscid-incompressible case, we can write the Bernoulli equation across the abrupt enlargement (see Figure 2.24) as

$$\left(\frac{p_2 - p_1}{\rho}\right)_{\text{ideal}} = \frac{V_1^2 - V_2^2}{2g_c}. \tag{2.52}$$

Note that this one-dimensional approach is valid in spite of the fact that there are strong streamline curvatures in the free mixing region between inlet and exit, simply because at axial stations 1 and 2 the streamlines *are* straight and parallel.

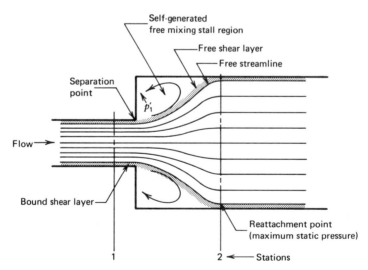

Figure 2.24 Abrupt enlargement, showing notation.

By continuity we have

$$\frac{V_2}{V_1} = \frac{A_1}{A_2} = \beta^2. \tag{2.53}$$

Combining (2.52) and (2.53) gives

$$\left(\frac{p_2 - p_1}{\rho}\right)_{\text{ideal}} = \frac{V_1^2}{2g_c}(1 - \beta^4), \tag{2.54}$$

which expresses the ideal pressure recovery across the abrupt enlargement in terms of the inlet velocity head ($V_1^2/2g_c$) and the diameter ratio (β).

For a diameter increase of 2 to 1, that is, for $\beta = 0.5$, for example, the ideal pressure recovery is seen to be 15/16 of the inlet velocity head. Later, for comparison, this same problem will be done for the case where losses are admitted (See Example 7.5).

REFERENCES

1. R. P. Benedict, "Analog simulation," *Electro-Technol.*, Science and Engineering Series No. 60, December 1963, p. 73.
2. G. Kirchhoff, *Poggendorfs Ann. Phys.*, Vol. 64, 1845, p. 497.
3. W. W. Soroka, *Analog Methods in Computation and Simulation*, McGraw-Hill, 1954.
4. W. J. Karplus, *Analog Simulation*, McGraw-Hill, 1958.
5. W. Li and S. Lam, *Principles of Fluid Mechanics*, Addison-Wesley, 1964, p. 112.
6. V. L. Streeter and E. B. Wylie, *Fluid Mechanics*, 6th ed., McGraw-Hill, 1975, p. 412.
7. W. B. Brower, "The application of the electrical analogy to two-dimensional problems in aeronautics," Rensselaer Polytechnic Institute Report, Department of Aeronautical Engineering, TRAE 5406, 1953.
8. V. K. Zworykin, et al, *Electron Optics and the Electron Microscope*, John Wiley, 1945.
9. H. Baumann, "The determination of temperature distribution in gas turbine rotor bodies and cylinders by the electrolytic tank method," *Brown-Boveri Rev.*, May-June 1953.
10. R. P. Benedict and C. A. Meyer, "Electrolytic tank analog for studying fluid flow fields within turbomachinery," *ASME Pap.* 57-A-120, December 1957.
11. R. V. Southwell, *Relaxation Methods in Engineering Science*, Oxford University Press, 1949.
12. D. N. deG. Allen, *Relaxation Methods*, McGraw-Hill, 1954.
13. H. W. Emmons, "The numerical solution of heat conduction problems," *Trans. ASME*, Vol. 65, August 1943, p. 607.

14. G. M. Dussinberre, "Numerical methods for transient heat flow," *Trans. ASME*, Vol. 67, November 1945, p. 703.
15. R. P. Benedict, "Transient heat flow," *Electro-Technol.*, Science and Engineering Series No. 36, December 1961, p. 94.
16. R. P. Benedict, "Two-dimensional transient heat flow," *Electro-Technol.*, May 1962.
17. R. H. Gallagher, J. T. Oden, C. Taylor, and O. C. Zienkiewicz, *Finite Elements in Fluids*, Vols. 1 and 2, John Wiley, 1975.
18. R. P. Benedict and J. S. Wyler, "Analytical and experimental studies of ASME flow nozzles," *Trans. ASME, J. Fluids Eng.*, September 1978, p. 265.
19. H. S. Bean, Ed. *Fluid Meters—Their Theory and Application*, Report of the ASME Research Committee on Fluid Meters, 6th ed., 1971.
20. R. P. Benedict, "Generalized contraction coefficient of an orifice for subsonic and supercritical flows," *Trans. ASME, J. Basic Eng.*, June 1971, p. 99.
21. D. A. Jobson, "On the flow of a compressible fluid through orifices," *Proc. Inst. Mech. Eng.*, Vol. 169, No. 73, 1955, p. 767.
22. R. P. Benedict, N. A. Carlucci, and S. D. Swetz, "Flow losses in abrupt enlargements and contractions," *Trans. ASME, J. Eng. Power*, January 1966, P. 73.

NOMENCLATURE

Roman

a acceleration
A area
C_c contraction coefficient
D pipe diameter
E total energy,
electric field intensity
f function
F force
g local gravitational acceleration
g_c gravitational constant
H temperature gradient
i electric current
K flow coefficient
l length
$\dot{m}$ mass flow rate
M mass

N distance
p static pressure
p_t total pressure
Q volumetric flow rate
r general radius
R outside radius
S distance
u velocity component
V directed velocity
V_e voltage
W external work
x, y, z Cartesian coordinates
Z vertical height

Greek

β diameter ratio
γ electrical conductivity
Δ finite difference
θ angle
λ characteristic function
ρ fluid density
φ potential
ψ stream function

Mathematical Symbols

∇ operator
∂ partial derivative
d total derivative

Subscripts

$1, 2, 3$ coordinates
r radial direction
x axial direction
θ circumferential direction

3
Solutions to Ideal-Compressible Pipe Flow

> ... thump, thump, thump, goes the pump in the chick in the egg on the second day after hen and rooster came to a decision... —*Gustav Eckstein*

3.1 GENERAL REMARKS

There is a striking difference between the compressible and the incompressible flow of fluids in pipes. In this chapter we apply the conservation equations of Chapter 1 to the flow in pipes of a hypothetical ideal-compressible fluid. Here too we review certain thermodynamic concepts as required for analyzing compressible pipe flow. Once more, our purpose is to gain insight into the basic behavior of pipe flow, now for compressible fluids. In other words, we examine here solutions to inviscid, variable density gas flow in pipes and through various piping components.

Since we are neglecting all frictional effects under the inviscid assumption, we will obtain only approximate solutions to gas flow in pipes.

3.2 THERMODYNAMIC CONCEPTS

Thermodynamics is a discipline dealing mainly with thermal energy, its transformations into other forms of energy, and the laws that govern such transformations. Basically, thermodynamics takes a *macroscopic* viewpoint of matter wherein the complicated substratum of molecular motions is consistently ignored. In thermodynamics the quantities heat, temperature, internal energy, and pressure, for example, receive no physical explanations in terms of bouncing billiard balls or dumbbell type molecules translating, rotating, or vibrating. Indeed, classical thermodynamics, having something positive to say only when statistical equilibrium reigns, and viewing matter on the macroscopic level only, might better be termed *thermostatics*.

The discipline of thermodynamics is built on certain *laws* which are simply axioms or postulates serving as the basis of study and taken as true. First, we will trace the logical development of these laws while forming no hypothesis as to the origins or meanings of the quantities involved in thermodynamics [1]. Some of the more important definitions used in this section are summarized in Table 3.1.

The concepts of pressure and work, so important in thermodynamics, are borrowed openly from mechanics, that is, from the study dealing

Table 3.1 Some Thermodynamic Definitions

Process: any change of state, that is, any change in any property. If the process can be graphed, it is the path of successive states through which the working substance passes.

System: a specific region (container) of real or imaginary boundaries, rigid or not, whose contents are to be analyzed as to transformations of energy within the boundaries and as to the passage of energy or matter across the boundaries.

Surroundings: the region outside boundaries of the specific system under study.

Isolated system: one in which no matter or energy crosses the boundaries, that is, the sum of all forms of energy within the system is conserved.

Closed system: one that is impermeable to matter, and in which only mechanical and thermal energy can cross the boundaries (this implies a *nonflow system*).

Adiabatic system: one that is isolated thermally from its surroundings, and in which only matter and mechanical energy can cross the boundaries (this implies *perfectly insulated real boundaries*).

Diathermic system: one that is isolated mechanically from its surroundings, and in which only matter and thermal energy can cross the boundaries (this implies *rigid boundaries*).

68 Solutions to Ideal-Compressible Pipe Flow

mainly with forces, their applications, and the laws that govern them. Very briefly, we take *pressure* as a macroscopic measure of the average compressive stress on the working substance. The *working substance* we define as the fluid under consideration, taken to be of a homogeneous and invariable chemical composition. Pressure may be expressed *empirically* in terms of an arbitrary base, such as gage or vacuum pressure, or it may be expressed *absolutely* in terms of a fixed zero base (i.e., with respect to a perfect vacuum). Pressure is thus a descriptive characteristic of the working substance, called, in thermodynamics, a *property*. Any two independent properties (volume is another example) establish the condition or *state* of the working substance.

By *work*, we mean an interaction between a system and its surroundings such that the sole effect on the surroundings can be expressed in terms of the displacement of a weight:

$$\mathbf{W} = \int W \cos\beta \, dZ, \tag{3.1}$$

where β is the angle between the force and the displacement.

It follows that the boundaries of the system must deform if work is involved. Furthermore, this work can be exerted in two directions: it can operate *on* the system, or it can be performed *by* the system. Specifically, work is said to be done by the system if the sole effect on the surroundings can be expressed in terms of the raising of a weight.

A differential amount of work is expressed as $\delta \mathbf{W}$ rather than $d\mathbf{W}$ to remind us that the inexact differential must be considered. We recall that the integral of an exact differential is independent of the path of integration; however, the work involved when the working substance passes from one state to another not only is a function of the end states (points) of the process but also, in general, is strongly dependent on the path followed. Thus work is, in general, a path function as opposed to a point function:

$$\int_1^2 \delta \mathbf{W} = \mathbf{W}_{1,2}. \tag{3.2}$$

It is meaningless to write $(\mathbf{W}_2 - \mathbf{W}_1)$ for the above integral since work represents energy in transition only. We do not speak of the work in a system at state 1 or 2, but only of how much work is involved in passing from state 1 to state 2; and, as we have pointed out, this depends on the specific path followed.

3.2.1 Zero-th Law

The concepts of thermal equilibrium and empirical temperature are basic to thermodynamics. These are clarified in the zero-th law. If there is a thermal interaction between system and surroundings, there generally will be a change of state of the working substance. However, it is a matter of experience that:

> If no observable change takes place in the pressure or volume of either the system or its surroundings when each is individually in contact with an auxiliary diathermic-closed system, the system and its surroundings are said to be in thermal equilibrium with the auxiliary system and with each other.

This generalization of experience introduces the concept of thermal equilibrium, and is the essence of what we will call the zero-th law.

By the zero-th law, thermometers (auxiliary diathermic-closed systems) are feasible. This law asserts the existence of a conserved quantity, *empirical temperature*—empirical because it is measured from an arbitrary base, as Celsius or Fahrenheit temperature. The law ensures that each particular state of thermal equilibrium (i.e., each *isotherm*) between the system and its surroundings can be labeled, although the assignment of numerical values to these isotherms is entirely arbitrary.

3.2.2 First Law

The concepts of internal energy and external heat are clarified by the first law. In addition, the concepts of friction and work are best examined in connection with the first law and its corollaries.

If there is a mechanical interaction between system and surroundings, there generally will be a change of state of the working substance. The work involved will, in general, be strongly dependent on the process. However, it is a matter of experience that:

> For any closed, adiabatic system, the work done is fixed by the end states of the working substance.

This generalization of experience, which we could not have expected in view of the general observation that work is a path function, is the essence of what we will call the first law of thermodynamics, following the general reasoning of Caratheodory [2–4].

The first law asserts the existence of a unique point function, the *internal energy*; empirical (i.e., relative) since its base point is entirely arbitrary (the zero datum is conventionally at 32°F). The law ensures that each particular state of the working substance can be given a characteristic number with respect to an arbitrary base, that is, the increase in internal energy accompanying any specific change of state taking place in a closed, adiabatic system can be realized experimentally simply by evaluating the work done on the system during the process:

$$\delta(\mathbf{W}_{on})_{closed,\ adiabatic} = dU$$

or

$$(\mathbf{W}_{on})_{1,2,closed,\ adiabatic} = U_2 - U_1, \qquad (3.3)$$

where, in this section, the internal energy is represented by the conventional symbol, U.

Mathematically, dU is an exact differential, meaning that its integral is determined solely by the end states of the process, independently of the process. Physically, whenever we mean, recognize, observe, or encounter $\delta(\mathbf{W}_{on})_{closed,\ adiabatic}$, we will think, say, or write dU instead, for convenience only. In keeping with this viewpoint, note that Kelvin first called the internal energy the mechanical energy [5].

Once the internal energy is realized in practice for a process that takes place within the restrictive closed, adiabatic system, we can write by the first law:

$$0 = dU - \delta(\mathbf{W}_{on})_{closed,\ adiabatic}. \qquad (3.4)$$

However, the adiabatic system is not generally encountered. The difference between the increase in internal energy of the working substance and the work done on the working substance in a closed system where the thermal-isolation restriction is removed (i.e., a *diabatic* system) need not equal zero but may have a definite numerical value, that is, in general,

$$0 \neq dU - \delta(\mathbf{W}_{on})_{closed}. \qquad (3.5)$$

This difference which may exist in a closed system because of an interaction between the system and its surroundings we define to be the *external heat transferred* or, more simply, the *heat*. This heat can be transferred across the system boundaries in two directions, that is, heat can be absorbed or rejected by the system. Specifically, the *heat absorbed* is defined as the increase in internal energy plus the work done *by* the closed system:

$$\delta Q_{added} = dU + \delta(\mathbf{W}_{by})_{closed}. \qquad (3.6)$$

Equation 3.6 defines heat: it may be taken as a corollary of the first law, and represents the complete *closed-system energy equation*. As a generalization of experience, there must be an empirical temperature difference between the system and its surroundings for an external heat transfer to be involved, and, of course, the system cannot be adiabatic.

A differential amount of heat is expressed as δQ rather that dQ to remind us that the inexact differential must be considered. This is so because the heat involved when the working substance passes from one state to another not only is a function of the end states of the process but also, in general, depends strongly on the path followed. Thus heat, like work, is a path function, and we write

$$\int_1^2 \delta Q = Q_{1,2}. \tag{3.7}$$

From the discussion of path functions, we know that it is meaningless to write $(Q_2 - Q_1)$ for the above integral. It is also incorrect to speak of the heat in a system at state 1 or 2 because heat, like work, is a measure of energy in transition only.

Since the concepts of internal energy and external heat are both defined in terms of the work done on a closed system, as indicated by (3.3) and (3.6), it is evident that the question of work and its evaluation has again become pertinent to this study. Work must be defined in such a way as to ensure the concreteness of the first law and its corollaries. Additional definitions are required first.

Wholly reversible process: a quasi-static change of state in the absence of mechanical interference, fluid mixing, turbulence, and unrestrained expansions in the system and its surroundings. Note that the quasi-static restriction as used here signifies a succession of statistical equilibrium states in which all properties of all parts of the system and its surroundings are constant with time, and thus implies the absence of fluid-viscosity effects, any finite unbalanced forces, and any finite empirical temperature differences within and between the system and its surroundings. A wholly reversible process is an idealization that, contrary to our experience, can proceed equally well in either direction from a given state, without external effects.

Internally reversible process: a quasi-static change of state within the system such that any irreversibilities are confined to the surroundings. The system, however, behaves as though the process is not reversible in the natural sense since permanent changes must be produced somewhere in the surroundings if the process is to proceed in either direction from a given state.

For any internally reversible process, the work done by a closed system is defined from mechanics, through (3.1), as

$$\delta(W_{by})_{closed,\ reversible} = p\,dV, \tag{3.8}$$

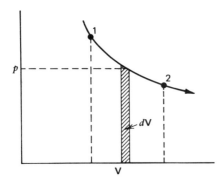

Figure 3.1 Area under pV curve is work $\left[(W_{by})_{\text{closed, reversible}} = \int_1^2 p\, dV \right]$.

where the volume must increase whenever work is done by the system, and the pressure must be absolute. This particular work term we call the *elastic work* since the boundaries of the closed system must deform if it is to exist. Like all work terms, the elastic work is in general a path function strongly dependent on the specific process involved.

An internally reversible process (and also a wholly reversible one) always can be drawn as a continuous curve on a property diagram, and the work involved in a closed system is always $\int p\, dV$, that is, the area under the p versus V curve in Figure 3.1. Conversely, an internally irreversible process never can be drawn as a continuous curve on a property diagram, and the work involved in a closed system may or may not be well approximated by $\int p\, dV$ (of course, it is meaningless to speak of the area under the p versus V curve since there is no curve). Thus, if a process is drawn as a continuous curve on a property diagram, internal reversibility (at least) is implied. See Figures 3.2 and 3.3.

Once the elastic work is realized in practice for the restrictive, internally reversible change of state of the working substance in a closed system, we have by (3.8)

$$0 = p\, dV - \delta(W_{by})_{\text{closed, reversible}}. \tag{3.9}$$

However, even the internally reversible process is not generally encountered. The difference between the elastic work done by a closed system and the work done by the same system when the internal reversible restriction is removed will not equal zero, in general, but will have a definite numerical value:

$$0 \neq p\, dV - \delta(W_{by})_{\text{closed}}. \tag{3.10}$$

Thermodynamic Concepts 73

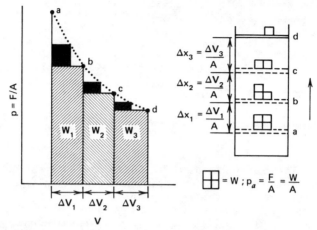

Figure 3.2 The reversible process as a limit. Imagine a weight, divided into four units, on a piston initially at rest at state a. By successively removing one unit of weight, with equilibrium established between each step, states b, c and d result, where the actual works involved are

$$W_1 = F_1 \Delta x_1 = \left(\tfrac{3}{4}W\right)\left(\frac{\Delta V_1}{A}\right) = p_b \Delta V_1,$$

$$W_2 = F_2 \Delta x_2 = \left(\tfrac{1}{2}W\right)\left(\frac{\Delta V_2}{A}\right) = p_c \Delta V_2,$$

$$W_3 = F_3 \Delta x_3 = \left(\tfrac{1}{4}W\right)\left(\frac{\Delta V_3}{A}\right) = p_d \Delta V_3,$$

as indicated by shaded areas on graph.

If the experiment is repeated, removing only one-half unit of weight per step, the darkened areas on the graph represent additional work involved.

If the experiment were repeated, removing an infinitesimal fraction of the weight per step, a plot of the states involved would approach a continuous curve as indicated by the dots from a to d. Such quasi-static changes of state represent a reversible process. Since the actual work approaches the area under the smooth curve, we note that the actual work approaches the reversible work as a limit.

This difference which will exist in a closed system in the presence of mechanical energy transfer or elastic work we define to be the internal heat generated or, more simply, the *friction*. Thus

$$\delta F = p\, dV - \delta(W_{by})_{closed}, \tag{3.11}$$

where the friction term is influenced by such factors within the system as mechanical interference between moving parts, shearing effect arising

74 Solutions to Ideal-Compressible Pipe Flow

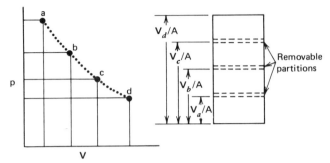

Figure 3.3 The irreversible process. Imagine a series of unrestrained expansions from initial equilibrium states, as indicated above by states b, c and d, obtained by successively removing the partitions.

If the experiment were repeated, using a large number of partitions, a plot of the states involved would again appear to approach a continuous curve as indicated by the dots from a to d. However, such unrestrained expansions are highly irreversible.

The mathematical tendency to pass a smooth curve through the many state points must be resisted, since the actual work (*on* or *by* the system) is zero, and the series of states approaches no meaningful thermodynamic limit.

because of a nonstatic process involving a viscous fluid, fluid mixing, turbulence, and unrestrained expansions. All-inclusive as the friction term appears, however, it is not the criterion of reversibility because, even in the constant-volume, closed-diathermic system where F is necessarily zero, there may be irreversibilities in the presence of thermal energy transfer. These will be discussed more fully under second law considerations. Thus the absence of friction is a necessary but not at all sufficient condition for reversibility.

As a generalization of experience, friction is a one-way street: either it is absent during a process, or it is positive, and friction involves happenings *within* system boundaries as opposed to work and heat, which are concerned with energy transfers *across* system boundaries in either direction. Like work and heat, friction is inherently a path function; hence it is instructive to note that (3.11) can be written also as

$$d\mathbf{V} = \frac{\delta F + \delta(\mathbf{W}_{by})_{closed}}{p}, \qquad (3.12)$$

where volume is inherently a point function. In other words, absolute pressure plays the role of an integrating factor in (3.12), mathematically transforming the inexact quantity, $(\delta F + \delta \mathbf{W})$, to one that is exact.

An important corollary of the first law, and a useful form of the energy equation for a closed (nonflow) system, can be obtained by combining

(3.6) and (3.11) to obtain

$$\delta Q + \delta F = dU + p\,d\mathbf{V}. \qquad (3.13)$$

Equation 3.13 brings together all the important concepts of thermodynamics so far discussed, namely, heat, friction, internal energy, and elastic work. Note that δF, in spite of its limitations, is the only term concerning irreversibility that must be considered in any of the energy equations. Equation 3.13 indicates that work can be converted, by means of internal heat generated (i.e., friction), to produce the same thermal effects on the system as an external heat transfer. In keeping with this viewpoint, Joule stated, "The most frequent way in which limiting force (work) is converted into heat is by means of friction..." [6].

3.2.3 General Energy

We have discussed the work done on or by a closed system and its relation to the nonflow energy equation. However, in pipe flow we are interested in open systems, and we inquire next as to the work involved and its relation to the energy equation for the more general open system under *steady flow* conditions, that is, when neither the mass nor the total energy of the working substance changes with time.

As already mentioned, work can be done on or by the working substance in any system only by boundary deformation. In a closed system, work crossing system boundaries is accounted for solely by friction and elastic work, as indicated by (3.11). In the more general piping system, however, additional forms of energy may be needed to account for work crossing system boundaries. The three energy terms that we will consider in connection with a moving fluid are flow work, kinetic energy, and potential energy. All of these are point functions.

Flow work: a work term arising solely because the working substance crosses system boundaries. From fluid mechanics a change in flow work involves only the fluid pressure and volume at system boundaries:

$$d\mathbf{W}_{\text{flow}} = d(p\mathbf{V}). \qquad (3.14)$$

(Note: In the nonflow system, this term may also exist, but it has no significance as a work term.)

Kinetic energy: an energy term arising solely because the working substance exhibits directed motion with respect to the real boundaries of the system. From mechanics a change in kinetic energy involves only the

76 Solutions to Ideal-Compressible Pipe Flow

fluid mass and its acceleration within system boundaries:

$$d(\text{K.E.}) = d\left(\frac{MV^2}{2g_c}\right). \tag{3.15}$$

Potential energy: an energy term arising solely because the working substance exhibits vertical displacement with respect to an arbitrary datum. From mechanics a change in potential energy involves only the fluid mass and its position within system boundaries:

$$d(\text{P.E.}) = \frac{d(MgZ)}{g_c}. \tag{3.16}$$

[*See also* (1.18)].

Thus the general equation for work done on the working substance is

$$\delta W_{on} = \delta(W_{on})_{closed} + dW_{flow} + d(\text{K.E.}) + d(\text{P.E.}) \tag{3.17}$$

or, upon substitution,

$$\delta W_{on} = \delta F - p\,dV + d(pV) + d\left(\frac{MV^2}{2g_c}\right) + \frac{d(MgZ)}{g_c}, \tag{3.18}$$

where any or all of the terms may be present in a given situation.

When (3.18), which represents a general accounting of mechanical energies, is combined with (3.13), which represents a general accounting of thermal energies, we obtain on a *per pound mass basis*

$$(\delta Q_{added} + \delta W_{on})_{\substack{external \\ effects}} = \left[du + d(pv) + \frac{V\,dV}{g_c} + \frac{g}{g_c}\,dZ\right]_{\substack{internal \\ effects}}, \tag{3.19}$$

where lower case u and v signify, respectively, internal energy per pound mass and volume per pound mass [see also (1.30)]. Equation 3.19 is called, with good reason, the *steady flow general energy equation* since it applies equally well to reversible and irreversible processes within piping systems. No friction term appears in (3.19) (i.e., internal irreversibilities in the presence of work transfer or elastic work are not recognized), simply because friction at the same time sets up counterbalancing work and heat terms, as shown by (3.13) and (3.18). (See Figure 3.4.)

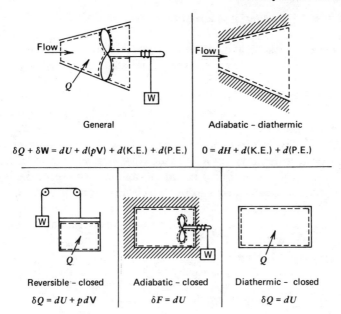

Figure 3.4 Thermodynamic system classifications.

3.2.4 Second Law

The concepts of absolute temperature and entropy are clarified by the second law. In addition, the dual nature of friction (now a work effect, now a heat effect) and its unidirectionality (the one-way-street idea), as met in the first law corollaries, are clarified by the second law.

Note that the story of thermodynamics has several loose ends at this point in the development. For instance, why didn't we encounter an absolute temperature function? The analogous absolute pressure was introduced, and actually was seen to be essential to thermodynamics. And why didn't we note an integrating factor for the inherent path function ($\delta Q + \delta F$) while we were doing so for the similar ($\delta W + \delta F$)? Recall, by the way, that absolute pressure was just such an integrating factor for the work and friction term. And, whereas we observed that some processes (namely, the irreversible ones) could not be represented by continuous curves on property maps, we came across no thermodynamics notion which indicated that some processes might not even be possible! In other words, the first law and its corollaries helped us only with our bookkeeping accounts concerning energy in its various forms; they did not help us to judge which processes were useful, or whether any limitations existed as to the useful effect of heat.* However, it is a matter of experience that:

For any adiabatic system, thermodynamic states exist that cannot be reached by any process from a given initial state.

This generalization of experience is the essence of what we will call the second law of thermodynamics, again following the general reasoning of Caratheodory [8].

The second law asserts the existence of two point functions of basic importance in thermodynamics, namely, the *entropy* and the *absolute temperature*, but the assertion is not at all obvious. The second law also ties up some of the loose ends by stating categorically that not all processes are possible. But, although we are forewarned, we are not yet armed. We cannot tell by looking at the law just which states are inaccessible.

To be more specific, we first note that internal energy, being a property, can be expressed in terms of any other two independent properties such as p and V. Thus we can rewrite (3.13) as

$$\delta Q + \delta F = dU + p\,dV = N\,d\sigma, \tag{3.20}$$

where N and σ are both functions of p and V alone. We can do this because we are assured mathematically that an integrating factor N exists for any two-variable linear differential equation; the resulting exact differential is then $d\sigma$.

However, N and σ are not yet identified. If the system is made up of more than one part, then the two variables, p and V, are no longer sufficient to define the state of the system. Mathematically, we should expect the existence of no integrating factor for the counterpart of (3.20) when it involves more than two variables. It is a profound generalization of experience that, if all parts of a system are temperature coupled (i.e., at a uniform empirical temperature), there will always be an integrating factor, which is necessarily a function of the empirical temperature and must have a fixed zero base, that is, an *absolute temperature*. Thus we meet the missing absolute temperature function (with the conventional symbol T), and at the same time we identify it with the missing integrating factor N. Note that the absolute temperature plays the same role for the inexact

---

*As early as 1824, Carnot [7] introduced the cycle concept, with his famous reversible engine operating between two isotherms and two adiabats, suggested that the way to judge the merits of any heat engine was to compare the maximum work obtainable with the heat absorbed, anticipated the first law by claiming that the maximum work was equivalent to the net heat available, and noted that this efficiency, $(\delta Q_2 - \delta Q_1)/\delta Q_1$, could be a function of the two *empirical* temperatures only, independently of the working substance.

$(\delta Q + \delta F)$ as absolute pressure does for the inexact $(\delta W + \delta F)$. The resulting exact differential, $d\sigma$, we define to be the differential *entropy* (with the conventional symbol dS). Thus

$$dS = \frac{\delta Q + \delta F}{T}. \tag{3.21}$$

Alternatively, entropy can be defined as

$$dS = dS_{\substack{\text{transferred} \\ \text{across boundaries}}} + dS_{\substack{\text{produced} \\ \text{within boundaries}}}, \tag{3.22}$$

where

$$dS_{\substack{\text{transferred} \\ \text{across boundaries}}} = \left(\frac{\delta Q}{T}\right)_{\substack{\text{evaluated at} \\ \text{boundaries}}}. \tag{3.23}$$

The entropy produced in the general case can only be given as the difference between the total entropy change and the entropy transferred. The often-quoted inequality first given by Clausius follows directly from (3.22) and (3.23). Thus

$$dS \geq \left(\frac{\delta Q}{T}\right)_{\substack{\text{evaluated at} \\ \text{boundaries}}}. \tag{3.24}$$

Neither (3.21) nor (3.22) is entirely satisfactory as a definition of entropy because the path of integration in the general irreversible process cannot be defined. However, since entropy is a point function, we must not overlook the important definition

$$dS = \left(\frac{\delta Q}{T}\right)_{\substack{\text{any reversible} \\ \text{path}}}. \tag{3.25}$$

Equation 3.25 ensures that the entropy change for any process (reversible or not) can be determined readily by evaluating the entropy change for any reversible process or series of processes between the actual end states. Although it is true that

$$\Delta S_{\text{irreversible}} = \Delta S_{\text{reversible}}$$

between the same end states, the implications are quite different. The irreversible entropy change includes irreparable entropy production, whereas the reversible entropy change represents entropy transfer only (e.g., if a system undergoing a reversible process exhibits an entropy increase, its surroundings will exhibit a corresponding entropy decrease).

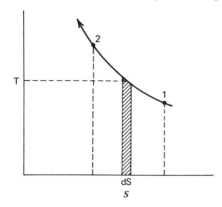

Figure 3.5 Area under TS curve is heat $\left[(Q_{added})_{reversible} = \int_1^2 T\,dS \right]$.

The very ease of computing ΔS, as indicated by (3.25), may cause us to lose sight of the irreversibilities that are involved in the usual process. To avoid this, entropy is best defined in combination as

$$dS = \frac{\delta Q + \delta F}{T} = dS_{transferred} + dS_{produced} = \left(\frac{\delta Q}{T}\right)_{\substack{\text{any reversible} \\ \text{path}}} \quad (3.26)$$

since (3.21), (3.22), and (3.25) are simply different viewpoints of the same concept. (See Figure 3.5.)

The entropy function, as given by (3.26), embodies the very concept we sought to indicate which processes are or are not possible. For while the second law claims only that there will be inaccessible states in an adiabatic system, consideration of the entropy production indicates precisely which states are excluded. In other words, since $dS_{produced}$ can be positive or zero but cannot be negative because of the one-wayness of irreversibilities, the second law can be restated more specifically in terms of S as

In any adiabatic system all processes requiring a decrease in entropy are impossible,

or (what amounts to the same thing)

If a process is irreversible for any reason, entropy will be produced.

The Basic Identity. A most important identity (i.e., statement that holds true for all working substances conforming to the laws of thermodynamics) is obtained by combining (3.13) and (3.21) and summarizes this entire section:

$$T\,dS = \delta Q + \delta F = du + p\,dv, \qquad (3.27)$$

where all terms are expressed as energy per pound mass. Observe that the first law portion of (3.27) disclaims the production or destruction of energy, but admits the possibility of energy transformations only, while the second law portion of (3.27) admits the impossibility of entropy destruction, but claims that entropy can be produced freely. In effect, the first law says that no kind of energy is ever lost in any process, while the second law says that the opportunity to convert all the supplied heat energy to useful work is lost forever in any real process.

3.2.5 The Perfect Gas

The laws of thermodynamics are often applied in the study of fluid processes. However, the characteristics of real fluids are often so complicated as to preclude direct analysis by these laws. Hence it is natural and necessary to seek a simplified fluid model that can be treated readily. A simplified macroscopic gas model will now be developed for this purpose.

The Boyle-Mariotte Law. In England, Boyle, in 1662 cautiously observed that over a limited pressure range the product of pressure and volume of a fixed mass of a real gas was essentially a constant, independent of pressure level, under isothermal conditions. On the continent of Europe, Mariotte announced this same observation in 1676. The Boyle-Mariotte law can be given as

$$(pv)_t = K_t, \qquad (3.28)$$

where the subscript t signifies that changes of state are allowed under conditions of constant temperature only; K_t indicates that although the isothermal pv product remains constant, its value changes as temperature changes. Temperature, of course, can be sensed by any empirical thermometer. We know today that Boyle's caution was quite in order, for no real gas satisfies (3.28) exactly.

The Charles–Gay-Lussac Law. Charles, in 1787, and Gay-Lussac, in 1802, found that equal volumes of real gases (oxygen, nitrogen, hydrogen, carbon dioxide, and air) would expand the same amount for a given temperature increase under isobaric conditions. The Charles–Gay-Lussac

82 Solutions to Ideal-Compressible Pipe Flow

law can be given as

$$\frac{1}{v_0}\left(\frac{v-v_0}{t-t_0}\right)_p = \alpha_{0p}, \tag{3.29}$$

where the subscript p signifies that changes of state are allowed under conditions of constant pressure only, the subscript 0 signifies that the variable is taken with respect to a definite reference state, and α_{0p} indicates that, although the isobaric cubical coefficient of expansion of any gas remains a constant, it value changes as the reference state or the pressure level changes. The symbol t represents temperature as measured on any empirical scale.

Similarly, when the volume of a given amount of any gas is held constant, the change in pressure is proportional to the change in temperature. Thus the Charles–Gay-Lussac law can be given as

$$\frac{1}{p_0}\left(\frac{p-p_0}{t-t_0}\right)_v = \alpha_{0v}, \tag{3.30}$$

where the subscript v signifies that changes of state are allowed under conditions of constant volume only, and α_{0v} indicates that, although the isochoric pressure coefficient of any gas remains constant, its value changes as the reference state or the volume changes. We know today that no real gas follows the Charles–Gay-Lussac laws exactly.

Clapeyron's Equation of State. Clapeyron, in 1834, first combined the Boyle-Mariotte and Charles–Gay-Lussac laws to give the equation of state of a gas as

$$pv = \bar{R}\left(t - t_0 + \frac{1}{\alpha_0}\right), \tag{3.31}$$

where $\bar{R}$ is a constant of integration, which can be evaluated at the reference state as

$$\bar{R} = p_0 v_0 \alpha_0, \tag{3.32}$$

and where the parenthetical expression in (3.31) is known as *temperature from the zero of the air thermometer*, for note that the quantity $(t - t_0 + 1/\alpha_0)$ is equivalent to expressing temperature on a new scale, whose zero is lower than that of the empirical scale of t by the quantity $(1/\alpha_0 - t_0)$, but whose unit temperature interval (i.e., degree) is the same as that of the empirical scale of t. We know today that no real gas satisfies (3.31) exactly.

Regnault's Ideal Gas. Regnault, in 1845, found that the mean cubical coefficient of expansion of any real gas is approximately $\frac{1}{273}$ per Celsius degree, in contrast to Gay-Lussac's evaluation of $\frac{1}{267}$ per Celsius degree.

Regnault realized, however, that the cubical coefficients of expansion of all permanent gases are only approximately equal. For simplicity he proposed an imaginary substance that perfectly fulfilled all conditions of the Boyle-Mariotte and Charles–Gay-Lussac laws, that is, he defined an *ideal* gas whose thermodynamic state satisfied Clapeyron's equation of state. The equation of state of Regnault's ideal gas is

$$pv = \bar{R}\left(t - t_0 + \frac{1}{\alpha}\right), \tag{3.33}$$

where the expression in parentheses represents temperature on the ideal-gas absolute-temperature scale. This is similar to the air-thermometer scale in that its zero is lower than that of the empirical scale of t by the constant $(1/\alpha - t_0)$, and its degree is the same size as that of the empirical scale of t. Today we know that no real gas satisfies the requirements of Regnault's ideal gas exactly.

Kelvin's Ideal Gas. William Thomson (later Lord Kelvin), in 1848, recognized that Sadi Carnot's analysis, in 1824, of a reversible heat engine operating between two isotherms and two adiabats provided a basis for defining an absolute thermometric scale since the efficiency of the Carnot engine was only a function of the two empirical temperatures, being independent of the working substance. His proposed *absolute thermodynamic temperature function* (θ) can be given in terms of the reversible Carnot heats as

$$\frac{\delta Q}{Q} = \frac{d\theta}{\theta} = \phi(t)\,dt \tag{3.34}$$

where θ is any arbitrary function of the empirical temperature, t.

Kelvin adopted Joule's suggestion and patterned the new temperature function after temperature from the zero of the air thermometer. By further relating the fundamental temperature interval of his scale to that of the empirical scale (i.e., by taking $\theta_{\text{steam}} - \theta_{\text{ice}} = t_{\text{steam}} - t_{\text{ice}}$), Kelvin succeeded in completely defining the absolute thermodynamic temperature scale. To experimentally realize temperatures on the θ scale, it was natural for Kelvin to turn to a gas thermometer:

$$\frac{\theta_2}{\theta_1} \simeq \left(\frac{p_2}{p_1}\right)_v \simeq \left(\frac{v_2}{v_1}\right)_p. \tag{3.35}$$

84 Solutions to Ideal-Compressible Pipe Flow

The difference in the α_0 factors for all real gases, however, means that no two gas thermometers can give the same temperature label to a common isotherm. Thus, although in theory Kelvin defined a thermodynamic temperature scale, independent of the thermometric substance, in effect he simply redefined an *ideal* gas whose thermodynamic state was now given by

$$pv = \bar{R}\theta. \tag{3.36}$$

Today we know that no real gas follows (3.36) exactly.

Real Gas. In *real* gases it has been shown by all existing experimental data that absolute temperature is proportional not simply to the pv product, as in Kelvin's ideal gas, but to the limiting value of the pv product as p approaches zero along the respective isotherm. Thus

$$\frac{T_2}{T_1} = \lim_{p \to 0} \frac{(pv)_2}{(pv)_1} = \frac{(pv)_2^0}{(pv)_1^0} = \frac{\theta_2}{\theta_1}, \tag{3.37}$$

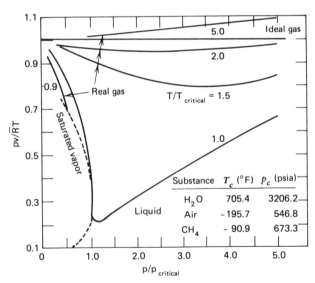

Figure 3.6 Generalized compressibility chart. This chart applies to any gas or gas mixture, and indicates departures from the ideal gas ($pv = \bar{R}T$) as a function of pressure and temperature, both referred to their critical values. At zero pressure, all gases act as the ideal gas, and again at the higher temperatures, with respect to the critical temperature, all gases approach the ideal gas.

where the superscript 0 indicates the zero pressure intercept. Equation 3.37, together with the relation $dT = dt$, serves completely to define absolute temperature today [9].

Note that the ideal gas expression, as given by (3.36), serves as an increasingly exact relation for all real gases as the pressure approaches zero along the respective isotherm (see Figure 3.6). Thus, when we describe a real gas by (3.36), we imply that both the size of the molecules and the intermolecular forces are negligible. It follows that the transport properties such as viscosity and thermal conductivity, which depend on molecular size and interaction, likewise must be considered negligible. In reality, however, the gas flow need not be inviscid or adiabatic when applying (3.36), the ideal gas equation of state, as long as these effects do not cause the state of the gas to deviate appreciably from that represented by (3.36).

The Perfect Gas. It is often convenient to go a step beyond (3.36) by defining the *perfect gas* of thermodynamics. The perfect gas has the equation of state of the ideal gas with the additional simplifying assumption of constant specific heat capacities (see Figure 3.7). Some characteris-

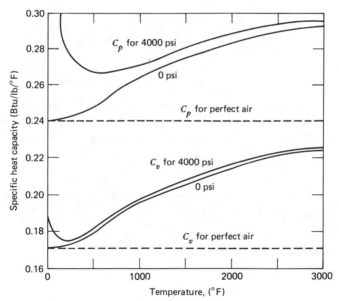

Figure 3.7 Comparison of specific heat capacities for air. At zero pressure, air acts essentially as an ideal gas (i.e., $C_p - C_v \approx$ constant). The higher the temperature, the more nearly this is true. Perfect air exhibits $C_p - C_v =$ constant. C_p and C_v for perfect air are based on $p = 0$ psi, $T = 0°F$.

tic equations of the perfect gas are

$$pv = \bar{R}T$$

and

$$C_p = \text{constant}, \tag{3.38}$$

along with

$$\left.\begin{array}{c} \dfrac{C_p}{C_v} = k, \\[6pt] C_p - C_v = \dfrac{\bar{R}}{J}, \\[6pt] C_p = \dfrac{k\bar{R}}{(k-1)J}, \\[6pt] C_v = \dfrac{\bar{R}}{(k-1)J}, \\[6pt] du = C_v\,dT = \dfrac{d(pv)}{(k-1)J}, \\[6pt] dh = C_p\,dT = \dfrac{k\,d(pv)}{(k-1)J}. \end{array}\right\} \tag{3.39}$$

In (3.39), k is the ratio of specific heats, $\bar{R}$ is the specific gas constant, J is Joule's mechanical equivalent of heat (i.e., $\simeq 778$ ft-lbf/Btu), and h is the enthalpy, a quantity that conveniently groups certain properties, that is,

$$h = u + pv. \tag{3.40}$$

Of course, as consistently followed in this study, the pound mass-pound force system is implied.

3.2.6 Thermodynamic Processes

The laws of thermodynamics can be applied to real fluids and, with greater simplicity, to the perfect gas of thermodynamics, as the working substances undergo changes of state. The relations that will allow evaluation of such performance quantities as work and heat will now be developed.

The Polytropic Process. The relation between p and v during an actual change of state may often be represented approximately by a member of the family

$$pv^n = K \text{ (a constant)}. \tag{3.41}$$

It is conventional to let (3.41) exactly represent all changes of state that take place reversibly in any system, and to call such processes *polytropic*. Curves representing these reversible changes of state are called *polytropes*.

The exponent n in (3.41) is a constant whose value is determined by the nature of the process. The significance of n is seen by rectifying (3.41), that is, by adjusting the scales of p and v so as to make the polytropes plot as straight lines. Taking the logarithm (to any base) of both sides of (3.41) will accomplish this, for note that

$$\log p = -n \log v + \log K \tag{3.42}$$

is in the form of a straight line ($y = Ax + B$), with the (3.42) counterparts of x, y, A and B clearly apparent (see Figure 3.8). In particular, n represents the negative slope of the polytrope when expressed on a log-log basis. It follows from a consideration of (3.42) that, whereas two given states may be joined by many different combinations of reversible processes, only one polytrope can connect given end states. Thus polytropes are very special reversible changes of state.

Polytropic Processes for the Perfect Gas. Referring to (3.41) and (3.42) or to the differentiated form of (3.42):

$$\frac{dp}{p} = -n \frac{dv}{v}, \tag{3.43}$$

we note the following simple reversible processes.

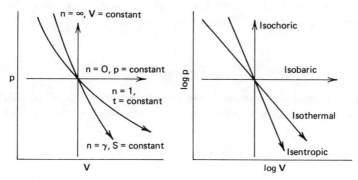

Figure 3.8 Rectifying several polytropic processes.

Isobaric. If we have a constant pressure process, then $dp=0$; but from (3.43), since $dv \neq 0$, n must $=0$. Alternatively, if $n=0$, this signifies a constant pressure process, which is represented on the pv diagram by a straight horizontal line (the log-log slope $=0$).

Isochoric. If we have a constant volume process, then $dv=0$; but from (3.43), since $dp \neq 0$, n must $=\infty$. Alternatively, if $n=\infty$, this signifies a constant volume process, which is represented on the pv diagram by a straight vertical line (the log-log slope $=\infty$).

Isothermal. If we have a constant temperature process, then, for the perfect gas, $pv=$ constant; and, from (3.41), n must $=1$. Alternatively, if $n=1$, this signifies a constant temperature process, which is represented on the pv diagram by an equilateral hyperbola (the log-log slope $=-1$).

Isentropic. If we have a process that is both adiabatic and reversible, we will arbitrarily denote this by setting $n=\gamma$ (the significance of γ is examined below). Alternatively, if $n=\gamma$, this signifies an isentropic process, which is represented on the pv diagram by a curve cutting across the isotherms (the log-log slope $=-\gamma$, with $\gamma > 1$).

pv-T Relation. For the perfect gas of (3.38) and (3.39) undergoing any reversible process described by (3.41), we obtain, by combination, the useful polytropic relations:

$$\frac{T_2}{T_1} = \left(\frac{p_2}{p_1}\right)^{(n-1)/n} = \left(\frac{v_1}{v_2}\right)^{n-1}. \tag{3.44}$$

The Significance of γ. Laplace, in 1816, while reviewing Newton's evaluation of the velocity of sound waves in air, that is,

$$a^2_{\text{Newton}} = g_c \left(\frac{\partial p}{\partial \rho}\right)_{\text{isothermal}} = \frac{g_c p}{\rho}, \tag{3.45}$$

correctly assumed that the rarefactions and compressions of air took place adiabatically (and reversibly) rather than isothermally. He obtained

$$a^2 = g_c \left(\frac{\partial p}{\partial \rho}\right)_{\text{adiabatic, reversible}} = g_c \frac{kp}{\rho}, \tag{3.46}$$

where k, the ratio of specific heats, was assumed to be a constant and has the additional significance

$$k = \frac{c_p}{c_v} = \frac{(\partial p/\partial \rho)_{\text{adiabatic, reversible}}}{(\partial p/\partial \rho)_{\text{isothermal}}}, \qquad (3.47)$$

as can be seen by comparing (3.45) and (3.46). Integration of (3.46) yields, for this isentropic case,

$$pv^k = K \text{(a constant)}. \qquad (3.48)$$

Since (3.46), (3.47), and (3.48) were obtained years before the perfect gas, the gas constant, the first law, Joule's law, and the thermodynamic temperature function were defined, it follows that these equations are independent of thermodynamic principles, as Maxwell had already noted in 1872.

From the later thermodynamic viewpoint the isentropic relation, (3.48), can be derived as follows. By the first law, (3.13), for the adiabatic-reversible case, we have

$$0 = du + p\,dv. \qquad (3.49)$$

This can be expressed in terms of the perfect gas as

$$0 = c_p p\,dv + c_v v\,dp, \qquad (3.50)$$

which, of course, integrates to give (3.48) once more. Thus, in either the historical or the thermodynamic development of (3.48), we note that k (the ratio of specific heats) actually represents, for the case of a perfect gas, the isentropic-process exponent γ as well.

3.2.7 Stagnation States

When a fluid is stopped *isentropically* (i.e., by a constant entropy process), the general energy equation of (3.19) becomes

$$0 = dh + \frac{V\,dV}{g_c} + \frac{g}{g_c}\,dZ. \qquad (3.51)$$

It is convenient in gas flow work to neglect the potential energy term in (3.51). This is justified by noting that the enthalpy and kinetic energy terms almost always dominate the energy balance. Thus (3.51) becomes for a gas:

$$0 = dh + \frac{V\,dV}{g_c}, \qquad (3.52)$$

which, when integrated between static and total states, yields

$$0 = \int_s^t dh + \int_V^0 \frac{V\,dV}{g_c}$$

or

$$h_t = h + \frac{V^2}{2g_c}. \qquad (3.53)$$

The *static* state means the undisturbed or free stream state, which is the actual state of the fluid. The *total* state means the state that results from stagnating the fluid by an isentropic process. The total term derives its name from the fact that it represents the sum of the static and dynamic terms, as seen in (3.53).

Since enthalpy and temperature are related through (3.39), it follows from (3.53) that

$$T_t = T + \frac{V^2}{2Jg_c C_p}, \qquad (3.54)$$

where conventional units of C_p, the specific heat capacity at constant pressure, are British thermal units per pound mass per degree Rankine (Btu/lbm°R).

An important relation between pressure and temperature already has been given in (3.44) for the general polytropic process. This can be

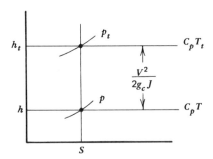

Figure 3.9 Various thermodynamic parameters.

expressed again in terms of an isentropic process between static and total states as

$$\frac{T_t}{T} = \left(\frac{p_t}{p}\right)^{(k-1)/k}. \qquad (3.55)$$

These concepts and terms involving the stagnation state are illustrated in Figure 3.9.

3.3 THE MACH NUMBER

The continuity equation of (1.14) can be differentiated logarithmically to yield

$$\frac{dA}{A} = -\frac{dV}{V} - \frac{d\rho}{\rho}$$

or, upon slight rearrangement,

$$\frac{dA}{A} = -\frac{dV}{V}\left[1 + \left(\frac{V^2}{V\,dV}\right)\frac{d\rho}{\rho}\right].$$

When this is combined with energy in the form of (3.60), namely, $\rho V\,dV = -g_c\,dp$, there results

$$\frac{dA}{A} = \frac{dV}{V}\left[\frac{V^2}{g_c(dp/d\rho)_s} - 1\right]. \qquad (3.56)$$

The quantity $g_c(dp/d\rho)_s$ is recognized from (3.46) as the square of the acoustic velocity, that is, of the velocity of propagation of a small pressure disturbance in a compressible fluid. For the perfect gas of (3.38), this can also be expressed as

$$a = \sqrt{\gamma g_c \bar{R} T}. \qquad (3.57)$$

In (3.56) the value of $(V/a)^2$ with respect to unity is seen to be critical, that is, as long as V/a is less than 1, the area must *decrease* (as in a nozzle) for the velocity to *increase*. Conversely, velocity can increase only if area increases when V/a exceeds unity. The velocity ratio (V/a) is so important in gas dynamics that it gets a special symbol (**M**) and is called the

92 Solutions to Ideal-Compressible Pipe Flow

Mach number, after the Viennese physicist, Ernst Mach, that is,

$$\mathbf{M} \equiv \frac{V}{a}. \tag{3.58}$$

Note that, if $\mathbf{M} < 1$, we speak of the *subsonic* regime; if $\mathbf{M} > 1$, we speak of the *supersonic* regime; while $\mathbf{M} = 1$ signifies a critical flow situation. Much more will be said and done with the Mach number as we continue this study.

3.4 THE ONE-DIMENSIONAL BERNOULLI ENERGY EQUATION

As in constant density flow, the one-dimensional analysis will apply up to and including complete pipe cross-sectional areas, whenever the streamlines are without curvature. However, in the compressible case the Bernoulli equation of (1.33) must be used in its differential form:

$$\frac{dp}{\rho} + \frac{V\,dV}{g_c} + \frac{g}{g_c}\,dZ = 0. \tag{3.59}$$

Consistently with the argument given under (3.51), the potential energy term in (3.59) is usually neglected in gas flow work, and (3.59) becomes

$$0 = \frac{dp}{\rho} + \frac{V\,dV}{g_c}. \tag{3.60}$$

This is seen to be a special form of the general energy equation of (3.19) where, in addition to overlooking potential energy changes, we neglect losses and work terms.

To apply (3.60), even in this ideal situation, we must be able to express density as a function of pressure before the required integration can be accomplished. By (3.48), for an isentropic process of a perfect gas, we have

$$p = K\rho^\gamma,$$

which, upon differentiation, becomes

$$dp = K\gamma\rho^{\gamma-1}\,d\rho. \tag{3.61}$$

With (3.61) applied to (3.60) we obtain, on integration,

$$\frac{V_2^2 - V_1^2}{2g_c} = \left(\frac{\gamma}{\gamma-1}\right)\left(\frac{p_1}{\rho_1} - \frac{p_2}{\rho_2}\right). \tag{3.62}$$

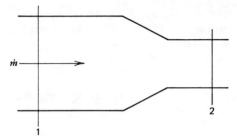

Figure 3.10 Reducer of Example 3.1.

Example 3.1. Estimate by compressible relations the pressure drop across the reducer of Figure 3.10, if air flows at a rate of 5 lbm/sec, when the inlet pressure is 20 psia, the inlet temperature is 80°F, and, geometrically, $D_1 = 12$ in. and $D_2 = 6$ in.

Solution: Assume an isentropic process between 1 and 2, with air taken to be a perfect gas at $\gamma = 1.4$, $\bar{R} = 53.35$ ft-lbf/°R lbm, and $C_p = 0.24$ Btu/lbm°R.

By continuity (1.14): $\dot{m} = \rho_1 A_1 V_1 = \rho_2 A_2 V_2$, where $A_1 = \pi/4$ ft² and $A_2 = \pi/16$ ft².

By perfect gas (3.38):

$$\rho_1 = \frac{p_1}{\bar{R} T_1} = \frac{20 \times 144}{53.35 \times (80 + 460)} = 0.09997 \text{ lbm/ft}^3.$$

Then

$$V_1 = \frac{\dot{m}}{\rho_1 A_1} = \frac{5}{0.09997 \times \pi/4} = 63.681 \text{ ft/sec}.$$

By isentropic relations (3.54), (3.55), and (3.62):

$$T_t = T_1 + \frac{V_1^2}{2 J g_c C_p} = 540 + \frac{63.681^2}{2 \times 778 \times 32.174 \times 0.24} = 540.3375 °R,$$

$$p_t = p_1 \left(\frac{T_t}{T_1}\right)^{\gamma/(\gamma-1)} = 20 \left(\frac{540.3375}{540}\right)^{3.5} = 20.043784 \text{ psia},$$

and

$$\frac{V_2^2 - V_1^2}{2 g_c} = \left(\frac{\gamma}{\gamma - 1}\right)\left(\frac{p_1}{\rho_1} - \frac{p_2}{\rho_2}\right).$$

94 Solutions to Ideal-Compressible Pipe Flow

But here we find three unknowns, namely, V_2, p_2, ρ_2, so an iterative (i.e., trial and error) solution method is introduced to obtain p_2 and hence the required pressure drop. The steps are, briefly, as follows:

1. Guess p_2.
2. Get T_2 from (3.55), since T_t and p_t are conserved in an isentropic process.
3. Get ρ_2 from (3.38).
4. Get V_{2c} from (3.62), where the subscript c signifies a calculated quantity.
5. Check whether the p_2 guessed is correct by comparing V_{2c} with V_2 determined from (1.14).

A good guess of p_2 is the incompressible value, which is obtained easily as follows:

$$(V_2)_{inc} = \frac{A_1}{A_2} V_1 = 4V_1 = 254.724 \text{ ft/sec.}$$

$$(p_{20})_{inc} = p_t - \frac{\rho_1 V_2^2}{2g_c \times 144} = 20.04378 - \frac{0.09997 \times 254.724^2}{2g_c \times 144} = 19.3438 \text{ psia.}$$

The subscript 0 signifies that this is our initial guess.

$$T_{20} = \frac{T_t}{(p_t/p_2)^{(\gamma-1)/\gamma}} = \frac{540.3375}{(20.04378/19.3438)^{0.28571}} = 534.8775°R,$$

$$\rho_{20} = \frac{p_2}{RT_2} = \frac{19.3438 \times 144}{53.35 \times 534.8775} = 0.097615 \text{ lbm/ft}^3,$$

and

$$V_{2c0} = \sqrt{\frac{2g_c \gamma \times 144}{\gamma - 1}\left(\frac{p_1}{\rho_1} - \frac{p_2}{\rho_2}\right) + V_1^2} = 256.00537 \text{ ft/sec.}$$

but

$$V_{20} = \frac{\dot{m}}{\rho_2 A_2} = \frac{5}{0.097615 \times \pi/16} = 260.8696 \text{ ft/sec.}$$

The velocity error of $\varepsilon_0 = V_{2c0} - V_{20} = -4.864$ shows us that p_2 was not guessed correctly. A second guess of p_2 is made arbitrarily as

$$p_{21} = 19.3189 \text{ psia.}$$

The subscript 1 signifies that this is our next guess. Following identical calculation steps, we obtain:

$$T_{21} = 534.6801°R,$$
$$\rho_{21} = 0.097525 \text{ lbm/ft}^3,$$
$$V_{2c1} = 260.5540 \text{ ft/sec},$$
$$V_{21} = 261.1104 \text{ ft/sec}.$$

The velocity error is now $\varepsilon_1 = V_{2c1} - V_{21} = -0.556$, which shows us that again p_2 was not guessed correctly. A third and final guess of p_2 is obtained by a straight line interpolation between the first two guesses (this is called the Newton-Raphson method), according to the relation

$$p_2 = p_{21} - \varepsilon_1 \left(\frac{p_{20} - p_{21}}{\varepsilon_0 - \varepsilon_1} \right)$$
$$= 19.3189 + 0.556 \left(\frac{19.3438 - 19.3189}{-4.864 + 0.556} \right).$$

There results

$$p_2 = 19.3157 \text{ psia},$$
$$T_2 = 534.655° R,$$
$$\rho_2 = 0.097514 \text{ lbm/ft}^3,$$
$$V_{2c} = 261.2045 \text{ ft/sec},$$
$$V_2 = 261.1400 \text{ ft/sec}.$$

Although the solution has not closed completely, the error is now reduced to $+0.064$, which is judged to be negligible. The pressure drop across the reducer is thus

$$\Delta p_{comp} = 20 - 19.3157 = 0.684 \text{ psi}.$$

It is instructive to compare this compressible result with the incompressible pressure drop, which is

$$\Delta p_{inc} = 20 - 19.3438 = 0.656 \text{ psi}.$$

This is seen to be not far off in this low velocity case where compressibility in unimportant. When we have covered generalized methods in Part III, this same problem will be done again in much simpler fashion (see Example 8.18).

3.5 ONE-DIMENSIONAL COMPRESSIBLE PRESSURE EQUATIONS

Two expressions that usefully relate the static and total pressures in compressible flows are developed in this section. The first one we owe to Newton's binomial expansion, and it can be developed as follows.

The total/static temperature relation of (3.54) can be written in terms of the Mach number through (3.57) and (3.58) as

$$\frac{T_t}{T} = 1 + \frac{\gamma - 1}{2} M^2, \qquad (3.63)$$

where we have made use of $\overline{R}/Jc_p = (\gamma - 1)/\gamma$ from (3.39). From the isentropic relations (3.55) and (3.63), we can write

$$\frac{p_t}{p} = \left(1 + \frac{\gamma - 1}{2} M^2\right)^{\gamma/(\gamma - 1)}. \qquad (3.64)$$

Equation 3.64 can be operated on according to the binomial expansion

$$(a+b)^N = a^N + \frac{N}{1} a^{N-1} b + \frac{N(N-1)}{1 \times 2} a^{N-2} b^2$$
$$+ \frac{N(N-1)(N-2)}{1 \times 2 \times 3} a^{N-3} b^3 + \ldots,$$

where, in (3.64),

$$a = 1, \quad b = \left(\frac{\gamma - 1}{2}\right) M^2, \quad N = \frac{\gamma}{\gamma - 1},$$
$$N - 1 = \frac{1}{\gamma - 1}, \quad N - 2 = \frac{2 - \gamma}{\gamma - 1},$$

to obtain

$$p_t = p + \left(\frac{\gamma p M^2}{2}\right) + \left(\frac{\gamma p M^2}{2}\right) \frac{M^2}{4} + \left(\frac{\gamma p M^2}{2}\right)\left(\frac{2 - \gamma}{24}\right) M^4 + \ldots. \qquad (3.65)$$

But $\gamma p M^2/2$ for a perfect gas is equivalent to $\rho V^2/2g_c$; hence (3.65) can be given as the familiar compressible counterpart of (2.5), that is,

$$p_t = p + \frac{\rho V^2}{2g_c}\left[1 + \frac{M^2}{4} + \frac{(2 - \gamma)}{24} M^4 + \ldots\right]. \qquad (3.66)$$

Note that the dynamic pressure of (3.66), that is, $(p_t - p)$, approaches the incompressible dynamic pressure, that is, $\rho V^2/2g_c$, as the Mach number approaches zero, that is, as compressible effects become less important.

Another expression for total pressure in compressible flow can be obtained by first defining a dimensionless flow number. The isentropic velocity relation of (3.62) is rewritten at a point in terms of the stagnation state where $V_1 = 0$, $p_1 = p_t$, and $\rho_1 = \rho_t$ as

$$\frac{V^2}{2g_c} = \left(\frac{\gamma}{\gamma-1}\right)\left(\frac{p_t}{\rho_t} - \frac{p}{\rho}\right)$$

or

$$V = \left\{\left(\frac{2\gamma}{\gamma-1}\right)\left(\frac{p_t g_c}{\rho_t}\right)\left[1 - \left(\frac{p}{p_t}\right)^{(\gamma-1)/\gamma}\right]\right\}^{1/2}. \qquad (3.67)$$

Expressing continuity (i.e., $\dot{m} = \rho A V$) in terms of (3.67), we have

$$\left(\frac{\dot{m}}{p_t A}\right)\left(\frac{\bar{R} T_t}{g_c}\right)^{1/2} = \left\{\left(\frac{2\gamma}{\gamma-1}\right)\left(\frac{p}{p_t}\right)^{2/\gamma}\left[1 - \left(\frac{p}{p_t}\right)^{(\gamma-1)/\gamma}\right]\right\}^{1/2}, \quad (3.68)$$

which is called the *total flow number* [10], to be discussed again in Section 3.6 on comparisons. Squaring both sides of (3.68), introducing the point pressure ratio $(R = p/p_t)$, and rearranging slightly yields

$$\left(\frac{\dot{m}}{Ap}\right)^2\left(\frac{\bar{R} T_t}{g_c}\right) = \left(\frac{2\gamma}{\gamma-1}\right)\{R^{2(1-\gamma)/\gamma} - R^{(1-\gamma)/\gamma}\},$$

which is seen to be a simple quadratic of the form

$$x^2 - x - C = 0,$$

where

$$x = R^{(1-\gamma)/\gamma} \quad \text{and} \quad C = \left(\frac{\gamma-1}{2\gamma}\right)\left(\frac{\dot{m}}{Ap}\right)^2\left(\frac{\bar{R} T_t}{g_c}\right).$$

Solving by the usual, $x = \frac{1}{2} \pm \sqrt{\frac{1}{4} + C}$, we obtain on substitution

$$p_t = p\left[\frac{1}{2} + \sqrt{\frac{1}{4} + \left(\frac{\gamma-1}{2\gamma}\right)\left(\frac{\dot{m}}{Ap}\right)^2\left(\frac{\bar{R} T_t}{g_c}\right)}\right]^{\gamma/(\gamma-1)}. \qquad (3.69)$$

98 Solutions to Ideal-Compressible Pipe Flow

Note that the specific gas constant has been denoted by $\overline{R}$ to distinguish it from the point pressure ratio (R). Equation 3.69 is especially useful for determining the effective total pressure across a passage (even when the flow is not one-dimensional) when it is not convenient or practical to install and traverse the pipe with a Pitot tube.

Example 3.2. In Example 3.1 the total pressure was determined by an exact isentropic relation to be 20.043784. Check this result by the use of (3.66).

Solution: To apply (3.66), it is clear that we need the Mach number at 1. By (3.57) and (3.58):

$$M_1 = \frac{V_1}{a_1} = \frac{63.681}{\sqrt{1.4 \times 32.174 \times 53.35 \times 540}} = \frac{63.681}{1139.1493} = 0.0559.$$

Then by (3.66):

$$p_t = 20 + \frac{0.09997 \times 63.681^2}{2 \times 32.174 \times 144}\left[1 + \frac{0.0559^2}{4} + \frac{0.6}{24}(0.0559)^4 + \cdots\right]$$

$$= 20.043786 \text{ psia}$$

which checks the more exact result very closely in this low Mach number case.

Example 3.3. Check the total pressure result of Example 3.1 by (3.69).

Solution: Using the values given in Example 3.1, we have upon substitution in (3.69)

$$p_t = 20\left[\frac{1}{2} + \sqrt{\frac{1}{4} + \left(\frac{0.4}{2.8}\right)\left(\frac{5}{\pi/4 \times 20 \times 144}\right)^2\left(\frac{53.35 \times 540.3375}{32.174}\right)}\right]^{1.4/0.4}$$

$$p_t = 20(1.00218929) = 20.043786 \text{ psia},$$

which again checks the other isentropic solution, and will check even at high Mach numbers.

3.6 COMPARISONS BETWEEN COMPRESSIBLE AND INCOMPRESSIBLE TREATMENTS

Often we prefer to treat a compressible fluid as though it were incompressible. We should, and usually do, have misgivings about this simplification.

Comparisons between Compressible and Incompressible Treatments 99

Although we are not always sure about the degree of approximation tacitly introduced, we nevertheless proceed to employ the more easily managed constant density fluid relations in working toward solutions of highly compressible flows.

3.6.1 Flow Numbers and Expansion Factors

In this section we contrast these two treatments in terms of several common parameters so that we can tell at a glance whether we can or cannot get away with the simplistic incompressible treatment. First, the compressible relations are reviewed. Next, their counterparts for the incompressible regime are developed. Useful relations between these two treatments are then developed in the form of expansion factors. Three dimensionless variables are used: $r = p_2/p_1$, the static pressure ratio; β, the diameter ratio; and γ, the isentropic exponent. Rapid solutions to representative compressible problems are given based on compressible graphs and tables, on the simplified incompressible graphs and tables, and on the simplified incompressible treatment together with the use of the expansion factors.

Compressible Treatment. From continuity, (1.14), we have

$$V_1 = \left(\frac{\rho_2}{\rho_1}\right)\left(\frac{A_2}{A_1}\right)V_2,$$

which combines with the isentropic relation, (3.48), and the geometric relation, $\beta = D_2/D_1$, to yield

$$V_1 = r^{1/\gamma}\beta^2 V_2. \tag{3.70}$$

Similarly, the Mach number at 1 can be given as

$$M_1 = r^{(\gamma+1)/2\gamma}\beta^2 M_2. \tag{3.71}$$

In terms of (3.62) and (3.70), we can express the velocity at 2 as

$$V_2 = \left\{ \frac{[2\gamma/(\gamma-1)]\left(\dfrac{p_1 g_c}{\rho_1}\right)\left[1 - r^{(\gamma-1)/\gamma}\right]}{1 - r^{2/\gamma}\beta^4} \right\}^{1/2}, \tag{3.72}$$

and by (3.44), (3.57), (3.58), and (3.71), the Mach number at 2 is

$$M_2 = \left\{ \frac{2r^{(1-\gamma)/\gamma}\left[1 - r^{(\gamma-1)/\gamma}\right]}{(\gamma-1)(1 - r^{2/\gamma}\beta^4)} \right\}^{1/2}. \tag{3.73}$$

100 Solutions to Ideal-Compressible Pipe Flow

Another expression for the Mach number at 2 can be obtained from the velocity expression of (3.67), based on the static/total pressure ratio, as

$$M_2 = \left\{ \left(\frac{2}{\gamma - 1} \right) R_2^{(1-\gamma)/\gamma} \left[1 - R_2^{(\gamma-1)/\gamma} \right] \right\}^{1/2}. \tag{3.74}$$

When these two Mach number relations, (3.73) and (3.74), are equated, we get the important static/total pressure ratio at 1 as

$$\frac{p_t}{p_1} = \left[\frac{1 - r^{(\gamma+1)/\gamma} \beta^4}{1 - r^{2/\gamma} \beta^4} \right]^{\gamma/(\gamma-1)}, \tag{3.75}$$

and, of course,

$$\frac{p_t}{p_2} = \left(\frac{p_t}{p_1} \right) \left(\frac{p_1}{p_2} \right). \tag{3.76}$$

A *static* flow number at 2 can be obtained from continuity in terms of (3.72) and (3.44) as

$$FN_2 = \left(\frac{\dot{m}}{p_2 A_2} \right) \left(\frac{\bar{R} T_2}{g_c} \right)^{1/2} = \left\{ \frac{2\gamma r^{(1-\gamma)/\gamma} \left[1 - r^{(\gamma-1)/\gamma} \right]}{(\gamma - 1)(1 - r^{2/\gamma} \beta^4)} \right\}^{1/2}. \tag{3.77}$$

This is similar to the *total* flow number derived in Section 3.5 as (3.68). Analogously to the V_1 and M_1 expressions of (3.70) and (3.71), we define the static flow number at 1 as

$$FN_1 = \left(\frac{\dot{m}}{p_1 A_1} \right) \left(\frac{\bar{R} T_1}{g_c} \right)^{1/2} = r^{(\gamma+1)/2\gamma} \beta^2 FN_2. \tag{3.78}$$

We have now arrived at three general expressions: M, p_t/p, and FN, defined in terms of the three dimensionless parameters, r, β, and γ. These are exact relations for any *isentropic* process of a *perfect* gas between any two arbitrary states.

The question that next attracts our attention is: What degree of approximation is made when these same parameters (M, p_t/p, and FN) are determined instead from incompressible relations? We proceed by developing equations, similar to those already given, for an ideal constant density fluid.

Incompressible Treatment. Continuity, under the constant density approximation, is simply

$$(V_1)_{\text{inc}} = \beta^2 (V_2)_{\text{inc}}. \tag{3.79}$$

Comparisons between Compressible and Incompressible Treatments

The Mach number at 1, assuming constant density, can be given as

$$(M_1)_{inc} = r^{(\gamma-1)/2\gamma} \beta^2 (M_2)_{inc}. \tag{3.80}$$

When (3.60) is integrated, under the constant density assumption, there results

$$\frac{V_2^2 - V_1^2}{2g_c} = \frac{p_1}{\rho} - \frac{p_2}{\rho}. \tag{3.81}$$

Combining (3.79) and (3.81), we obtain

$$(V_2)_{inc} = \left[\frac{2(p_1 g_c/\rho_1)(1-r)}{1-\beta^4} \right]^{1/2}, \tag{3.82}$$

and by (3.44), (3.57), (3.58), and (3.82) the Mach number at 2 is

$$(M_2)_{inc} = \left[\frac{2r^{(1-\gamma)/\gamma}(1-r)}{\gamma(1-\beta^4)} \right]^{1/2}. \tag{3.83}$$

Another expression for the Mach number at 2 can be obtained from the velocity expression of (3.81), based now on the total conditions at 1 (where $V_1 = 0$ and $p_1 = p_t$) and on the *inlet* density, as

$$(M_2)_{inc} = \left[\frac{2r^{1/2}(1-R_2)}{\gamma R_2} \right]^{1/2}. \tag{3.84}$$

When the two Mach numbers relations, (3.83) and (3.84), are equated, we get the static/total pressure ratio at 1 as

$$\left(\frac{p_t}{p_1}\right)_{inc} = \frac{1-r\beta^4}{1-\beta^4}, \tag{3.85}$$

and, of course,

$$\left(\frac{p_t}{p_2}\right)_{inc} = \left(\frac{p_t}{p_1}\right)_{inc} \left(\frac{p_1}{p_2}\right). \tag{3.86}$$

A static flow number at 2 can be obtained from continuity in terms of (3.82) and (3.44), based on the *inlet* density, as

$$(FN_2)_{inc} = \left(\frac{\dot{m}}{p_2 A_2}\right) \left(\frac{\bar{R} T_2}{g_c}\right)^{1/2} = \left[\frac{2r^{-(\gamma+1)/\gamma}(1-r)}{1-\beta^4} \right]^{1/2}. \tag{3.87}$$

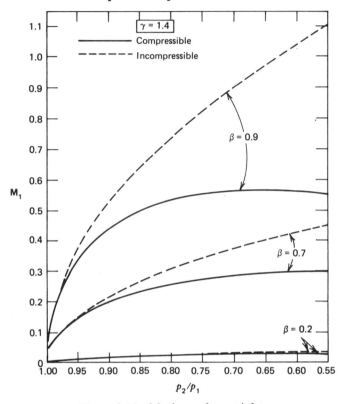

Figure 3.11 Mach number at inlet.

Analogously to the V_1 and M_1 expressions of (3.79) and (3.80), we define the static flow number at 1 as

$$(FN_1)_{inc} = \left(\frac{\dot{m}}{p_1 A_1}\right)\left(\frac{\overline{R}T_1}{g_c}\right)^{1/2} = r^{(\gamma+1)/2\gamma}\beta^2 (FN_2)_{inc}. \tag{3.88}$$

These isentropic relations, (3.79) through (3.88), are compared directly with their compressible counterparts, (3.70) through (3.78), in Figures 3.11 through 3.16, to note the degree of approximation made in computing dimensionless velocities, dimensionless total pressures, and dimensionless flow rates of compressible fluids through the use of incompressible relations.

We consider next certain general relations that *link* the compressible and incompressible treatments. These treatments are related through a family

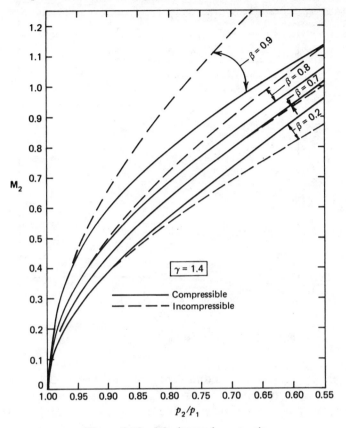

Figure 3.12 Mach number at exit.

of expansion factors defined in general as

$$P_{comp} = (EF) P_{inc} \qquad (3.89)$$

where P is a general parameter, and EF is the expansion factor.

Thus the isentropic velocity expansion factor at 2 is defined through (3.72) and (3.82) as

$$EF_{V2} = \frac{V_{comp,2}}{V_{inc,2}} = \left\{ \frac{\gamma[1 - r^{(\gamma-1)/\gamma}](1 - \beta^4)}{(\gamma - 1)(1 - r)(1 - r^{2/\gamma}\beta^4)} \right\}^{1/2}, \qquad (3.90)$$

and, from (3.70) and (3.79),

$$EF_{V1} = r^{1/\gamma} EF_{V2}. \qquad (3.91)$$

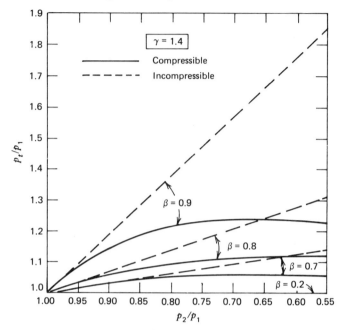

Figure 3.13 Total/static pressure ratio at inlet.

These velocity expansion factors are given in Tables 3.2 and 3.3 and plotted generally in Figures 3.17 and 3.18. They are used to correct velocities computed by incompressible equations to compressible velocities.

The isentropic pressure expansion factor is defined through (3.75) and (3.85) as

$$EF_{p_t} = \frac{(p_t/p_1)_{\text{comp}}}{(p_t/p_1)_{\text{inc}}} = \left(\frac{1-\beta^4}{1-r\beta^4}\right)\left[\frac{1-r^{(\gamma+1)/\gamma}\beta^4}{1-r^{2/\gamma}\beta^4}\right]^{\gamma/(\gamma-1)}. \quad (3.92)$$

There is only one pressure expansion factor since $p_{t1} = p_{t2}$ in the isentropic case. It is given in Table 3.4 and plotted in Figure 3.19. It is used to correct total pressures computed by incompressible equations to compressible total pressures.

The isentropic flow expansion factor is defined through (3.77) and (3.87) as

$$EF_{\dot{m}} = \frac{(FN_2)_{\text{comp}}}{(FN_2)_{\text{inc}}} = \left\{\frac{\gamma r^{2/\gamma}[1-r^{(\gamma-1)/\gamma}](1-\beta^4)}{(\gamma-1)(1-r)(1-r^{2/\gamma}\beta^4)}\right\}^{1/2}. \quad (3.93)$$

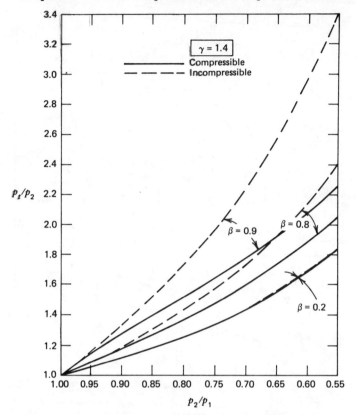

Figure 3.14 Total/static pressure ratio at exit.

Several comments can be made concerning (3.93). First, it is clear that $EF_{\dot{m}} = EF_{V_1}$ since both are based on the inlet density. Second, there is only one flow expansion factor since $\dot{m}_1 = \dot{m}_2 =$ constant. Finally, the *flow* expansion factor, of this family of expansion factors, is to be recognized as the so-called adiabatic expansion factor of fluid metering [11] (see Part IV). The flow expansion factor of (3.93) is given in Table 3.2 and plotted in Figure 3.17. It is used to correct flow rates computed by incompressible equations to compressible flow rates.

Example 3.4. Air flows isentropically in a pipe between two axial stations, the diameter ratio of which is $\beta = 0.6$. Static pressure measurements indicate that $p_2/p_1 = r = 0.8$. Find the Mach numbers, the flow numbers, and the static/total pressure ratios at stations 1 and 2.

106 Solutions to Ideal-Compressible Pipe Flow

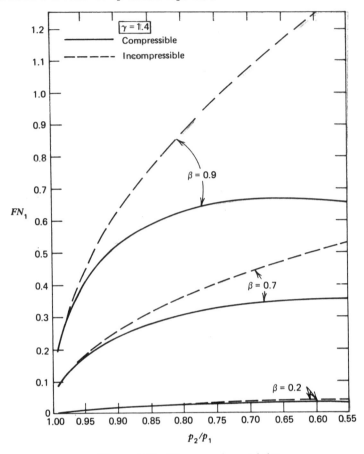

Figure 3.15 Flow number at inlet.

Solution: *Directly*, by the solid curves on the graphs indicated:

$$M_1 = 0.17924 \quad \text{(Figure 3.11)},$$
$$M_2 = 0.60283 \quad \text{(Figure 3.12)},$$
$$FN_1 = 0.21208 \quad \text{(Figure 3.15)},$$
$$FN_2 = 0.71327 \quad \text{(Figure 3.16)},$$
$$p_t/p_1 = 1.02267 \quad \text{(Figure 3.13)},$$
$$p_t/p_2 = 1.27834 \quad \text{(Figure 3.14)}.$$

Comparisons between Compressible and Incompressible Treatments 107

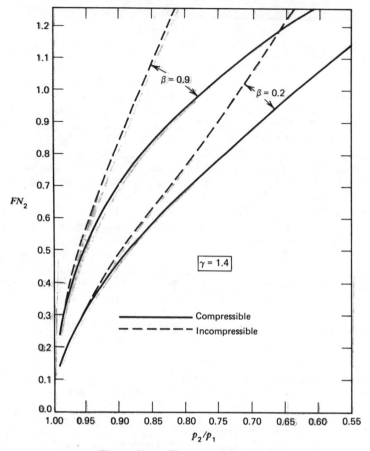

Figure 3.16 Flow number at exit.

Indirectly, by the pertinent expansion factors and the dotted curves on the graphs indicated:

$M_1 = EF_{V1} \times (M_1)_{inc} = 0.86900 \times 0.20626 = 0.17924$ (Figures 3.11, 3.17),

$FN_1 = EF_{\dot{m}} \times (FN_1)_{inc} = 0.86900 \times 0.24405 = 0.21208$ (Figures 3.15, 3.17),

$p_t/p_1 = EF_{pt} \times (p_t/p_1)_{inc} = 0.99310 \times 1.02978 = 1.02267$ (Figures 3.13, 3.19).

3.6.2 Hydraulic-Gas Analogy [12]

Gas systems also can be studied by means of water systems if a valid *analogy* can be shown to exist between the two systems [13–15]. This

Table 3.2 Flow Expansion Factor and Velocity Expansion Factor at 1 ($\gamma = 1.4$)

p_2/p_1	Beta								
	0.1	0.2	0.3	0.4	0.5	0.6	0.7	0.8	0.9
0.99	0.99463	0.99462	0.99457	0.99444	0.99415	0.99357	0.99239	0.98975	0.98137
0.98	0.98923	0.98920	0.98911	0.98886	0.98829	0.98714	0.98481	0.97961	0.96343
0.97	0.98380	0.98376	0.98363	0.98325	0.98240	0.98069	0.97725	0.96958	0.94612
0.96	0.97833	0.97829	0.97811	0.97761	0.97650	0.97424	0.96970	0.95966	0.92940
0.95	0.97284	0.97279	0.97257	0.97194	0.97056	0.96777	0.96217	0.94984	0.91324
0.94	0.96732	0.96726	0.96699	0.96625	0.96461	0.96129	0.95465	0.94013	0.89760
0.93	0.96177	0.96170	0.96139	0.96053	0.95863	0.95480	0.94715	0.93051	0.88244
0.92	0.95618	0.95610	0.95575	0.95478	0.95263	0.94829	0.93966	0.92098	0.86774
0.91	0.95056	0.95047	0.95008	0.94900	0.94660	0.94177	0.93219	0.91155	0.85347
0.90	0.94491	0.94481	0.94438	0.94319	0.94055	0.93524	0.92473	0.90220	0.83961
0.85	0.91615	0.91600	0.91538	0.91367	0.90989	0.90234	0.88756	0.85665	0.77560
0.80	0.88647	0.88629	0.88550	0.88332	0.87853	0.86900	0.85056	0.81285	0.71885
0.75	0.85580	0.85559	0.85464	0.85205	0.84636	0.83512	0.81360	0.77048	0.66769
0.70	0.82405	0.82380	0.82272	0.81978	0.81331	0.80061	0.77654	0.72924	0.62087
0.65	0.79109	0.79082	0.78963	0.78637	0.77926	0.76535	0.73925	0.68887	0.57747
0.60	0.75679	0.75650	0.75522	0.75172	0.74408	0.72922	0.70158	0.64912	0.53673
0.55	0.72100	0.72069	0.71934	0.71564	0.70760	0.69204	0.66334	0.60971	0.49806

Table 3.3 Velocity Expansion Factor at 2 ($\gamma = 1.4$)

p_2/p_1	Beta								
	0.1	0.2	0.3	0.4	0.5	0.6	0.7	0.8	0.9
0.99	1.00179	1.00178	1.00173	1.00161	1.00132	1.00073	0.99954	0.99688	0.98844
0.98	1.00360	1.00358	1.00349	1.00323	1.00266	1.00149	0.99913	0.99385	0.97743
0.97	1.00543	1.00540	1.00526	1.00487	1.00401	1.00226	0.99874	0.99091	0.96693
0.96	1.00728	1.00724	1.00705	1.00654	1.00539	1.00306	0.99839	0.98806	0.95690
0.95	1.00915	1.00909	1.00886	1.00822	1.00678	1.00388	0.99807	0.98529	0.94732
0.94	1.01103	1.01097	1.01069	1.00991	1.00820	1.00473	0.99779	0.98261	0.93816
0.93	1.01294	1.01286	1.01253	1.01163	1.00963	1.00560	0.99754	0.98002	0.92939
0.92	1.01486	1.01477	1.01440	1.01337	1.01109	1.00649	0.99733	0.97750	0.92099
0.91	1.01680	1.01671	1.01629	1.01513	1.01257	1.00740	0.99715	0.97507	0.91295
0.90	1.01877	1.01866	1.01820	1.01691	1.01406	1.00834	0.99700	0.97272	0.90523
0.85	1.02892	1.02876	1.02806	1.02614	1.02189	1.01341	0.99682	0.96210	0.87107
0.80	1.03965	1.03943	1.03850	1.03595	1.03033	1.01916	0.99753	0.95331	0.84307
0.75	1.05103	1.05077	1.04961	1.04643	1.03944	1.02563	0.99920	0.94624	0.82000
0.70	1.06315	1.06284	1.06145	1.05764	1.04930	1.03292	1.00186	0.94084	0.80102
0.65	1.07612	1.07574	1.07413	1.06970	1.06002	1.04111	1.00560	0.93707	0.78553
0.60	1.09004	1.08961	1.08777	1.08272	1.07172	1.05032	1.01051	0.93495	0.77308
0.55	1.10507	1.10460	1.10252	1.09686	1.08454	1.06068	1.01670	0.93450	0.76338

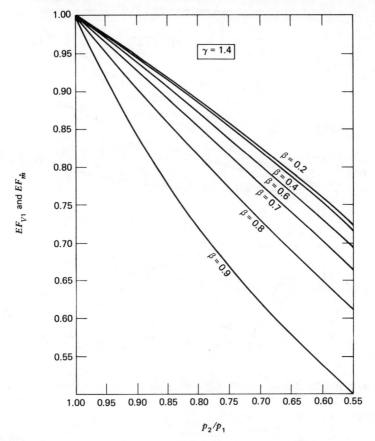

Figure 3.17 Isentropic velocity and flow expansion factors at inlet in terms of r, β, γ.

means that there must be a resemblance between the differential equations that characterize the two fields of study. If such is true, we can draw quantitative mathematical and experimental conclusions concerning phenomena in gas flow in terms of the analogous phenomena in water flow. To exploit the analogy, an *analog* (which in this case is an apparatus called a water table) must be available [16, 17].

To get down to specifics, high speed gas flows are costly to produce and control, and the required measuring schemes are necessarily esoteric and temperamental. However, a reliable, simplified, slow-motion water table analog can be used to study such high speed gas flows.

The analogous expressions, which relate the two-dimensional inviscid flows of a perfect gas and a constant density liquid, are next examined.

110 Solutions to Ideal-Compressible Pipe Flow

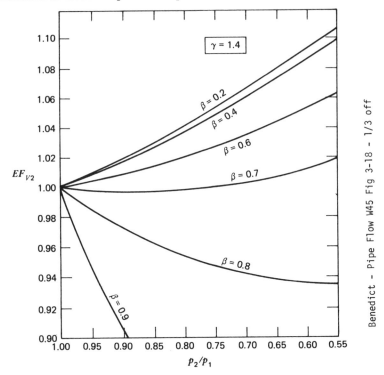

Figure 3.18 Isentropic velocity expansion factor at exit in terms of r, β, γ.

Equation of State. For the gas:

$$p = \rho \bar{R} T. \tag{3.94}$$

For the liquid:

$$p = \rho g / g_c (H - Z), \tag{3.95}$$

Here we have assumed that the liquid flow is in a channel as shown in Figure 3.20, and that vertical accelerations therein are negligible compared with the acceleration of gravity. Of course, (3.95) requires that

$$\frac{\partial p}{\partial x} = \rho \frac{g}{g_c} \frac{\partial H}{\partial x} \tag{3.96}$$

and similarly for $\partial p / \partial y$. As long as the free surface curvature of the liquid does not significantly influence the hydrostatic pressure distribution of (3.95), we can say that the water flow is quasi-two-dimensional.

Table 3.4 Pressure Expansion Factor ($\gamma = 1.4$)

					Beta				
p_2/p_1	0.1	0.2	0.3	0.4	0.5	0.6	0.7	0.8	0.9
0.99	1.00000	1.00000	1.00000	1.00000	0.99999	0.99998	0.99996	0.99988	0.99943
0.98	1.00000	1.00000	1.00000	0.99999	0.99997	0.99993	0.99982	0.99951	0.99779
0.97	1.00000	1.00000	0.99999	0.99997	0.99993	0.99984	0.99961	0.99891	0.99522
0.96	1.00000	1.00000	0.99999	0.99995	0.99988	0.99971	0.99930	0.99809	0.99180
0.95	1.00000	1.00000	0.99998	0.99993	0.99981	0.99955	0.99892	0.99706	0.98763
0.94	1.00000	0.99999	0.99997	0.99990	0.99973	0.99935	0.99846	0.99582	0.98280
0.93	1.00000	0.99999	0.99996	0.99986	0.99963	0.99912	0.99791	0.99439	0.97739
0.92	1.00000	0.99999	0.99994	0.99982	0.99952	0.99885	0.99729	0.99277	0.97145
0.91	1.00000	0.99999	0.99993	0.99977	0.99939	0.99855	0.99659	0.99097	0.96506
0.90	1.00000	0.99998	0.99991	0.99971	0.99925	0.99822	0.99582	0.98900	0.95826
0.85	1.00000	0.99996	0.99980	0.99936	0.99832	0.99606	0.99087	0.97682	0.91978
0.80	1.00000	0.99993	0.99965	0.99886	0.99703	0.99310	0.98427	0.96133	0.87687
0.75	0.99999	0.99989	0.99945	0.99822	0.99540	0.98937	0.87616	0.94321	0.83230
0.70	0.99999	0.99985	0.99921	0.99745	0.99342	0.98493	0.96670	0.92302	0.78772
0.65	0.99999	0.99979	0.99893	0.99653	0.99111	0.97980	0.95603	0.90122	0.74409
0.60	0.99998	0.99973	0.99860	0.99549	0.98848	0.97402	0.94426	0.87819	0.70196
0.55	0.99998	0.99965	0.99823	0.99431	0.98554	0.96762	0.93153	0.85426	0.66164

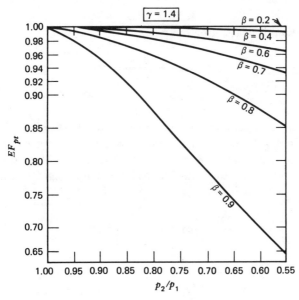

Figure 3.19 Isentropic total pressure expansion factor in terms of r, β, γ.

112 Solutions to Ideal-Compressible Pipe Flow

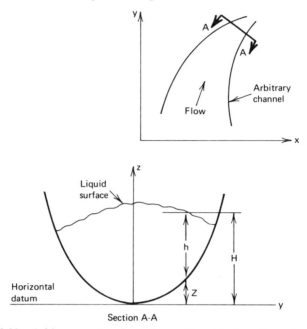

Figure 3.20 Arbitrary water channel considered in two-dimensional analysis.

Conservation of Mass. For the gas:

$$\frac{\partial}{\partial x}(\rho u_1) + \frac{\partial}{\partial y}(\rho u_2) = 0; \qquad (3.97)$$

this is obtained from (1.10).

For the liquid: we cannot use the ready-made (1.10) since it is applicable only to a fixed volume cubical element, which fails to account for variations in liquid depth. Instead, we consider an element of fluid as shown in Figure 3.21, for which the two-dimensional steady state continuity equation is

$$\frac{\partial}{\partial h}(hu_1) + \frac{\partial}{\partial y}(hu_2) = 0. \qquad (3.98)$$

There is an apparent paradox here since we started with a water channel of *arbitrary* geometry, whereas the two-dimensional gas flow requires a flow channel bounded by parallel plates. This is resolved when the depth (h) of

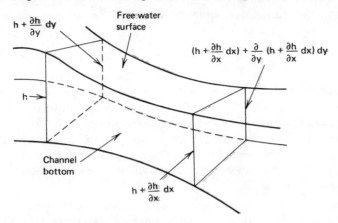

Figure 3.21 Liquid element used for continuity in two-dimensional flow.

(3.98) is replaced by $(H-Z)$ to obtain

$$h\frac{\partial u_1}{\partial x} + u_1\left(\frac{\partial H}{\partial x} - \frac{\partial Z}{\partial x}\right) + h\frac{\partial u_2}{\partial y} + u_2\left(\frac{\partial H}{\partial y} - \frac{\partial Z}{\partial y}\right) = 0 \qquad (3.99)$$

and when (3.97) is expanded to

$$\rho\frac{\partial u_1}{\partial x} + u_1\frac{\partial \rho}{\partial x} + \rho\frac{\partial u_2}{\partial y} + u_2\frac{\partial \rho}{\partial y} = 0. \qquad (3.100)$$

Now we see that a direct correspondence of (3.99) and (3.100) is possible only if $\partial Z/\partial x = \partial Z/\partial y = 0$. This requires a horizontal, flat-bottom water channel with vertical sides, so that $h = H$. With these restrictions observed, we note that h_{liquid} is analogous to ρ_{gas}.

Conservation of Momentum. Since we are limiting this inquiry to inviscid fluids, it is the two-dimensional Euler equations of (1.21) that apply.
For the gas these are

$$-g_c\frac{\partial p}{\partial x} = \rho\left(u_1\frac{\partial u_1}{\partial x} + u_2\frac{\partial u_2}{\partial y}\right) \qquad (3.101)$$

and similarly for $\partial p/\partial y$. The thermodynamic property (p) always can be expressed in terms of ρ and S as

$$\frac{\partial p}{\partial x} = \left(\frac{\partial p}{\partial \rho}\right)_S\left(\frac{\partial \rho}{\partial x}\right) + \left(\frac{\partial p}{\partial S}\right)_\rho\left(\frac{\partial S}{\partial x}\right) \qquad (3.102)$$

and similarly for $\partial p/\partial y$.

114 Solutions to Ideal-Compressible Pipe Flow

For the constant density liquid, we have for the Euler equations

$$-(gh)\frac{\partial h}{\partial x} = h\left(u_1\frac{\partial u_1}{\partial x} + u_2\frac{\partial u_1}{\partial y}\right) \qquad (3.103)$$

and similarly for $\partial h/\partial y$, where we have made use of relation (3.96).

A comparison of (3.101) and (3.102) with (3.103) indicates that direct correspondence between the gas and water flows can exist only if $\partial S/\partial x_k = 0$ in the gas flow. But this is the isentropic restriction which we are observing here. Thus again we see that ρ_{gas} is analogous to h_{liquid}. In addition, we note the important correspondence between $g_c(\partial p/\partial \rho)_S$ in the gas and gh in the liquid.

By (3.46), $\sqrt{g_c(\partial p/\partial \rho)_S} = a$, the acoustic velocity in the gas, which is the velocity of a *pressure wave*, of small amplitude, propagated by the elastic properties (i.e., density variations) of the gas. Similarly, *long gravity waves* are propagated in a liquid by the elastic properties (i.e., depth variations) of a free surface liquid at a wave velocity of $\sqrt{gh} = c$.

Conservation of Energy. As already discussed, energy relations are redundant when momentum has been applied. Nevertheless, we can view the analogy through the familiar concepts of thermodynamics and hydraulics if we make use of the general energy equation.

For the gas (3.19) yields

$$\frac{T_t}{T} = 1 + (\gamma - 1)\frac{M^2}{2}. \qquad (3.63)$$

For the liquid the Bernoulli equation of (2.1) can be given as

$$E_t = \frac{g}{g_c}h + \frac{V^2}{2g_c} \qquad (3.104)$$

by again using (3.95), where E_t is the total liquid energy per pound mass. This can be rearranged to

$$\frac{E_t}{(g/g_c)h} = 1 + \frac{Fr^2}{2} \qquad (3.105)$$

by defining the Froude number of a liquid as $Fr = V/c$.

A comparison of (3.63) and (3.105) indicates the direct correspondence of T_t and E_t, of T and $(g/g_c)h$, and of the Mach and Froude numbers, provided that $\gamma_{gas} = 2$. It further follows that $(T_2/T_1)_{gas}$ must equal $(h_2/h_1)_{liquid}$.

Comparisons between Compressible and Incompressible Treatments

Table 3.5 Analogous Terms for the Hydraulic-Gas Analogy

Perfect Gas	Constant Density Liquid
T_t/T	$E_t/h(g/g_c)$
T_2/T_1	h_2/h_1
ρ_2/ρ_1	h_2/h_1
p_2/p_1	$(h_2/h_1)^2$
M	Fr
V_2/V_1	V_2/V_1
$a^2 = \gamma p g_c/\rho$	$c^2 = gh$

In summary: A constant density, inviscid liquid flowing in a shallow, horizontal, open channel of rectangular cross section, where the free surface curvature is small compared to the depth, represents the adiabatic flow of an inviscid perfect gas of specific heat ratio, $\gamma = 2$, in a two-dimensional channel of rectangular cross section when there are no discontinuities in the flow. The exact analogous relations are given in Table 3.5.

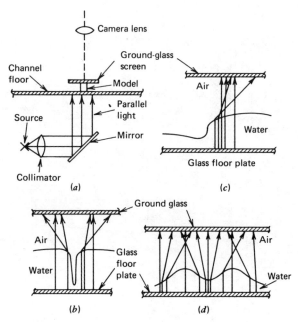

Figure 3.22 Analysis of method for photographing free surface of water table analog. (*a*) Shadowgraph ray diagram. (*b*) Vertex. (*c*) Hydraulic jump. (*d*) Capillary wave.

116 Solutions to Ideal-Compressible Pipe Flow

Often the value of γ varies between 1.1 (as for wet steam) and 1.4 (as for air). In such cases, interpretation of the results from the analogy developed for $\gamma = 2$ presents obvious difficulties. However, when it is possible to consider the gas flow to be *one-dimensional*, values of γ other than 2 can be simulated exactly by redesigning the water table channel bottom. For example, for $\gamma = 1.5$ a triangular channel is used, while for $\gamma = 1.4$ a parabolic channel is required. The references cited in this section should be consulted for these refinements to the basic analogy.

Having tabulated the various analogous terms and equations for the hydraulic-gas analogy, we have only to make visual observations, photographic recordings, or depth measurements in the water table to obtain gas flow solutions. The analog consists of a thin sheet of water (on the order of $\frac{1}{2}$ in. deep) flowing over a smooth, horizontal glass plate, with appropriate solid boundaries, mounted on the glass bottom, that are geometrically similar to those of the simulated gas flow models. Measurable depth variations in the liquid correspond to the required, but inaccessible, temperature, pressure, or density variations in the gas. Figure 3.22 shows how visual or photographic results are obtained. A schematic layout of a suitable water table analog is given in Figure 3.23.

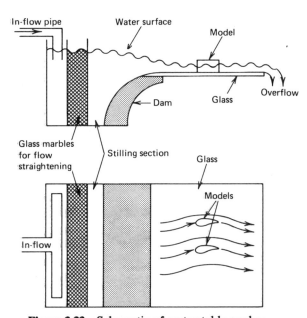

Figure 3.23 Schematic of water table analog.

3.7 IDEALIZED SOLUTIONS

To illustrate some of the ideas involved in working with compressible flows, ideal solutions to several typical piping element flow problems are given in this section. A combined continuity-energy equation is developed in a generalized form to make possible the easy solution of compressible flow problems with area change, as in a nozzle. The orifice contraction solution of Chapter 2 is extended to include subsonic and supercritical coefficients in terms of a generalized momentum equation.

3.7.1 Several Nozzle Solutions

In isentropic flow, since total pressure and total temperature are conserved, only an area change can cause a change in the thermodynamic state of a compressible fluid. It is instructive to develop an equation that combines the continuity and energy concepts, as an aid in solving isentropic problems that involve area change.

From continuity and the general velocity expression of (3.67) we get

$$\dot{m} = \rho_1 A_1 V_1 = \rho_2 A_2 V_2$$

or

$$\left(\frac{p_1}{RT_1}\right) A_1 \left\{ \left(\frac{2\gamma}{\gamma-1}\right)\left(\frac{p_t g_c}{\rho_t}\right)\left[1 - R_1^{(\gamma-1)/\gamma}\right] \right\}^{1/2}$$

$$= \left(\frac{p_2}{RT_2}\right) A_2 \left\{ \left(\frac{2\gamma}{\gamma-1}\right)\left(\frac{p_t g_c}{\rho_t}\right)\left[1 - R_2^{(\gamma-1)/\gamma}\right] \right\}^{1/2}. \quad (3.106)$$

Canceling where possible, and using the isentropic relation of (3.41), we find that this reduces to

$$\left(\frac{T_2}{T_1}\right)\left(\frac{p_1}{p_2}\right)\left(\frac{A_1}{A_2}\right)\left[1 - R_1^{(\gamma-1)/\gamma}\right]^{1/2} = \left[1 - R_2^{(\gamma-1)/\gamma}\right]^{1/2}. \quad (3.107)$$

By algebra we have

$$\left(\frac{A_2}{A_1}\right)\left(\frac{A_1}{A^*}\right) = \frac{A_2}{A^*}. \quad (3.108)$$

118 Solutions to Ideal-Compressible Pipe Flow

Here A^* is the area at the reference state, where the Mach number is unity. Combining (3.107) and (3.108), where state 2 is arbitrarily taken as the reference state, we obtain

$$\frac{A}{A^*} = \frac{[(\gamma-1)/(\gamma+1)]^{1/2}[2/(\gamma+1)]^{1/(\gamma-1)}}{R^{1/\gamma}[1-R^{(\gamma-1)/\gamma}]^{1/2}}. \qquad (3.109)$$

This area ratio relation is of such general usefulness that values for it are given in Table 3.6 for $\gamma = 1.4$ in terms of pressure ratio (R) of (3.109); Mach number (M) of (3.64); and temperature ratio (T/T_t) of (3.55) [18].

Example 3.5. If air flows through a nozzle that has an inlet area of 25 in.² and a throat area of 6 in.², find the maximum velocity at inlet if $T_1 = 80°F$.

Solution: Consider air to be a perfect gas at $\gamma = 1.4$. The maximum velocity at 1 will occur when the throat velocity is at a maximum. This will occur at $M_2 = 1$ and $A_2/A^* = 1$.

By (3.108):

$$\frac{A_1}{A^*} = \left(\frac{A_2}{A^*}\right)\left(\frac{A_1}{A_2}\right) = (1)\left(\frac{25}{6}\right) = 4.16667.$$

Table 3.6 Generalized Isentropic Compressible Flow Table ($\gamma = 1.4$)

Pressure Ratio, p/p_t	Temperature Ratio, T/T_t	Mach Number M	Area Ratio, A^*/A
1.00000	1.00000	0	0
0.98000	0.99424	0.17013	0.28894
0.96000	0.98840	0.24220	0.40412
0.94000	0.98248	0.29863	0.48938
0.92000	0.97646	0.34720	0.55858
0.90000	0.97035	0.39090	0.61715
0.88000	0.96414	0.43127	0.66789
0.86000	0.95782	0.46922	0.71249
0.84000	0.95141	0.50536	0.75203
0.82000	0.94488	0.54009	0.78729
0.80000	0.93823	0.57372	0.81880
0.78000	0.93147	0.60650	0.84701
0.76000	0.92458	0.63862	0.87222
0.74000	0.91757	0.67022	0.89469
0.72000	0.91041	0.70144	0.91464

Table 3.6 Continued

Pressure Ratio, p/p_t	Temperature Ratio, T/T_t	Mach Number M	Area Ratio, A^*/A
0.70000	0.90311	0.73239	0.93222
0.68000	0.89566	0.76318	0.94756
0.66000	0.88806	0.79389	0.96079
0.64000	0.88028	0.82461	0.97199
0.62000	0.87234	0.85542	0.98123
0.60000	0.86420	0.88639	0.98858
0.58000	0.85587	0.91761	0.99409
0.56000	0.84733	0.94914	0.99778
0.54000	0.83857	0.98107	0.99970
0.52828	0.83333	1.00000	1.00000
0.52000	0.82958	1.01348	0.99985
0.50000	0.82034	1.04645	0.99825
0.48000	0.81082	1.08008	0.99490
0.46000	0.80102	1.11446	0.98979
0.44000	0.79091	1.14969	0.98291
0.42000	0.78047	1.18591	0.97424
0.40000	0.76967	1.22324	0.96375
0.38000	0.75847	1.26183	0.95139
0.36000	0.74684	1.30186	0.93712
0.34000	0.73475	1.34353	0.92088
0.32000	0.72213	1.38707	0.90258
0.30000	0.70893	1.43277	0.88214
0.28000	0.69510	1.48096	0.85945
0.26000	0.68053	1.53205	0.83438
0.24000	0.66515	1.58655	0.80677
0.22000	0.64882	1.64510	0.77642
0.20000	0.63139	1.70853	0.74311
0.18000	0.61266	1.77795	0.70652
0.16000	0.59239	1.85484	0.66630
0.14000	0.57021	1.94130	0.62194
0.12000	0.54564	2.04046	0.57280
0.10000	0.51795	2.15719	0.51795
0.08000	0.48596	2.29978	0.45606
0.06000	0.44761	2.48403	0.38495
0.04000	0.39865	2.74634	0.30066
0.02000	0.32703	3.20769	0.19386
0.	0.	∞	0.

120 Solutions to Ideal-Compressible Pipe Flow

Linear interpolation in Table 3.6 yields $R_1 \simeq 0.9834$ and $\mathbf{M}_1 \simeq 0.1413$. More exact results, as by (3.109), are

$$R_1 = 0.9863, \quad \mathbf{M}_1 = 0.14054.$$

Hence $(V_1)_{\max} = \mathbf{M}_1 a = 0.14054 \sqrt{1.4 \times 32.174 \times 53.35 \times 540} = 160.1$ ft/sec. Equation 3.108 and Table 3.6 can be used to solve most but not all isentropic compressible flow problems. For example, if flow from a plenum inlet is considered, we see that $A_2/A_1 \simeq 0$, and (3.108) is useless. In such cases the total flow number of (3.68) provides an easy solution.

Example 3.6. Find the required flow area at 2 for isentropic flow if air flows at 1 lbm/sec from a plenum inlet at $p_1 = 20$ psia, $T_1 = 80°$F, and $p_2 = 15$ psia.

Solution: Since we have a plenum inlet, $p_1 = p_t =$ constant, $T_1 = T_t =$ constant, and $R_2 = p_2/p_t = 15/20 = 0.75$.
From (3.68):

$$FN_2 = \left(\frac{\dot{m}}{p_t A_2}\right)\left(\frac{\overline{R}T_t}{g_c}\right)^{1/2} = \left\{\left(\frac{2\gamma}{\gamma-1}\right)R_2^{2/\gamma}\left[1 - R_2^{(\gamma-1)/\gamma}\right]\right\}^{1/2}$$

$$= [7 \times 0.663004 \times 0.078908]^{1/2} = 0.605155.$$

Solving for A_2 gives

$$A_2 = \left(\frac{\dot{m}}{p_t}\right)\left(\frac{\overline{R}T_t}{g_c}\right)^{1/2}\left(\frac{1}{0.605155}\right) = 2.47238 \text{ in.}^2$$

3.7.2 Generalized Contraction Coefficient

In Section 2.6.2 we considered the contraction coefficient of an orifice with constant density fluid flow. In this section we develop a similar expression for ideal compressible fluid flow through an orifice [19].

Several authors [20–22] consider that the fluid forces on the walls of the orifice perpendicular to the flow would be the same for both liquid and gas flows. S. L. Bragg [23], of Rolls Royce, however, modified Jobson's plenum analysis by introducing an assumption that specifies a velocity distribution upstream of the orifice which is different for compressible, as compared to incompressible, flow.

Starting with (2.45), we include now the back pressure (p_b) (see Figure 3.24), since the jet pressure (p_3) need not correspond to the back pressure

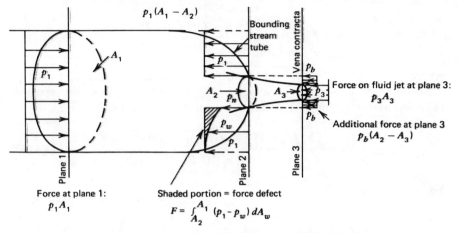

Figure 3.24 Generalized orifice for compressible flow.

for compressible flow. We obtain in general

$$p_1 A_1 - p_1(A_1 - A_2) + \int_{A_2}^{A_1}(p_1 - p_w)dA_w$$
$$- p_b(A_2 - A_3) - p_3 A_2 = \frac{\dot{m}(V_3 - V_1)}{g_c}, \quad (3.110)$$

which reduces to

$$(p_1 - p_b)A_2 - (p_3 - p_b)C_c A_2 + F_D = \frac{\dot{m}(V_3 - V_1)}{g_c}, \quad (3.111)$$

where $C_c = A_3/A_2$ according to (2.42), and F_D is the force defect as defined by the integral term of (3.110), just as described under (2.47). The first term of (3.111) can be written as

$$(p_1 - p_b)A_2 = p_1 A_2(1 - r_b), \quad (3.112)$$

where $r_b = p_b/p_1$.

The second term of (3.111) is

$$(p_3 - p_b)C_c A_2 = p_1 A_2(r_3 - r_b)C_c, \quad (3.113)$$

where $r_3 = p_3/p_1$, and is the significant static pressure ratio across the

122 Solutions to Ideal-Compressible Pipe Flow

orifice. The third term of (3.111) is first given, as in (2.49), as

$$F_D = f\left(\frac{\dot{m}^2}{\rho_1 A_2 g_c}\right) = \frac{f\rho_3^2 C_c^2 A_2 V_3^2}{\rho_1 g_c}, \qquad (3.114)$$

where f is a dimensionless force coefficient. To further evaluate (3.114), it is necessary to define the jet velocity (V_3). In terms of the static pressure ratio (r_3) across the orifice, V_3 can be given in terms of (3.62) and (3.70) as

$$V_3^2 = \frac{[2\gamma/(\gamma-1)](p_1 g_c/\rho_1)\left[1-r_3^{(\gamma-1)/\gamma}\right]}{1-r_3^{2/\gamma}C_c^2\beta^4}. \qquad (3.115)$$

Thus (3.114) becomes

$$F_D = p_1 A_2 \left\{ \frac{fC_c^2 r_3^{2/\gamma}\left[1-r_3^{(\gamma-1)/\gamma}\right]}{1-r_3^{2/\gamma}C_c^2\beta^4} \left(\frac{2\gamma}{\gamma-1}\right) \right\}. \qquad (3.116)$$

The fourth term of (3.111) is

$$\frac{\dot{m}(V_3 - V_1)}{g_c} = p_1 A_2 \left\{ \frac{C_c r_3^{1/\gamma}\left[1-r_3^{(\gamma-1)/\gamma}\right]}{1+r_3^{1/\gamma}C_c\beta^2} \left(\frac{2\gamma}{\gamma-1}\right) \right\}. \qquad (3.117)$$

Equations 3.112 through 3.117 combine according to (3.111) to yield the generalized cubic in C_c:

$$\left[\frac{Z(r_3-r_b)r_3^{1/\gamma}\beta^4}{2(1-r_b)}\right]C_c^3 + \left[r_3^{1/\gamma}\left(f+\beta^2 - Z\frac{\beta^4}{2}\right)\right]C_c^2$$

$$-\left[1+\frac{Z(r_3-r_b)}{2(1-r_b)r_3^{1/\gamma}}\right]C_c + \frac{Z}{2r_3^{1/\gamma}} = 0, \qquad (3.118)$$

where

$$Z = \frac{(1-r_b)(\gamma-1)}{\gamma\left[1-r_3^{(\gamma-1)/\gamma}\right]}. \qquad (3.119)$$

In (3.118) it is the compressible force coefficient (f) that remains the unknown.

Toward a solution, Bragg [23] recognized that the upstream fluid is compressed significantly under certain conditions (as when $C_{ci} > 0.65$ and when $r_3 < 0.7$). Hence he modified the uncompressed assumption by stating that it is the *mass velocity* pattern, rather than the velocity pattern, which is independent of the flow. He particularized the mass flow distribution on the stream tube boundary upstream of the orifice for plenum flow by specifying that

$$\rho_W V_W = \frac{k_0 \dot{m}}{A_W}. \tag{3.120}$$

In other words, Bragg introduced the assumption that the mass velocity (ρV) at any given radius in the boundary was proportional to the through-mass flow, while at a given flow rate the boundary mass velocity was inversely proportional to the square of the radius. The constant of proportionality (k_0), which is the *link* to be carried over to compressible flow, was

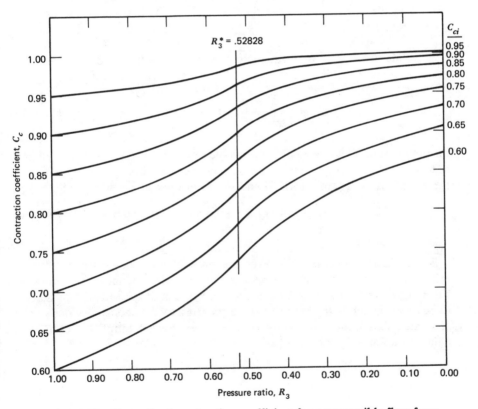

Figure 3.25 Generalized contraction coefficient for compressible flow from plenum inlet.

given as

$$k_0^2 = 2f_{i0}, \qquad (3.121)$$

where the subscript i signifies *incompressible*, and the subscript 0 indicates the $\beta = 0$ condition. The net result is that the compressible force coefficient for plenum inlet flow is

$$f_0 = f_{i0}\left\{\frac{2}{R_N^{1/\gamma}} - \frac{(\gamma-1)(1-R_N)}{\gamma R_N^{2/\gamma}[1-R_N^{(\gamma-1)/\gamma}]}\right\}, \qquad (3.122)$$

where R_N represents the static/total pressure ratio at a radius directly at the inner edge of the orifice opening. We see that (3.122) defines a compressible force coefficient in terms of the incompressible plenum force coefficient, modified by a compressibility correction factor. Note that (3.122) yields $f_0 = f_{i0}$ as R_N approaches unity, which corresponds to the incompressible case.

Subsonic Solution. In the subsonic case, r_3 necessarily equals r_b, and the general cubic of (3.118) reduces to a quadratic having the solution

$$C_{c0} = \frac{1}{2R_3^{1/\gamma}f_0}\left[1-(1-2Zf_0)^{1/2}\right] \qquad (3.123)$$

for the plenum inlet case. The inaccessible pressure ratio (R_N) at the orifice edge can be given in terms of the readily specified pressure ratio (R_3) at the vena contracta, from (3.120) and the total flow number of (3.68), as

$$R_N^{2/\gamma}\left[1-R_N^{(\gamma-1)/\gamma}\right] = k_0^2 C_c^2 R_3^{2/\gamma}\left[1-R_3^{(\gamma-1)/\gamma}\right]. \qquad (3.124)$$

Supercritical Solution. Whenever the pressure (p_b) downstream of the orifice is less than the critical pressure (p^*), the fluid at the vena contract will not yet have expanded to the back pressure, but instead will be taken to be at p^*, which is the pressure corresponding to the acoustic velocity (a) of (3.46). This limiting pressure ratio is given by the familiar thermodynamic relation

$$R^* = \frac{p^*}{p_t} = \left(\frac{2}{\gamma+1}\right)^{\gamma/(\gamma-1)}. \qquad (3.125)$$

Under such conditions the general cubic of (3.118) reduces to a quadratic having the solution

$$C_{c0}^* = \frac{1}{2R_*^{1/\gamma}f_0}\left\{\left[1+\frac{(R_*-R_b)}{\gamma R_*}\right]\right.$$
$$\left. -\left\{\left[1+\frac{(R_*-R_b)}{\gamma R_*}\right]^2 - \frac{2f_0(\gamma+1)(1-R_b)}{\gamma}\right\}^{1/2}\right\}, \quad (3.126)$$

where R_N is calculated from (3.124), with R_3 replaced by R^*.

The compressible contraction coefficients of (3.123) and (3.126) have been calculated, for $\gamma = 1.4$, for C_{ci} from 0.6 to 0.95, and for R_3 from 1 to 0. These analytical results are shown in Figure 3.25 and in Table 3.7. Typical comparisons between the generalized contraction coefficient and experimental results are shown in Figure 3.26 for $\beta = 0.2405$.

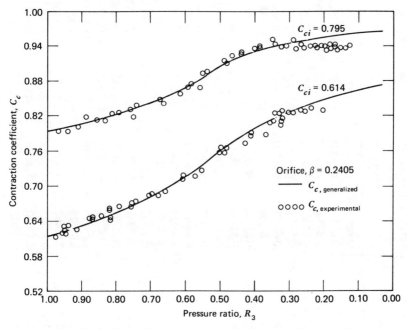

Figure 3.26 Comparisons between generalized contraction coefficient, analytical versus experimental.

Table 3.7 Generalized Compressible Contractions for Various Incompressible Contractions for Plenum Inlet Flow

R_3	Generalized				C_c	($\gamma=1.4, \beta=0$)			
1.00	0.60000	0.65000	0.70000	0.75000	0.80000	0.85000	0.90000	0.95000
0.98	0.60349	0.65331	0.70305	0.75273	0.80233	0.85186	0.90131	0.95069
0.96	0.60709	0.65673	0.70622	0.75556	0.80475	0.85379	0.90267	0.95141
0.94	0.61083	0.66028	0.70950	0.75849	0.80726	0.85579	0.90409	0.95216
0.92	0.61469	0.66395	0.71290	0.76154	0.80987	0.85788	0.90557	0.95295
0.90	0.61870	0.66777	0.71644	0.76472	0.81259	0.86006	0.90712	0.95376
0.88	0.62286	0.67173	0.72012	0.76802	0.81543	0.86233	0.90874	0.95462
0.86	0.62718	0.67585	0.72395	0.77146	0.81838	0.86471	0.91043	0.95552
0.84	0.63167	0.68014	0.72794	0.77505	0.82148	0.86720	0.91220	0.95647
0.82	0.63634	0.68460	0.73209	0.77880	0.82471	0.86980	0.91406	0.95746
0.80	0.64120	0.68925	0.73643	0.78272	0.82810	0.87254	0.91602	0.95852
0.78	0.64627	0.69411	0.74097	0.78682	0.83165	0.87541	0.91809	0.95963
0.76	0.65155	0.69918	0.74571	0.79112	0.83538	0.87844	0.92027	0.96080
0.74	0.65707	0.70449	0.75069	0.79564	0.83930	0.88164	0.92258	0.96206
0.72	0.66285	0.71004	0.75590	0.80038	0.84343	0.88501	0.92503	0.96340
0.70	0.66889	0.71587	0.76137	0.80537	0.84780	0.88859	0.92765	0.96484
0.68	0.67523	0.72198	0.76713	0.81063	0.85241	0.89238	0.93043	0.96638
0.66	0.68189	0.72841	0.77320	0.81618	0.85729	0.89642	0.93341	0.96805
0.64	0.68888	0.73518	0.77959	0.82206	0.86248	0.90072	0.93661	0.96986
0.62	0.69625	0.74231	0.78635	0.82828	0.86799	0.90532	0.94006	0.97184
0.60	0.70402	0.74985	0.79351	0.83489	0.87386	0.91025	0.94379	0.97402
0.58	0.71223	0.75783	0.80110	0.84191	0.88014	0.91556	0.94784	0.97644
0.56	0.72092	0.76629	0.80916	0.84940	0.88685	0.92127	0.95227	0.97915
0.54	0.73013	0.77528	0.81775	0.85740	0.89406	0.92745	0.95712	0.98221

0.52	0.73982	0.78473	0.82679	0.86584	0.90169	0.93402	0.96232	0.98557
0.50	0.74907	0.79369	0.83528	0.87368	0.90866	0.93988	0.96677	0.98819
0.48	0.75773	0.80201	0.84307	0.88075	0.91483	0.94492	0.97042	0.99010
0.46	0.76585	0.80973	0.85022	0.88717	0.92031	0.94928	0.97343	0.99153
0.44	0.77348	0.81692	0.85681	0.89299	0.92521	0.95306	0.97594	0.99263
0.42	0.78064	0.82362	0.86290	0.89831	0.92959	0.95638	0.97806	0.99349
0.40	0.78740	0.82988	0.86853	0.90316	0.93354	0.95930	0.97987	0.99418
0.38	0.79377	0.83575	0.87375	0.90761	0.93711	0.96189	0.98142	0.99475
0.36	0.79978	0.84124	0.87860	0.91170	0.94034	0.96419	0.98276	0.99522
0.34	0.80547	0.84640	0.88312	0.91547	0.94328	0.96625	0.98393	0.99562
0.32	0.81086	0.85126	0.88733	0.91895	0.94596	0.96810	0.98497	0.99596
0.30	0.81597	0.85583	0.89127	0.92218	0.94842	0.96976	0.98588	0.99624
0.28	0.82083	0.86014	0.89495	0.92517	0.95067	0.97127	0.98669	0.99650
0.26	0.82544	0.86422	0.89841	0.92796	0.95275	0.97265	0.98742	0.99672
0.24	0.82983	0.86808	0.90166	0.93055	0.95467	0.97390	0.98808	0.99692
0.22	0.83401	0.87173	0.90472	0.93297	0.95644	0.97505	0.98867	0.99709
0.20	0.83800	0.87519	0.90760	0.93524	0.95809	0.97611	0.98921	0.99725
0.18	0.84180	0.87848	0.91032	0.93737	0.95963	0.97708	0.98970	0.99739
0.16	0.84544	0.88160	0.91289	0.93937	0.96106	0.97799	0.99015	0.99751
0.14	0.84891	0.88458	0.91533	0.94124	0.96239	0.97882	0.99056	0.99763
0.12	0.85224	0.88741	0.91764	0.94301	0.96364	0.97960	0.99094	0.99774
0.10	0.85543	0.89011	0.91982	0.94468	0.96481	0.98032	0.99130	0.99783
0.08	0.85849	0.89269	0.92190	0.94626	0.96592	0.98099	0.99162	0.99792
0.06	0.86142	0.89516	0.92388	0.94776	0.96695	0.98162	0.99192	0.99800
0.04	0.86424	0.89752	0.92576	0.94917	0.96793	0.98222	0.99221	0.99808
0.02	0.86695	0.89977	0.92756	0.95052	0.96885	0.98277	0.99247	0.99815
0.00	0.86955	0.90194	0.92927	0.95179	0.96973	0.98329	0.99272	0.99821

REFERENCES

1. R. P. Benedict, "Essentials of thermodynamics," *Electro-Technol.*, Science and Engineering Series No. 43, July 1962, p. 108.
2. C. Caratheodory, "Investigations into the foundations of thermodynamics," *Math. Ann.*, Vol. 67, 1909, p. 355.
3. P. T. Lansberg, "Foundations of thermodynamics," *Rev. Mod. Phys.*, Vol. 28, 1956, p. 363.
4. J. N. Bronsted, *Principles and Problems in Energetics*, Interscience Publishers, 1955, pp. 40–61.
5. W. Thompson, "On the dynamical theory of heat, with numerical results deduced from Mr. Joule's equivalent of a thermal unit, and M. Regnault's observations in steam," Part 2, *Trans. R. Soc. Edinb.*, March 1851.
6. J. P. Joule and W. Thomson, "On the thermal effects of fluids in motion," Part 2, *Phil. Trans. R. Soc. London*, Vol. 144, June 15, 1854, p. 350.
7. S. Carnot, *Reflections on the Motive Power of Fire*, Dover Publications, New York, 1960.
8. J. A. Goff, *Thermodynamic Notes*, 4th ed., University of Pennsylvania Press, 1947.
9. R. P. Benedict, *Fundamentals of Temperature, Pressure, and Flow Measurements*, 2nd ed., John Wiley, 1977, p. 21.
10. R. P. Benedict, "Some comparisons between compressible and incompressible treatments of compressible fluids," *Trans. ASME, J. Basic Eng.*, September 1964, p. 527.
11. H. S. Bean, Ed., *Fluid Meters—Their Theory and Application*, Report of the ASME Research Committee on Fluid Meters, 6th ed., 1971.
12. R. P. Benedict, "Analog simulation," *Electro-Technol.*, Science and Engineering Series No. 60, December 1963, p. 73.
13. E. Preiswerk, "Application of the methods of gas dynamics to water flows with free surface," *NACA TM 934*, 1940.
14. R. A. A. Bryant, "The one-dimensional and two-dimensional gas dynamics analogies," *Austr. J. Appl. Sci.*, Vol. 7, No. 4, 1956, p. 296.
15. J. W. Hoyt, "The hydraulic analogy for compressible gas flow," *Appl. Mech. Rev.*, Vol. 15, June 1962.
16. A. H. Shapiro, "Free surface water table," in *High Speed Aerodynamics and Jet Propulsion*, Vol. IX, Princeton University Press, 1954.
17. W. H. T. Loh, "Hydraulic analogue for one-dimensional unsteady gas dynamics," *J. Franklin Inst.*, January 1960, p. 43.
18. R. P. Benedict and W. G. Steltz, *Handbook of Generalized Gas Dynamics*, Plenum Press, 1966.
19. R. P. Benedict, "Generalized contraction coefficient of an orifice for subsonic and supercritical flows," *Trans. ASME, J. Basic Eng.*, June 1971, p. 99.
20. E. Buckingham, "Note on contraction coefficients of jets of gas," *J. Res. Natl. Bur. Stand.*, Vol. 6, 1931, p. 765.
21. R. G. Cunningham, "Orifice meters with supercritical compressible flow," *Trans. ASME*, Vol. 73, July 1951, p. 625.

22. D. A. Jobson, "On the flow of a compressible fluid through orifices," *Proc. Inst. Mech. Eng.*, Vol. 169, No. 73, 1955, p. 767.
23. S. L. Bragg, "Effect of compressibility on the discharge coefficients of orifices and convergent nozzles," *J. Mech. Eng. Sci.*, Vol. 2, No. 1, 1960, p. 35.

NOMENCLATURE

Roman

- a acoustic velocity
- A area
- A^* area where Mach number is unity
- C_c contraction coefficient
- C_p specific heat at constant pressure
- C_v specific heat at constant volume
- EF expansion factor
- E_t total energy
- f force coefficient
- F loss term (as friction)
- F_D force defect
- Fr Froude number
- FN flow number
- g local gravity
- g_c gravitational constant
- h depth, specific enthalpy
- H depth
- J Joule's constant
- k ratio of specific heats
- k_0 constant of proportionality
- K constant
- K.E.   kinetic energy
- $\dot{m}$ mass flow rate
- M mass
- **M**   Mach number
- n polytropic exponent
- N function of p, v

130 Solutions to Ideal-Compressible Pipe Flow

- p pressure
- P general parameter
- P.E.   potential energy
- Q heat
- r static/static pressure ratio
- R static/total pressure ratio
- $\bar{R}$ gas constant
- R^* pressure ratio when Mach number is unity
- S entropy
- t empirical temperature
- T absolute temperature
- u internal energy per pound-mass, velocity component
- U internal energy
- v specific volume
- V average fluid velocity
- $\mathbf{V}$ volume
- W weight
- $\mathbf{W}$ work
- x, y Cartesian coordinates
- Z vertical displacement, combination of variables

Greek

- α cubical coefficient of expansion
- β angle, diameter ratio
- γ isentropic exponent
- Δ finite difference
- ε error
- θ temperature function
- ρ fluid density
- σ factor of p, v

Subscript

- t total

II
REAL FLUID CONCEPTS

Here we treat the several perturbations that are superimposed on the ideal, potential flow of Part I. These include the boundary layer that is built up on any solid boundary when a real fluid, with viscosity, flows. The various regions within the boundary layer are considered as to the velocity profiles, the various boundary layer thicknesses, and the modifications to the continuity, momentum, and energy equations across the boundary layer. Finally, the loss parameters that characterize boundary layer flow are treated in some detail.

4

The Boundary Layer

...now contemplate the tip of your wiggling minimus, your little finger, and the thought of what goes on there may make you laugh outright...
—*Gustav Eckstein*

4.1 GENERAL REMARKS

The various mathematical relations leading up to the boundary layer concept and the boundary layer relations are first considered. The experimental and theoretical bases for the boundary layer concept are briefly given. The different types of boundary layer, the transition criteria, and the separation phenomenon are discussed. Boundary layer parameters used to describe the shape and thickness of the several boundary layer types are defined mathematically and graphically where possible. Certain simplified applications of these relations, concepts, and parameters toward solutions to flat plate flow problems are given to illustrate the basic ideas of a boundary layer solution. Finally, a stepping type numerical solution, useful for more complex axisymmetric flow problems, is described in enough detail to make possible its application toward piping and piping component solutions in terms of the boundary layer.

4.2 THE CONCEPT

The development of the boundary layer concept can be traced quite clearly from the first formal equations of Bernoulli and Euler for describing *ideal flow*, through the *overcomplex real flow* equations of Navier and Stokes, to the *empirical* relations of Hagen, Poiseuille, Darcy, and Weisbach, finally reaching the *grand boundary layer hypothesis* of Prandtl.

4.2.1 Ideal Relations

Daniel Bernoulli, in 1738 [1], and Leonhard Euler, shortly thereafter [2], as successive professors of mathematics at St. Petersburg, Russia, first expounded on the principles of *ideal flow*. Such ideal flows form the subject matter of *hydrodynamics*. The name of this discipline was coined by Bernoulli to cover frictionless, incompressible flow in which the effects of viscosity are completely neglected.

In such flows, only the pressure and inertia forces are taken to be significant, whereas the tangential forces which are associated with shear stresses (and hence viscosity) are considered to exert only a negligible effect on the flow. The absence of shear stresses in pipe flow means quite simply that there are no variations in velocity, in a direction perpendicular to the pipe axis, and that as a consequence the fluid *must slip* past the solid boundaries. From another viewpoint, since the viscosity is considered to be vanishingly small, it follows that the Reynolds number (i.e., VD/ν) must be taken as approaching infinity in ideal flow.

This idea of very large Reynolds number flow is not so far-fetched when we consider that the very fluids in most common usage do indeed exhibit very small viscosities. For example, at room temperature, $\nu_{air} \simeq 6 \times 10^{-3}$ ft²/sec and $\nu_{water} \simeq 10^{-5}$ ft²/sec. Hence the Reynolds numbers, in general, will be quite large, and ordinarily these fluids can be treated as ideal over most of the flow passage.

Other names for the *ideal* flow of hydrodynamics are the *reversible* flow of thermodynamics, the *potential* flow of mathematics, and the *loss-free* flow of hydraulics.

Briefly, the equations to be associated with ideal incompressible flow in two-dimensional pipe flow are as follows.

For continuity:

$$\frac{du}{dx} + \frac{dv}{dy} = 0. \tag{4.1}$$

For energy:

$$\frac{p}{\rho} + \frac{V^2}{2g_c} + \frac{g}{g_c} Z = \text{constant}, \tag{4.2}$$

where (4.2) is the so-called Bernoulli equation.

For momentum:

$$\frac{du}{dt} = -\frac{g_c}{\rho} \frac{\partial p}{\partial x} - g \frac{\partial Z}{\partial x}, \tag{4.3}$$

where (4.3) is the Euler equation, in the x direction only, and all terms have units in the pound force-pound mass system of feet per second per second. The du/dt term in (4.3) is clarified, in two-dimensional flow, by the general definition

$$\frac{dP}{dt} = u\frac{\partial P}{\partial x} + v\frac{\partial P}{\partial y} + \frac{\partial P}{\partial t}, \qquad (4.4)$$

where dP/dt is the *total derivative* of a parameter P.

Although ideal formulations, such as the Bernoulli-Euler equations given above, serve very well in describing many flow situations, there are other flow phenomena that are intrinsically viscosity oriented and hence cannot be handled by the ideal relations. For example, solutions based on the ideal flow equations fail completely to account for form and pressure drag of immersed bodies, wakes behind bluff bodies, total pressure drop across piping components, entropy-increasing processes, skin friction along pipe walls, unrestrained expansions across abrupt area changes, secondary flows in elbows, irreversible mixing losses, and the like. This is so because all real fluids set up and transmit *tangential* as well as normal stresses (i.e., shear as well as pressure stresses), especially in regions near solid boundaries. It is these very tangential forces in fluids (i.e., the forces that are directly related to viscosity) that are overlooked by the ideal relations.

Hence, whereas ideal fluids are presumed to slip at fluid-solid interfaces, it is an experimentally confirmed fact that real fluids *do not slip*, but instead adhere to the solid boundaries. Since the slip and no-slip conditions constitute the main difference between real and ideal fluids, it follows that the ideal flow relations must break down in the vicinity of the flow boundaries.

4.2.2 Exact Relations

A giant step toward correcting this idealization of hydrodynamics was taken in 1827 by Louis Marie Henri Navier [3], who presented a purely mathematical momentum balance in which, in addition to the Euler terms, he included hypothetical forces between adjacent molecules. For example, in the x direction, Navier's equation can be expressed as

$$\frac{du}{dt} = -\frac{g_c}{\rho}\frac{\partial p}{\partial x} - g\frac{\partial Z}{\partial x} + \epsilon \nabla^2 u, \qquad (4.5)$$

where ϵ was taken as an unknown function of molecular spacing to which he attached no physical significance [4], and where ∇^2 represented, in

136 The Boundary Layer

two-dimensional flow, the Laplacian operator

$$\nabla^2 = \frac{\partial^2}{\partial x^2} + \frac{\partial^2}{\partial y^2}. \tag{4.6}$$

In 1849, George Gabriel Stokes [5] rederived the Navier equations, but replaced the unknown ϵ function with the physically significant $\mu/\rho = \nu$, where ν is now called the kinematic viscosity. Of course the celebrated Navier-Stokes equation, that is, (4.5) with ν replacing ϵ, reduces to Euler's ideal equation, (4.3), for inviscid flow.

Although the Navier-Stokes equations are believed to be generally valid in describing real fluid flow, no general proof to this effect has been advanced. There are, however, several isolated simplified cases that bear out this belief. Thus, while the Navier-Stokes equations include the viscosity effects, the overall flow problem has not yet been solved in practice. The reason is that solutions to the Navier-Stokes equations are rare for most flow situations because of the complexity of the resulting system of nonlinear, partial differential equations.

4.2.3 Empirical Relations

One approach toward a satisfying solution to the flow problems of general interest is to use the ideal Bernoulli-Euler equations to describe the overall flow situation, and then to correct the results by means of *empirical* equations. This field of endeavor is commonly known as *hydraulics*. For example, the Hagen-Poiseuille law (to be discussed in greater detail), namely,

$$\Delta p = \frac{128 \mu Q L}{\pi D^4 g_c}, \tag{4.7}$$

can be used to compute the pressure drop caused by the viscosity (μ) in *laminar* flow, whereas the Darcy-Weisbach equation (also to be discussed in greater detail), namely,

$$\Delta p = f \frac{L}{D} \frac{\rho V^2}{2 g_c}, \tag{4.8}$$

can be used to compute the pressure drop in *laminar* or *turbulent* flow in terms of f, an empirical friction factor discussed in Chapter 6.

Incidentally, it has long been noted that the empirical Hagen-Poiseuille law for pressure drop in laminar pipe flow is one of those experimentally based laws that are in complete accord with the Navier-Stokes equations. As mentioned before, however, the agreement of this one law cannot be taken as proof of the general validity of the Navier-Stokes equations.

4.2.4 Basic Boundary Layer Concept

Still another approach was taken by Ludwig Prandtl [6] in 1904, when, as professor of mechanical engineering at the Polytechnic Institute of Hannover, he advanced the concept of a *boundary layer* to explain certain experimental observations and, as it turns out, to help predict others. His grand hypothesis was simply this. If the flow in the main was well described by the ideal Bernoulli-Euler relations, and if the ideal relations broke down near solid boundaries, he would for purposes of analysis divide the flow into two parts. One part would concern a thin region immediately adjacent to the solid boundary (this is the boundary layer) where, at large Reynolds numbers, *all* the viscous effects are confined. The other part would include all the flow outside the boundary layer (this is the ideal, potential, loss-free core flow) where, at large Reynolds numbers, all viscous effects are to be ignored.

In the boundary layer the viscous forces are comparable in magnitude to the inertia forces (and hence the ideal flow relations are invalid here). And, in this region, the tangential velocities vary smoothly across the boundary layer, from zero at the walls (the no-slip condition) to the ideal, potential velocity of the core flow. It is this sharp change in velocity with distance from the wall (i.e., the large velocity gradient), rather than the magnitude of the velocity itself (which can be very small), that causes the shear stress,

$$\tau = \mu \frac{\partial u}{\partial y}, \qquad (4.9)$$

to be so important. Thus it is that here, in the boundary layer, are to be found all the flow losses.

Prandtl, who after his 1904 paper was called to be a professor at Gottingen, where he founded the world-famed Kaiser Wilhelm Institute for Flow Research, applied the Navier-Stokes equations, with their viscous terms, to his very thin boundary layer and, making certain bold assumptions, obtained boundary layer relations which were simplified enough so that approximate solutions could be obtained. To the core flow, where the velocity gradient is necessarily small, Prandtl applied the Bernoulli-Euler equations and accepted the ideal, potential solution as establishing the flow pattern, the pressure distribution, and so on.

Thus it is seen that Prandtl's boundary layer theory spans the gap between the *ideal* approach of hydrodynamics and the *empirical* approach of hydraulics, making use of the *exact* relations of Navier-Stokes as the bridge.

4.2.5 Boundary Layer Relations

Prandtl's assumptions can be listed as follows.

1. In the very thin boundary layer region, pressure is a function of x only. As a consequence, the $\partial p / \partial x$ term in the Navier-Stokes equation becomes dp/dx. A further implication of this assumption is that the pressure in the boundary layer is constant in a direction perpendicular to the flow, at a value equal to that in the core flow. This means that the pressure across the boundary layer, from the pipe wall to the core, is the same as that determined by the ideal flow equations. In effect, this one sweeping assumption of Prandtl's means that the core flow *forces* its pressure on the boundary layer flow also.
2. In the thin boundary layer region the flow direction is parallel to the pipe walls. As a consequence, the $\partial^2 u / \partial x^2$ term in the Navier-Stokes equation is negligible compared to the $\partial^2 u / \partial y^2$ term. In other words, shear stresses in the boundary layer are important only in a direction perpendicular to the flow, according to (4.9).

Applying assumptions 1 and 2 to the Navier-Stokes equation, (4.5), and neglecting gravitational forces in the boundary layer, yields for two-dimensional pipe flow

$$\frac{du}{dt} = -\frac{g_c}{\rho}\frac{dp}{dx} + \nu \frac{\partial^2 u}{\partial y^2}. \tag{4.10}$$

On comparing Prandtl's boundary equation, (4.10), with Euler's, (4.3), we see that they differ only in the viscous term. On comparing (4.10) with Navier-Stokes's equation, (4.5), we note the simplifications that have accrued through the Prandtl assumptions.

Thus the boundary layer problem has been reduced to this: Given a solid boundary-large Reynolds number flow (as in a pipe flow), with density (ρ) and viscosity (μ) taken as constants for simplicity, find the velocity components (u and v), from the continuity equation, (4.1), and the momentum equation, (4.10), where the pressure [$p(x)$] is to be obtained from the ideal flow relations. The mathematical boundary conditions are as follows:

1. At $y=0$, $u=v=0$ (which is the no-slip condition).
2. At the outer edge of the boundary layer, $u=U$, where U is the core velocity of potential flow.

3. The relation between pressure and velocity in the core is, by the Bernoulli equation, (4.2), $p + \rho U^2/2g_c =$ constant.

In spite of these simplifications, there are only a very limited number of exact solutions for Prandtl's boundary layer relations. One of these, that for a flat plate, will be discussed next since it embodies the boundary layer concepts that will carry over to pipe flow.

4.2.6 The Blasius Solution

A student and later collaborator of Prandtl at Gottingen, Paul Richard Heinrich Blasius, first applied boundary layer equations (4.1) and (4.10) to one of the simplest of cases, a flat plate in a zero pressure gradient, and in 1907 he did indeed arrive at a complete solution. The Blasius solution is not in closed form, however, and thus cannot be given as a single equation, but instead is based on a complex series solution [7]. His numerical methods need not concern us here since more exact numerical results have since been obtained, notably by L. Howarth [8] in 1938. Blasius succeeded in integrating the differential boundary layer equations and thus in obtaining the velocity distribution for a laminar boundary layer on a flat plate at zero angle of incidence. His results, based, however, on Howarth's numerical values, are shown graphically in Figure 4.1. The coordinates, in accord with dimensional analysis, are u/U and $y\sqrt{U/\nu x}$, where u is the boundary layer velocity in the x direction parallel to the plate, U is

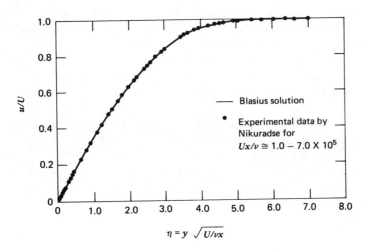

Figure 4.1 The first boundary layer solution—the Blasius velocity profile for a flat plate (after Li and Lam[9]).

the undisturbed free stream velocity outside the boundary layer, and y is the distance perpendicular to the plate and the flow direction. Really, the $y\sqrt{U/\nu x}$ parameter, called η by Blasius and Howarth, should be looked upon as proportional to the dimensionless variable y/δ, where δ is the y value when $u = 0.99 U$ and hence is a measure of the boundary layer thickness.

The velocity profile predicted by Blasius has been confirmed experimentally many times since he obtained his solution. In Figure 4.1, for example, the experimental points of Johann Nikuradse, a staff member at Gottingen, are shown, as indicated by Reference 9.

Other results based on the Blasius solution, such as δ^*, θ, C_f, and H, are discussed, more appropriately, in Sections 4.4 and 4.5 after these terms have been defined. Here, it is sufficient to note that Blasius' work was the first demonstration that the Prandtl boundary layer equations were indeed simplified enough to yield to mathematical analysis, and that at the same time such solutions were in agreement with the experimental facts.

That it was necessary for Blasius to limit his work to a *laminar* boundary layer implies that the other types of boundary layers and situations are possible. These will be discussed next.

4.3 BOUNDARY LAYER REGIMES

4.3.1 Laminar

As a result of microscopic viscous action between the various wall layers of a fluid, a boundary layer begins to build up onward from the leading edge of a flat plate or from the inlet to a pipe. This is normally a *laminar* boundary layer, so called because the fluid particles move in paths parallel to the boundaries, that is, since there are no radial velocity components, the individual fluid layers (or laminae) slide relative to each other with purely axial velocities.

In laminar flow the force balance is straightforward, there being no net inertial force to consider. In other words, the particles are accelerated by the pressure gradient in the flow direction, and retarded by the shear stresses set up by the velocity gradient in a direction perpendicular to the flow. The pressure drop that accompanies laminar flow varies linearly with velocity, as seen by (4.7), the Hagen-Poiseuille law.

Initially, the laminar boundary layer is very thin, but its thickness increases steadily as the viscous action extends further and further into the flow and retards more and more fluid.

4.3.2 Transition

In 1839, Gotthilf Heinrich Ludwig Hagen noted that laminar flow ceased when the *velocity* increased beyond a certain limit, and turbulent flow ensued. By 1854, Hagen had included *viscosity* as a variable affecting transition. Moreover, in 1869, Hagen concluded that transition from laminar to turbulent flow depended on three variables, V, μ, and D, the latter being the *pipe diameter* [10].

However, it was not until 1883 that Osborne Reynolds, in his classical work [11], found the unifying principle that others before him had sought. He showed, by dimensional analysis and by experiment, that *transition* from laminar to turbulent flow depends only on the single parameter that now bears his name: the Reynolds number, $R_D = VD/\nu$. Beyond a critical Reynolds number, on the order of 3000 for pipe flow, the laminar flow gives way abruptly to turbulent flow. Reynolds distinguished between these two very different flow regimes in terms of the path taken by fluid particles. In laminar flow the particles follow paths parallel to the flow boundaries. In turbulent flow the particles move in erratic, haphazard paths.

Transition also can be discussed in terms of the boundary layer concept. As the laminar boundary layer develops and thickens, after its inception, it becomes progressively more unstable, and eventually it transforms into a turbulent boundary layer. A stability theory has been given that explains the origin of turbulence in terms of arbitrary small disturbances which initiate instabilities in the laminar boundary layer and cause the breakdown of the laminar structure. Walter Gustav Johannes Tollmien [12], in his mathematical analysis of boundary layer stability, showed that stability was dependent on two parameters, δ^*/λ and $U\delta^*/\nu$, δ^* being the

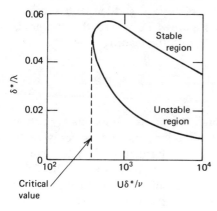

4.2 Stability of a laminar boundary layer on a flat plate (after Tollmien [12]).

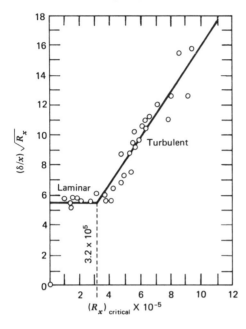

4.3 Boundary layer transition on a flat plate (after Schlichting [13], using Hansen's [14] points).

boundary layer displacement thickness (to be discussed), and λ being the wavelength of the small disturbances. Tollmien's results, given in Figure 4.2, indicate that, whenever the Reynolds number (based on δ^*) exceeds a certain critical value, the flow becomes unstable if disturbances of certain wavelengths are present. However, even though the laminar flow is in the unstable region, turbulent flow does not necessarily follow since the small disturbances must be amplified many times before transition takes place.

Experimentally, the abrupt transition from laminar to turbulent flow in the boundary layer is most clearly discernible by the sudden large increase in the boundary layer thickness (see Figure 4.3), where the Reynolds number (R_x) is based on the distance (x) from the leading edge of the flat plate.

4.3.3 Turbulent

As we have just seen, beyond certain high Reynolds numbers a transition takes place in the flow structure such that laminar flow gives way to turbulent flow. Perhaps turbulent flow is best characterized by the irregular secondary motion of the fluid particles at right angles to the flow, which is superimposed on the principal motion in the flow direction. This turbulent mixing (which is the macroscopic turbulent counterpart of the microscopic mixing of laminar flow):

Boundary Layer Regimes 143

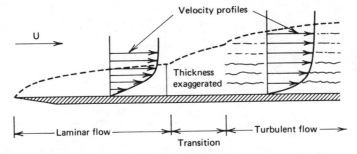

Figure 4.4 Schematic of boundary layer development on a flat plate.

1. Causes high energy momentum to be transferred into the boundary layer, with an attendant thickening of the boundary layer (see Figure 4.4).
2. Is accompanied by rapid fluctuations in velocity and pressure at every point, centered about steady mean values.
3. Causes the velocity distribution to be more uniform than in laminar flow, which simply means that the velocity gradient near the walls must be larger in turbulent flow.
4. Introduces apparent shear stresses, called Reynolds stresses, which can be many times larger than those caused by viscosity alone.
5. Causes the dissipation of large amounts of energy, so that the pressure drop that accompanies turbulent flow is much larger than that of laminar flow, being now proportional to the square of the velocity [see (4.8)].
6. Causes the effective viscosity and density to be very different so that their joint effect can no longer be lumped into ν, as in the laminar case.
7. Makes wall roughness of fundamental importance in regard to flow resistance, whereas in laminar flow it has no effect.

All in all, the turbulent boundary layer is considerably different from the laminar boundary layer, and special velocity profiles and friction factors will be developed in later chapters to handle this far more complex flow situation.

4.3.4 Separation

In certain piping components, such as diffusers, after abrupt area changes, and in flow over blunt bodies, the static pressure may *rise* in the direction of flow. In such cases, whereas the fluid in the outer layers need only face the rising pressure, the fluid near the walls must overcome the wall shear

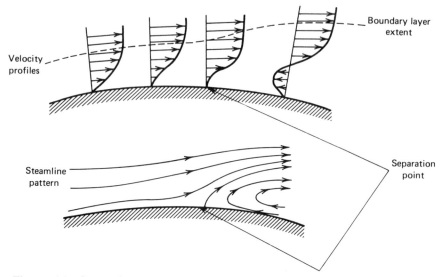

Figure 4.5 Separation: effect of an adverse pressure gradient on the boundary layer.

stress as well as the adverse pressure gradient. Hence the wall layers will be retarded more than the outer layers. Eventually, if these unbalanced forces continue to act over sufficient distances, the inner layer may come to rest at the surface, and then proceed to flow backwards.

At the point where there is no forward flow, the velocity gradient at the wall becomes zero, and hence the shear stress at the wall also becomes zero. Such a point is called the *separation point* because just beyond it, the reversed flow causes a rapid thickening of the boundary layer and the wall streamline is diverted outward and leaves the boundary (see Figure 4.5).

Large energy losses are to be associated with separated flow; the pressure distribution after separation no longer can be described by the ideal flow relations, nor can the flow region in the vicinity of the boundary be considered thin in the sense of the boundary layer approximations. Thus, although the skin friction effects are negligible after separation, the flow in no way can be considered ideal, and in addition all boundary layer solutions break down.

4.4 BOUNDARY LAYER PARAMETERS

In forming the Reynolds number, a reference size or length dimension must be used. In boundary layer work, along with the plate length (L) or the pipe diameter (D), a boundary layer thickness is often used. For example, we have already encountered $R_D = VD/\nu$, $R_L = VL/\nu$, and, in Tollmien's stability theory, $R_{\delta*} = V\delta*/\nu$. Now, as we apply the boundary layer equations, we will make use of $R_\theta = V\theta/\nu$. The latter two Reynolds numbers make use of $\delta*$, the boundary layer displacement thickness, and θ, the boundary layer momentum thickness. Thus it is clear that a variety of arbitrary measures of boundary layer thicknesses are used to describe certain dynamic attributes of the boundary layer.

4.4.1 Boundary Layer Thickness (δ)

The simplest measure of the boundary layer thickness at first seems to be the lateral or radial extent of the boundary layer. However, since the velocity (u) in the boundary layer approaches the core velocity (U) asymptotically, it is not at all obvious where to draw the line on the boundary layer extent. Certainly, the boundary layer thickness (δ) must be taken far enough from the wall to include *most* of the velocity changes, but just how far is enough remains an arbitrary judgment. Typical definitions for the boundary layer extent are as follows:

$$\delta_{99} = y \text{ where } u = 0.99\,U,$$

or

$$\delta_{995} = y \text{ where } u = 0.995\,U,$$

or even

$$\delta_\infty = y \text{ where } u = U.$$

The most commonly accepted definition of boundary layer extent is δ_{99}, and this is shown schematically in Figure 4.6.

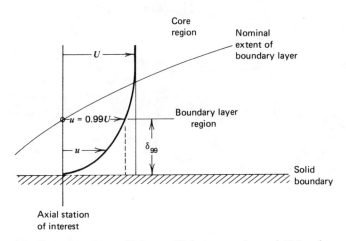

Figure 4.6 Boundary layer thickness (δ) in terms of u and U (u = boundary layer velocity, U = core velocity).

4.4.2 Displacement Thickness (δ^*)

In real flow, where viscous effects are admitted, there is a layer of retarded fluid adjacent to the solid boundaries within which, for axisymmetric pipe flow, the velocity, $u = f(r)$, necessarily varies from zero at the boundary to the free stream core velocity (U) at the outer extent of the boundary layer. This complex *real* flow situation, consisting of the boundary layer region and the inviscid core, can be transformed into an entirely equivalent hypothetical *ideal* flow situation, involving inviscid flow only, by introducing the concept of a displacement boundary layer thickness.

The general form used to describe mass flux (which is another name for mass flow rate) is $\dot{m} = \rho U A$. As applied to the boundary layer in a pipe we can write

$$\dot{m}_{\text{B.L.}} = 2\pi \rho U \int_{R-\delta}^{R} r\, dr - 2\pi \rho \int_{R-\delta}^{R} (U-u) r\, dr, \quad (4.11)$$

where the various terms and limits are shown in Figure 4.7. The first term in (4.11) represents the *ideal situation*, where the potential core velocity (U) is applied across the complete boundary layer. The second term in (4.11) is a *correction* introduced to account for the velocity defect ($U-u$) in the boundary layer.

Another expression for mass flux in the boundary layer can be written in terms of δ^*, a fictive boundary layer thickness such that, if the solid boundary were *displaced* inward by this amount, the potential core velocity alone could be used to determine mass flux over the remaining flow region. In other words,

$$\dot{m}_{\text{B.L.}} = 2\pi \rho U R (\delta - \delta^*). \quad (4.12)$$

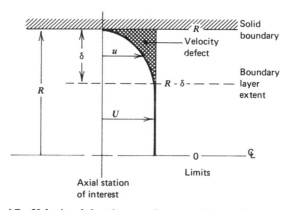

Figure 4.7 Velocity defect in an axisymmetric boundary layer.

Boundary Layer Parameters

The basis of (4.12) is clarified in Figure 4.8.

When (4.11) and (4.12) are equated, the defining equation for δ^* results [15]. Thus

$$\delta^* = \int_{R-\delta}^{R} \left(1 - \frac{u}{U}\right)\left(\frac{r}{R}\right) dr, \qquad (4.13)$$

which presumes only that $\delta^* \ll R$. A nondimensional form of (4.13) is

$$\left(\frac{\delta^*}{R}\right)_{inc} = \int_0^1 \left(1 - \frac{u}{U}\right)\left(\frac{r}{R}\right) d\left(\frac{r}{R}\right). \qquad (4.14)$$

A compressible form of (4.14) is

$$\left(\frac{\delta^*}{R}\right)_{comp} = \int_0^1 \left(1 - \frac{\rho u}{\rho_c U}\right)\left(\frac{r}{R}\right) d\left(\frac{r}{R}\right), \qquad (4.15)$$

where ρ_c is the density of the core fluid.

For reference in Chapter 14, we next derive an expression for mass flux, in terms of the displacement thickness, which applies over the *complete* flow passage:

$$\dot{m} = \int_A d\dot{m} = 2\pi\rho \int_0^R u r \, dr,$$

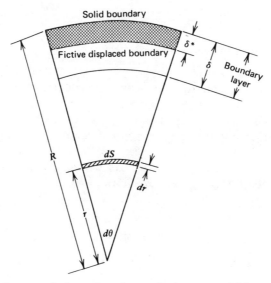

Figure 4.8 Axisymmetric boundary layer displacement thickness (δ^*). In general, $dA = dr\, dS = r\, dr\, d\theta$. For axial symmetry, $dA = 2\pi r\, dr, S = 2\pi R, A_\delta = 2\pi R\delta$.

or

$$\dot{m} = 2\pi\rho UR\left(\int_0^{R-\delta}\left(\frac{r}{R}\right)dr + \int_{R-\delta}^{R}\frac{u}{U}\frac{r}{R}dr\right),$$

which, combined with (4.13), becomes

$$\dot{m} = 2\pi\rho UR\left(\int_0^{R-\delta}\left(\frac{r}{R}\right)dr + \int_{R-\delta}^{R}\frac{r}{R}dr - \delta^*\right).$$

Finally, on simplification, we obtain [15]

$$\dot{m} = \rho(\pi R^2) U\left(1 - 2\frac{\delta^*}{R}\right). \tag{4.16}$$

The term $\rho(\pi R^2)U$ in (4.16) represents the *ideal* situation, where the potential core velocity (U) is applied across the complete flow passage. The second term, $[(1-2\delta^*/R)]$, is a *correction* that accounts for the presence of the boundary layer.

In summary: δ^*, the displacement thickness defined by (4.13), is the amount by which the solid boundary must, in principle, be displaced radially into the flow passage to account for the defect of mass flow in the boundary layer and still allow the fluid in the remaining region ($0 \leq r \leq R - \delta^*$) to be treated (*concerning mass flux*) as in potential flow at the free stream core velocity.

4.4.3 Momentum Thickness (θ)

The retardation of flow in the boundary layer causes a momentum flux defect in comparison to the ideal flow situation, in a manner exactly analogous to the mass flux defect just described. This complex real flow situation can be transformed into an equivalent hypothetical ideal situation by introducing the concept of a momentum boundary layer thickness.

The general form used to describe momentum flux is $\dot{M} = U\dot{m}$. As applied to the boundary layer, we can write the defect of momentum flux in the boundary layer for axisymmetric flow as

$$\dot{M}_{\text{defect}} = \int_{R-\delta}^{R} (U-u)(\rho u 2\pi r\, dr). \tag{4.17}$$

Another expression for this momentum flux defect in the boundary layer can be written in terms of θ, a fictive boundary layer thickness such that the potential core velocity alone can be used to determine the momentum flux defect, as

$$\dot{M}_{\text{defect}} = U(\rho U 2\pi R\theta). \tag{4.18}$$

When (4.17) and (4.18) are equated, the defining equation for θ results. Thus

$$\theta = \int_{R-\delta}^{R} \left(\frac{u}{U}\right)\left(1 - \frac{u}{U}\right)\left(\frac{r}{R}\right) dr. \tag{4.19}$$

A nondimensional form of (4.19) is

$$\left(\frac{\theta}{R}\right)_{\text{inc}} = \int_{0}^{1} \left(\frac{u}{U}\right)\left(1 - \frac{u}{U}\right)\left(\frac{r}{R}\right) d\left(\frac{r}{R}\right). \tag{4.20}$$

A compressible form of (4.20) is

$$\left(\frac{\theta}{R}\right)_{\text{comp}} = \int_{0}^{1} \left(\frac{\rho u}{\rho_c U}\right)\left(1 - \frac{u}{U}\right)\left(\frac{r}{R}\right) d\left(\frac{r}{R}\right). \tag{4.21}$$

We next derive an expression for momentum flux, in terms of momentum thickness, which applies over the complete flow passage:

$$\dot{M} = \int_{A} (MU) d\dot{m} = 2\pi\rho \int_{0}^{R} u^2 r \, dr$$

or

$$\dot{M} = 2\pi\rho U^2 R \left[\int_{0}^{R-\delta} \left(\frac{r}{R}\right) dr + \int_{R-\delta}^{R} \left(\frac{u}{U}\right)^2 \frac{r}{R} dr \right],$$

which, combined with (4.19), becomes

$$\dot{M} = 2\pi\rho U^2 R \left(\int_{0}^{R-\delta} \left(\frac{r}{R}\right) dr + \int_{R-\delta}^{R} \frac{u}{U} \frac{r}{R} dr - \theta \right),$$

and, upon introducing (4.13), becomes

$$\dot{M} = 2\pi\rho U^2 R \left(\int_{0}^{R} \frac{r}{R} dr - \delta^* - \theta \right).$$

On simplification we obtain

$$\dot{M} = \rho(\pi R^2) U^2 \left(1 - 2\frac{\delta^*}{R} - 2\frac{\theta}{R} \right).$$

150 The Boundary Layer

As before, the first term above represents the ideal momentum flux, while the second term is a correction to account for the boundary layer effects.

In summary: θ, the momentum thickness defined by (4.19), is the amount of the boundary layer thickness that must be used to account for the defect in momentum flux when the potential core velocity is to be used.

4.4.4 Energy Thickness (δ^{**})

In addition to the mass flux defect and the momentum flux defect in the boundary layer, there is an analogous kinetic energy flux defect, which we will account for by introducing the concept of an energy boundary layer thickness.

The general form used to describe kinetic energy flow is $\dot{K}.E. = (U^2/2g_c)\dot{m}$. As applied to the boundary layer, we can write the defect of kinetic energy flux in the boundary layer for axisymmetric flow as

$$\dot{K}.E._{\text{defect}} = \int_{R-\delta}^{R} \left(\frac{U^2 - u^2}{2g_c} \right)(\rho u 2\pi r\, dr). \tag{4.22}$$

Another expression for this energy flux defect in the boundary layer can be written in terms of δ^{**}, a fictive boundary layer thickness such that the potential core velocity alone can be used to determine the energy flux defect, as

$$\dot{K}.E._{\text{defect}} = \left(\frac{U^2}{2g_c} \right)(\rho U 2\pi R \delta^{**}). \tag{4.23}$$

When (4.22) and (4.23) are equated, the defining equation for δ^{**} results [15]. Thus

$$\delta^{**} = \int_{R-\delta}^{R} \left(\frac{u}{U}\right)\left(1 - \frac{u^2}{U^2}\right)\left(\frac{r}{R}\right) dr. \tag{4.24}$$

A nondimensional form of (4.24) is

$$\left(\frac{\delta^{**}}{R}\right)_{\text{inc}} = \int_{0}^{1} \left(\frac{u}{U}\right)\left(1 - \frac{u^2}{U^2}\right)\left(\frac{r}{R}\right) d\left(\frac{r}{R}\right). \tag{4.25}$$

A compressible form of (4.25) is

$$\left(\frac{\delta^{**}}{R}\right)_{\text{comp}} = \int_{0}^{1} \left(\frac{\rho u}{\rho_c U}\right)\left(1 - \frac{u^2}{U^2}\right)\left(\frac{r}{R}\right) d\left(\frac{r}{R}\right). \tag{4.26}$$

Boundary Layer Parameters 151

For reference in Chapter 14, we next derive an expression for the total energy flux lost over the length, $(x_2 - x_1)$. For axisymmetric flow, the energy flux loss at an axial position is given in terms of (4.22), (4.23), and (4.59) as

$$\frac{d}{dx}\left[\left(\frac{U^2}{2g_c}\right)(\rho U 2\pi R \delta^{**})\right] = \mu \int_{R-\delta}^{R}\left(\frac{\partial u}{\partial r}\right)^2 (2\pi r\, dr).$$

Strictly speaking, this energy equation applies only to laminar boundary layers, but the left-hand side (which we will use) applies also to turbulent boundary layers ([13], p. 125).

Separating variables and integrating over the length, $(x_2 - x_1)$, we obtain [16]

$$\left(\frac{2\pi\rho}{2g_c}\right)\int_1^2 d(RU^3\delta^{**}) = (2\pi\mu)\int_{x_1}^{x_2}\int_{R-\delta}^{R}\left(\frac{\partial u}{\partial r}\right)^2 r\, dr\, dx.$$

Hence the total energy flux lost between x_1 and x_2 is

$$(\dot{E}_{\text{loss}})_{1,2} = \left(\frac{2\pi\rho}{2g_c}\right)(R_2 U_2^3 \delta_2^{**} - R_1 U_1^3 \delta_1^{**}). \tag{4.27}$$

Also for future reference, we derive an expression for kinetic energy flux, in terms of energy thickness, that applies over the complete flow passage:

$$\dot{K}.E. = \int_A K.E.\, d\dot{m} = \frac{2\pi\rho}{2g_c}\int_0^R u^3 r\, dr$$

or

$$\dot{K}.E. = \frac{2\pi\rho U^3 R}{2g_c}\left(\int_0^{R-\delta}\left(\frac{r}{R}\right) dr + \int_{R-\delta}^{R}\frac{u^3}{U^3}\frac{r}{R}\, dr\right),$$

which, combined with (4.13) and (4.25), becomes

$$\dot{K}.E. = \frac{2\pi\rho U^3 R}{2g_c}\left(\int_0^R \frac{r}{R}\, dr - \delta^* - \delta^{**}\right).$$

On simplification, we obtain [15]

$$\dot{K}.E. = \frac{\rho\pi R^2 U^3}{2g_c}\left(1 - 2\frac{\delta^*}{R} - 2\frac{\delta^{**}}{R}\right). \tag{4.28}$$

152 The Boundary Layer

As before, the first term in (4.28) represents the ideal kinetic energy flux, while the second term is a correction to account for the boundary layer effects.

In summary: δ^{**}, the energy thickness defined by (4.24), is the amount of the boundary layer thickness that must be used to account for the defect in kinetic energy flux when the potential core velocity is to be used.

4.4.5 Comparisons

The various boundary layer thickness equations are summarized in Table 4.1.

Since it is seldom necessary to specify a detailed boundary layer velocity distribution to obtain useful results, we next illustrate the relative magnitudes of the various boundary layer thickness parameters by assuming a reasonable boundary layer velocity profile. For example, to describe the velocity profile in a laminar boundary layer on a flat plate, we can assume [9] that

$$\frac{u}{U} = \sin\left(\frac{\pi}{2}\frac{y}{\delta}\right). \tag{4.29}$$

This profile yields, according to Table 4.1, for the displacement thickness

$$\frac{\delta^*}{\delta} = \int_0^1 \left[1 - \sin\left(\frac{\pi}{2}\frac{y}{\delta}\right)\right] d\left(\frac{y}{\delta}\right),$$

for the momentum thickness

$$\frac{\theta}{\delta} = \int_0^1 \sin\left(\frac{\pi}{2}\frac{y}{\delta}\right) d\left(\frac{y}{\delta}\right) - \int_0^1 \sin^2\left(\frac{\pi}{2}\frac{y}{\delta}\right) d\left(\frac{y}{\delta}\right),$$

and for the energy thickness

$$\frac{\delta^{**}}{\delta} = \int_0^1 \sin\left(\frac{\pi}{2}\frac{y}{\delta}\right) d\left(\frac{y}{\delta}\right) - \int_0^1 \sin^3\left(\frac{\pi}{2}\frac{y}{\delta}\right) d\left(\frac{y}{\delta}\right).$$

Making use of the general forms

$$\int_0^1 \sin ax\, dx = -\frac{1}{a}\cos ax,$$

$$\int_0^1 \sin^2 ax\, dx = \frac{x}{2} - \frac{\sin 2ax}{4a}, \tag{4.30}$$

Table 4.1 Summary of Defining Equations for Various Boundary Layer Thickness Parameters

Boundary Layer Thickness Parameter	Flat Plate Incompressible	Application Axisymmetric Pipe Incompressible	Axisymmetric Pipe Compressible
Displacement	$\dfrac{\delta^*}{\delta} = \int_0^1 \left(1 - \dfrac{u}{U}\right) d\left(\dfrac{y}{\delta}\right)$	$\dfrac{\delta^*}{R} = \int_0^1 \left(1 - \dfrac{u}{U}\right)\left(\dfrac{r}{R}\right) d\left(\dfrac{r}{R}\right)$	$\dfrac{\delta^*}{R} = \int_0^1 \left(1 - \dfrac{\rho u}{\rho_c U}\right)\left(\dfrac{r}{R}\right) d\left(\dfrac{r}{R}\right)$
Momentum	$\dfrac{\theta}{\delta} = \int_0^1 \left(\dfrac{u}{U}\right)\left(1 - \dfrac{u}{U}\right) d\left(\dfrac{y}{\delta}\right)$	$\dfrac{\theta}{R} = \int_0^1 \left(\dfrac{u}{U}\right)\left(1 - \dfrac{u}{U}\right)\left(\dfrac{r}{R}\right) d\left(\dfrac{r}{R}\right)$	$\dfrac{\theta}{R} = \int_0^1 \left(\dfrac{\rho u}{\rho_c U}\right)\left(1 - \dfrac{u}{U}\right)\left(\dfrac{r}{R}\right) d\left(\dfrac{r}{R}\right)$
Energy	$\dfrac{\delta^{**}}{\delta} = \int_0^1 \left(\dfrac{u}{U}\right)\left(1 - \dfrac{u^2}{U^2}\right) d\left(\dfrac{y}{\delta}\right)$	$\dfrac{\delta^{**}}{R} = \int_0^1 \left(\dfrac{u}{U}\right)\left(1 - \dfrac{u^2}{U^2}\right)\left(\dfrac{r}{R}\right) d\left(\dfrac{r}{R}\right)$	$\dfrac{\delta^{**}}{R} = \int_0^1 \left(\dfrac{\rho u}{\rho_c U}\right)\left(1 - \dfrac{u^2}{U^2}\right)\left(\dfrac{r}{R}\right) d\left(\dfrac{r}{R}\right)$

and

$$\int_0^1 \sin^3 ax\, dx = -\frac{1}{a}\cos ax + \frac{1}{3a}\cos^3 ax.$$

yields

$$\frac{\delta^*}{\delta} = 0.36338,$$

$$\frac{\theta}{\delta} = 0.13662, \qquad (4.31)$$

and

$$\frac{\delta^{**}}{\delta} = 0.21221.$$

These results are compared graphically in Figure 4.9.

4.4.6 Shape Factors

Two ratios of the boundary layer thickness have been found useful to specify the shape of the boundary layer velocity profile. These are

$$H = \frac{\delta^{**}}{\theta} \qquad (4.32)$$

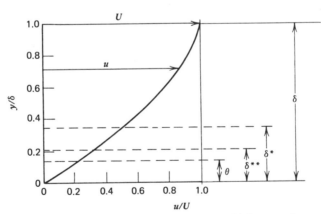

Figure 4.9 Comparison of various boundary layer thicknesses for an arbitrary laminar velocity profile on a flat plate.

and

$$H_{12} = \frac{\delta^*}{\theta}. \tag{4.33}$$

We will have occasion to discuss these shape factors in more detail in Section 4.6, when we consider more complex boundary layer solutions.

Here, as a brief illustration, we can compare the results from our approximate profile of (4.29) with those based on the more exact profile of the Blasius solution given in Figure 4.1. Many references (e.g., 9, 10, 13), indicate that the Blasius solution leads to the following values:

$$\left(\frac{\delta}{x}\right)_B = \frac{4.96}{\sqrt{R_x}}, \quad \left(\frac{\delta^*}{x}\right)_B = \frac{1.721}{\sqrt{R_x}}, \quad \left(\frac{\theta}{x}\right)_B = \frac{0.664}{\sqrt{R_x}}. \tag{4.34}$$

These can also be expressed, by simple ratioing, in the following forms:

$$\left(\frac{\delta^*}{\delta}\right)_B = 0.34698, \quad \left(\frac{\theta}{\delta}\right)_B = 0.13387, \quad (H_{12})_B = 2.592. \tag{4.35}$$

The approximate results given in (4.31) are seen to compare favorably with the more exact values of (4.35), confirming the assertion that the exact velocity profile need not be known to obtain useful information about the boundary layer.

The quantity $(\delta/x)_{\text{approx}}$, based on (4.29), will be obtained next, to allow further comparison, as we apply the boundary layer equations to a simplified flat plate example.

4.5 SIMPLIFIED APPLICATIONS

4.5.1 General Continuity Considerations

For steady flow through a control volume, the net mass flow crossing the control surface is zero. In equation form, for unit width, this is expressed as

$$\dot{m}_{\text{net}} = \Sigma \dot{m}_S = \oint \rho (q \cos \gamma) \, dS = 0, \tag{4.36}$$

where $q \cos \gamma$ is the velocity vector perpendicular to the differential surface (dS).

4.5.2 General Momentum Considerations

From Newton's second law the time rate of change of momentum equals the net external force. For steady flow through a control volume, this net force is simply the sum of the momentum fluxes crossing the control surface. In equation form, for unit width, this is expressed as

$$\Sigma F_x = \Sigma \dot{M}_s = \frac{1}{g_c} \oint_S u\rho (q \cos \gamma) \, dS, \qquad (4.37)$$

where u is the x component of the velocity.

4.5.3 General Flat Plate Considerations

As the basis of a simplified application of the boundary layer equations, consider next the control surface on a flat plate, as shown in Figure 4.10. Note that $\cos \gamma$ for the rectangular control surface will be -1 or $+1$, depending on whether the direction of the particular velocity is *into* or *out of* the control volume, respectively. Referring to Figure 4.10, and applying continuity and momentum expressions to each straight line section of the control surface, we obtain the following. From (4.36) we have

$$\dot{m}_{OA} = \int_0^\delta \rho(-U)\,dy = -\rho U \delta,$$

$$\dot{m}_{AB} = \int_0^x \rho(v)\,dx,$$

$$\dot{m}_{BC} = \int_0^\delta \rho(u)\,dy,$$

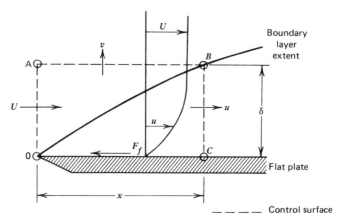

Figure 4.10 Control surface and notation for a flat plate.

and, summing around the complete surface,

$$\dot{m}_{AB} = -\dot{m}_{OA} - \dot{m}_{BC}$$

or

$$\dot{m}_{AB} = \rho U \delta - \int_0^\delta \rho u \, dy. \qquad (4.38)$$

From (4.37) we have

$$g_c \dot{M}_{OA} = \int_0^\delta \rho U(-U) \, dy = -\rho U^2 \delta,$$

$$g_c \dot{M}_{AB} = \int_0^x \rho U(v) \, dx = U \int_0^x \rho v \, dx,$$

$$g_c \dot{M}_{BC} = \int_0^\delta \rho u(u) \, dy = \rho \int_0^\delta u^2 \, dy,$$

where, from (4.38),

$$g_c \dot{M}_{AB} = \rho U^2 \delta - \rho U \int_0^\delta u \, dy,$$

in the pound force-pound mass system of units.

Since a uniform pressure is assumed to prevail over a flat plate, it is clear that the only force which opposes the inertial force set up by the net momentum flux is the force of friction (F_f), according to Figure 4.10. Thus

$$g_c \Sigma F_x = -g_c F_f = -\rho U^2 \delta + \left(\rho U^2 \delta - \rho U \int_0^\delta u \, dy \right) + \rho \int_0^\delta u^2 \, dy.$$

Simplifying, we obtain a general expression for the total *frictional force* exerted by a flat plate over the length x, namely,

$$F_f = \frac{\rho}{g_c} \left(U \int_0^\delta u \, dy - \int_0^\delta u^2 \, dy \right). \qquad (4.39)$$

Primarily, from a boundary layer analysis, we are interested in the wall shear stress (τ_0), the local skin friction coefficient (C_f), and the boundary layer thickness (δ). For unit width of a flat plate, general expressions for these quantities can be given as

$$\tau_0 = \frac{dF_f}{dx} \qquad (4.40)$$

and

$$C_f = \frac{\tau_0}{\rho U^2/2g_c}. \tag{4.41}$$

It remains to estimate δ/x, τ_0, and C_f, given reasonable approximations to the velocity profile, for both laminar and turbulent boundary layers.

4.5.4 Laminar Boundary Layer on a Flat Plate

Once again, it is convenient to assume the velocity profile of (4.29) for this case. Then (4.39) becomes

$$F_f = \frac{\rho U^2}{g_c}\left[\int_0^\delta \sin\left(\frac{\pi}{2}\frac{y}{\delta}\right)dy - \int_0^\delta \sin^2\left(\frac{\pi}{2}\frac{y}{\delta}\right)dy\right] = \left(\frac{4-\pi}{2\pi}\right)\frac{\rho U^2 \delta}{g_c}, \tag{4.42}$$

making use of the general forms of (4.30).

From (4.40) and (4.42), we obtain for the wall shear stress

$$\tau_0 = \left(\frac{4-\pi}{2\pi}\right)\frac{\rho U^2}{g_c}\frac{d\delta}{dx}, \tag{4.43}$$

while, from (4.9), we obtain as another expression

$$\tau_0 = \frac{\mu}{g_c}\left[U\cos\left(\frac{\pi}{2}\frac{y}{\delta}\right)_{y=0}\left(\frac{\pi}{2\delta}\right)\right] = \frac{\mu U \pi}{2g_c\delta}, \tag{4.44}$$

making use of the general form $d/dx(\sin u) = \cos u\, du/dx$.

Equating (4.43) and (4.44), separating the variables (δ and x), and integrating, we have

$$\left(\frac{\delta}{x}\right)_{\text{approx}} = \left[\left(\frac{2\pi^2}{4-\pi}\right)\left(\frac{\nu}{Ux}\right)\right]^{1/2} = \frac{4.795}{\sqrt{R_x}}, \tag{4.45}$$

which compares quite favorably with the Blasius result of (4.34), namely,

$$\left(\frac{\delta}{x}\right)_B = \frac{4.96}{\sqrt{R_x}}.$$

From (4.44) and (4.45) there results

$$(\tau_0)_{\text{approx}} = \frac{\mu U \pi}{2g_c(4.795x/\sqrt{R_x})} = \frac{0.328\rho U^2}{\sqrt{R_x}\, g_c}.$$

And finally, when this expression for τ_0 is combined with (4.41), we obtain

$$(C_f)_{\text{approx}} = \frac{0.655}{\sqrt{R_x}},$$

which compares favorably with the accepted Blasius result of

$$(C_f)_B = \frac{0.664}{\sqrt{R_x}}.$$

4.5.5 Turbulent Boundary Layer on a Flat Plate

For this case, it is convenient to assume the 1/7 power law velocity profile of (5.13), namely, $u = U(y/\delta)^{1/7}$. Then (4.39) leads to

$$F_f = \frac{\rho U^2}{g_c}\left[\int_0^\delta \left(\frac{y}{\delta}\right)^{1/7} dy - \int_0^\delta \left(\frac{y}{\delta}\right)^{2/7} dy\right] = \left(\frac{7}{72}\right)\frac{\rho U^2 \delta}{g_c}. \quad (4.46)$$

But, consistent with (5.13), as shown in Section 6.5, is the expression

$$\frac{U}{V^*} = 8.74\left(\frac{V^* \delta}{\nu}\right)^{1/7}, \quad (4.47)$$

from which we get

$$V^* = \frac{0.15 U}{(U\delta/\nu)^{1/8}},$$

where V^* is the *friction velocity* of (5.18), namely, $V^* = \sqrt{\tau_0 g_c/\rho}$. Combining these two expressions for V^*, we obtain

$$\tau_0 = \frac{0.02251 \rho U^2}{g_c(U\delta/\nu)^{1/4}}. \quad (4.48)$$

160 The Boundary Layer

Another expression for τ_0 is obtained through (4.40) and (4.46) as

$$\tau_0 = \left(\frac{7}{72}\right) \frac{\rho U^2 dS}{g_c \, dx}. \tag{4.49}$$

Equating (4.48) and (4.49), separating the variables (δ and x), and integrating, we have

$$\left(\frac{\delta}{x}\right)_{\text{approx}} = \left[\left(\frac{5}{4}\right)\left(\frac{72}{7}\right)(0.02251)\right]^{0.8} \left(\frac{\nu}{Ux}\right)^{1/5} = \frac{0.3709}{R_x^{1/5}}. \tag{4.50}$$

From (4.48) and (4.50) there results

$$(\tau_0)_{\text{approx}} = \frac{0.02251 \rho U^2}{g_c (U/\nu)^{1/4} (0.3709 x/R_x^{1/5})^{1/4}} = \frac{0.02884 \rho U^2}{g_c R_x^{1/5}}.$$

And finally, when this expression for τ_0 is combined with (4.41), we obtain

$$(C_f)_{\text{approx}} = \frac{0.05769}{R_x^{1/5}},$$

which can be compared with the accepted result [13, p. 432] for the local skin friction coefficient for a turbulent boundary layer on a flat plate:

$$C_f = \frac{0.0592}{R_x^{1/5}}.$$

The difference is caused by a slight adjustment in the latter expression to account for experimental observations.

4.6 MORE COMPLEX SOLUTION METHODS

4.6.1 Momentum Integral Equation

The boundary layer momentum equation of (4.10) can be integrated (i.e., averaged) across the entire boundary layer to obtain the two-dimensional, incompressible *momentum integral equation*, as first suggested by Theodor von Karman [17] in 1921. Briefly, the continuity equation of (4.1) is used in the form

$$v = -\int_0^y \frac{\partial u}{\partial x} dy$$

and the energy equation of (4.2) is used in the form

$$\frac{g_c}{\rho}\frac{dp}{dx} = -\frac{U\,dU}{dx}$$

to obtain

$$\int_0^h \left(u\frac{\partial u}{\partial x} - \frac{\partial u}{\partial y}\int_0^y \frac{\partial u}{\partial x}\,dy - \frac{U\,dU}{dx} \right) dy = \int_0^h \frac{\partial}{\partial y}\left(\frac{\mu\,\partial u/\partial y}{\rho}\right) dy. \quad (4.51)$$

The right-hand side of (4.51) can be reduced to $-\tau_0 g_c/\rho$ in the pound force-pound mass system, when the boundary conditions are applied (i.e., $\partial u/\partial y = 0$ for the upper limit, outside the boundary layer, and $\mu\,\partial u/\partial y = \tau_0$ for the lower limit). The second term on the left of (4.51) can be integrated by parts to yield

$$-\int_0^h \left(\frac{\partial u}{\partial y}\int_0^y \frac{\partial u}{\partial x}\,dy\right) dy = -U\int_0^h \frac{\partial u}{\partial x}\,dy + \int_0^h u\frac{\partial u}{\partial x}\,dy.$$

Thus (4.51) can be written in the equivalent forms

$$\int_0^h \left(U\frac{\partial u}{\partial x} + U\frac{dU}{dx} - 2u\frac{\partial u}{\partial x} \right) dy = \frac{\tau_0 g_c}{\rho}$$

and

$$\int_0^h \frac{\partial}{\partial x}[u(U-u)]\,dy + \frac{dU}{dx}\int_0^h (U-u)\,dy = \frac{\tau_0 g_c}{\rho}. \quad (4.52)$$

In 1931, just ten years after von Karman's integral method was given, E. Gruschwitz [18] expressed (4.52) in the form used today. He introduced θ, the momentum thickness; δ^*, the displacement thickness; and H_{12}, the shape factor. This development can be traced by noting that (4.19) can be rewritten as

$$\theta U^2 = \int_0^h [u(U-u)]\,dy \quad (4.53)$$

and that (4.13) can be rewritten as

$$\delta^* U = \int_0^h (U-u)\,dy. \quad (4.54)$$

It follows that, when (4.53), (4.54), and (4.33) are introduced into (4.52), there results

$$\frac{d}{dx}(\theta U^2) + \frac{dU}{dx}(\delta^* U) = \frac{\tau_0 g_c}{\rho}$$

or, rearranged slightly,

$$\frac{d\theta}{dx} + \frac{\theta}{U}\frac{dU}{dx}(2+H_{12}) = \frac{\tau_0 g_c}{\rho U^2}. \tag{4.55}$$

Equation 4.55 is the conventional momentum integral equation for a two-dimensional, incompressible boundary layer (see, e.g., [9], p. 246; [13], p. 467; [19], p. 657; [20], p. 117; [22], p. 88).

Note that in the absence of a pressure gradient, as on a flat plate, (4.55) can be expressed in terms of (4.41) in the reduced form

$$\frac{d\theta}{dx} = \frac{C_f}{2} = \frac{\tau_0 g_c}{\rho U^2}, \tag{4.56}$$

which indicates that in this case the shear stress and/or the skin friction coefficient is proportional to the rate of change of the momentum defect along the bounding surface.

A more general form of (4.55) was given by S. B. Au in 1972 (see [20], p. 166, and [21]) as

$$\frac{d\theta}{dx} = \frac{R}{R-\theta}\left\{\left[\frac{\theta^2}{2RU}(H_{12}^2+2) + \frac{\theta U}{R_g g_c T}\left(1-\frac{\theta}{2R}\right) - \frac{\theta}{U}(H_{12}+2)\right]\right.$$
$$\left. \times \frac{dU}{dx} - \frac{\theta}{R}\frac{dR}{dx} + \frac{\tau_0 g_c}{\rho U^2}\right\}, \tag{4.57}$$

which applies to laminar and turbulent boundary layers for compressible fluids in two-dimensional flow with axial symmetry. When the fluid can be considered incompressible, (4.57) reduces to

$$\frac{d\theta}{dx} = \frac{R}{R-\theta}\left\{\left[\frac{\theta^2}{2RU}(H_{12}^2+2) - \frac{\theta}{U}(H_{12}+2)\right]\frac{dU}{dx} - \frac{\theta}{R}\frac{dR}{dx} + \frac{\tau_0 g_c}{\rho U^2}\right\}. \tag{4.58}$$

Finally, when the radius approaches ∞, as in flat plate work, (4.58) reduces to (4.55), the two-dimensional von Karman momentum integral equation.

We must realize that having the momentum integral equation is not enough. Certain assumptions must still be made about the various boundary layer parameters involved. There are many ways to introduce the empirical information necessary to describe the shape parameter (H_{12}) and the wall shear stress (τ_0), and hence there are many methods for solving the momentum integral equations of (4.55), (4.57), and (4.58). One of the simplest and clearest methods available in the literature for obtaining boundary layer solutions, which at the same time leads to reliable results, was first given Alfred Walz [22] in 1966. In the rest of this chapter, we will concentrate on the Walz method of solution.

4.6.2 Energy Integral Equation

In addition to the momentum integral equation of (4.55), Walz makes use of an energy integral equation of similar form. Briefly, in 1948, K. Wieghardt [23] first averaged the boundary layer energy equation across the entire boundary layer to obtain the two-dimensional, incompressible *energy integral equation* as

$$\frac{\rho}{2}\frac{d}{dx}\int_0^h u(U^2-u^2)\,dy = \mu\int_0^h \left(\frac{\partial u}{\partial y}\right)^2 dy \tag{4.59}$$

(see Section 4.4.4). By introducing δ^{**}, the energy thickness of (4.24), in the two-dimensional form

$$\delta^{**}U^3 = \int_0^h u(U^2-u^2)\,dy, \tag{4.60}$$

(4.59) can be expressed as

$$\frac{d}{dx}(\delta^{**}U^3) = 2\nu\int_0^h \left(\frac{\partial u}{\partial y}\right)^2 dy$$

or, rearranged slightly,

$$\frac{d\delta^{**}}{dx} + 3\frac{\delta^{**}}{U}\frac{dU}{dx} = -\frac{2\tau_0 g_c}{\rho U^2}. \tag{4.61}$$

Equation 4.61 is the conventional energy integral equation for a two-dimensional, incompressible boundary layer (see, e.g., [9], p. 292 or [22], p. 91).

4.6.3 Walz Stepping Equations

Walz transformed both the momentum equation of (4.55) and the energy equation of (4.61) through the functions

$$Z = \theta R_\theta^N, \qquad (4.62)$$

a new thickness parameter, which is used in the momentum equation, and

$$H = \frac{\delta^{**}}{\theta},$$

the shape factor of (4.32), which is used in the energy equation. He obtained for momentum

$$\frac{dZ}{dx} + \frac{Z}{U}\frac{dU}{dx} F_1 - F_2 = 0 \qquad (4.63)$$

and for energy

$$\frac{dH}{dx} + \frac{H}{U}\frac{dU}{dx} F_3 - \frac{F_4}{Z} = 0. \qquad (4.64)$$

Thus the boundary layer problem has been resolved to solving simultaneously the two ordinary first-order differential equations, (4.63) and (4.64), in terms of two unknowns, H and Z.

The various quantities involved in (4.62) through (4.64) are given for ready reference in Table 4.2.

Walz next expressed these transformed momentum and energy equations in finite difference forms as

$$\frac{Z_{i+1}}{Z_i}\left(\frac{R_{i+1}}{R_i}\right)^{1+N} = A_Z + B_Z \bar{F}_2\left(\frac{\Delta x}{Z_i}\right)\left[1 + \frac{(R_{i+1}/R_i)^{1+N}}{2}\right] \qquad (4.65)$$

for the axisymmetric stepping form of momentum, and

$$\frac{H_{i+1}}{H_i} = A_H + \frac{B_H \bar{F}_4}{\bar{Z}}\left(\frac{\Delta x}{H_i}\right) \qquad (4.66)$$

for the stepping form of energy, independently of the axisymmetric condition. Note that average values of H and Z are used in the small Δx intervals, and that a linear change between specified values of U and R is used in the interval.

Table 4.2 Various Quantities Involved in the Walz Equations (4.63) and (4.64)[a]

$$R_\theta = \frac{U\theta}{\nu} = \left[R_D\left(\frac{U}{U_{ref}}\right)\left(\frac{Z}{d}\right)\right]^{1/(1+N)}$$

$F_1 = 2 + N + (1+N)H_{12}$

$F_2 = (1+N)\alpha$

$F_3 = 1 - H_{12}$

$F_4 = 2\beta R_\theta^{N-NN} - \alpha H$

$NN_L = 1$

$N_L = 1$

$NN_T = 0.2317H - 0.2644 - 87{,}000(2-H)^{20}$

$N_T = 0.268$

$\alpha_L = 1.7261(H-1.515)^{0.7158}$

$\alpha_T = 0.03894(H-1.515)^{0.7}$

$\beta_L = 0.1564 + 2.1921(H-1.515)^{1.7}$

$\beta_T = 0.00481 + 0.0822(H-1.5)^{4.81}$

$H_{12L} = 4.0306 - 4.2845(H-1.515)^{0.3886}$

$H_{12T} = 1 + 1.48(2-H) + 104(2-H)^{6.7}$

[a] After Walz [22].

The various quantities introduced by (4.65) and (4.66) are given for ready reference in Table 4.3. Other relations of interest are

$$C_f = \frac{\tau_0}{q} = \frac{2\alpha}{R_\theta^N} \qquad (4.67)$$

and

$$\frac{\theta}{D} = \left[\frac{Z}{D}\left(\frac{U_{ref}}{R_D U}\right)^N\right]^{1/(1+N)}. \qquad (4.68)$$

Table 4.3 Various Quantities Involved in the Walz Equations (4.65) and (4.66)[a]

$$\Delta X = X_{i+1} - X_i$$

$$\overline{H} = \frac{H_{i+1} + H_i}{2}$$

$$\overline{Z} = \frac{Z_{i+1} + Z_i}{2}$$

$$A_Z = \left(\frac{U_i}{U_{i+1}}\right)^{F_1}$$

$$B_Z = \frac{1 - A_Z(U_i/U_{i+1})}{(1+F_1)(1 - U_i/U_{i+1})}$$

$$A_H = \left(\frac{U_i}{U_{i+1}}\right)^{F_3}$$

$$B_H = \frac{1 - A_H(U_i/U_{i+1})}{(1+F_3)(1 - U_i/U_{i+1})}$$

[a]After Walz [22].

Starting with appropriate values of H and Z (normally, in flow from a plenum where the boundary layer has its inception, $H_{\text{initial}} = 1.572$, the flat plate value, and $Z_{\text{initial}} = 0$ at $x = 0$), one proceeds with the stepwise solution until either transition (see Section 4.3) or separation (also see Section 4.3) occurs. *Transition*, by the Walz method, is indicated when R_θ exceeds a critical value defined by

$$\log R_\theta^* = 2.42 + 24.2(H - 1.572). \qquad (4.69)$$

In the first element that shows transition, the solution is changed from a laminar boundary layer to a turbulent one by simply redefining Z as

$$Z_T = \theta R_\theta^{0.268}. \qquad (4.70)$$

From then on, one proceeds with the stepwise solution, using only turbulent relations. *Separation*, by the Walz method, is indicated when $H = 1.515$. Note that the *lower bound* of transition is indicated by (4.69). This is shown as the extreme left-hand value of R_θ in Figure 4.2. The *upper bound* of transition is represented by the laminar separation point.

4.6.4 Application of Walz Solution to a Nozzle

As an illustration of this more complex type of boundary layer solution, we next apply the Walz method to the common piping component, the flow nozzle [15, 16]. The flow through circular cylindrical pipes and through many piping components, such as the flow nozzle, can be considered as *three-dimensional with axial symmetry*. When we further limit our example to *incompressible* flow, we can apply the Walz stepping equations, (4.65) and (4.66), directly as given. The nozzle itself has been chosen for this example partly because its radius varies with length (as opposed to the simpler constant diameter pipe), but mainly because it is generally useful.

For the standard ASME nozzle profile shown in Figure 4.11, the first task is to form a table of given dimensionless values of the potential velocities (U/U_{ref}) and radii (R/d), as a function of nozzle profile position (S/d). For the velocity reference we choose the ideal potential velocity at the nozzle throat. For the reference length we choose the throat diameter.

While it is a matter of geometry only to ascertain R/d as a function of S/d, it is another matter to determine the potential velocities throughout the nozzle. These are obtained from a solution of the Laplace equation for velocity potential (see Section 2.6.1). In two-dimensional Cartesian coordinates, for example, this expression is simply

$$\frac{\partial^2 \phi}{\partial x^2} + \frac{\partial^2 \phi}{\partial y^2} = 0. \tag{4.71}$$

In the fluid field, the governing equation is

$$\dot{m} = -\rho A \frac{d\phi}{dS},$$

where the potential gradient is the fluid velocity, that is,

$$U = \frac{d\phi}{dS}. \tag{4.72}$$

Solution of the Laplace equation of (4.71), and hence the potential velocities of (4.72), can be obtained by numerical relaxation methods [24] or by electrical relaxation methods [25, 26]. Figure 4.12 presents typical potential velocity solutions for a plenum inlet installation of the nozzle of Figure 4.11, where the significant differences between straight two-dimensional flow and three-dimensional flow with axial symmetry are apparent. For this example we are interested in the axisymmetric results.

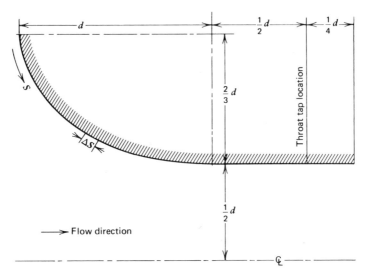

Figure 4.11 Geometry of ASME low β elliptical nozzle.

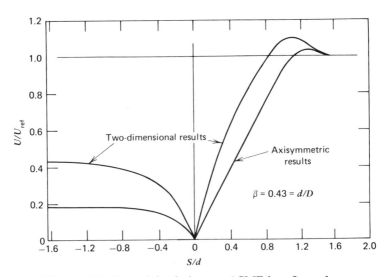

Figure 4.12 Potential solutions to ASME low β nozzle.

```
INCOMPRESSIBLE AXI-SYMMETRIC BOUNDARY LAYER SOLUTION,WALZ METHOD, THROAT REYNOLDS NO.= 1.0E+07
```

I	S/D	U/UT	R/D	Z/D	HBAR	RDEL	DEL2/D	TAU/Q	DEL1/D
1	0.000	.140	1.167	0.00000	1.572	0.0	0.	0.	0.
2	.010	.144	1.156	.00416	1.578	77.4	5.376E-05	6.202E-03	1.378E-04
3	.020	.148	1.146	.00790	1.584	108.1	7.307E-05	4.747E-03	1.833E-04
4	.025	.152	1.142	.00901	1.599	117.0	7.700E-05	5.054E-03	1.837E-04
5	.030	.156	1.135	.01022	1.604	126.2	8.092E-05	4.859E-03	1.905E-04
68	1.000	.910	.533	.11187	1.638	1009.0	1.109E-04	7.631E-04	2.365E-04
69	1.045	.937	.524	.12370	1.627	1076.6	1.149E-04	6.677E-04	2.531E-04
70	1.080	.965	.518	.12543	1.634	1100.2	1.140E-04	6.821E-04	2.462E-04
71	1.135	.993	.510	.14064	1.622	1181.7	1.190E-04	5.891E-04	2.659E-04
72	1.175	1.022	.508	.14063	1.630	1198.8	1.173E-04	6.141E-04	2.556E-04
73	1.200	1.040	.506	.14214	1.632	1215.9	1.169E-04	6.125E-04	2.534E-04

```
        LAMINAR-TURBULENT TRANSITION,S/D= 1.262,   RDEL=1347.2,   HBAR= 1.601
```

74	1.320	1.060	.500	.00184	1.761	2412.9	2.276E-04	3.624E-03	3.096E-04
75	1.400	1.040	.500	.00354	1.770	3987.4	3.834E-04	3.243E-03	5.161E-04
Point	S/d	U/U$_{ref}$	R/d	Z/d	$\bar{H}$	R_θ	θ/d	τ_0/q	δ^*/d

Figure 4.13 Portion of computer printout results of boundary layer analysis for ASME nozzle.

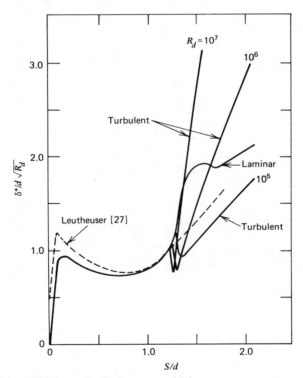

Figure 4.14 ASME nozzle displacement thickness versus contour position.

170 The Boundary Layer

Given the numerical results of Figure 4.12 and the geometry specifications of Figure 4.11, we can form the required table of dimensionless values of S/d, R/d, and U/U_{ref}, a portion of which is shown in the first three columns of Figure 4.13, and proceed with the boundary layer solution. Note that ΔS replaces Δx in the stepping equations when varying radii are involved.

Starting at the first element, at the nozzle inlet, with $Z_i = 0$ and $H_i = 1.572$, at $S = 0$, we solve simultaneously for Z_{i+1} and H_{i+1} through (4.65) and (4.66). Along with Z and H, we form the important boundary layer parameters, in dimensionless form: δ^*/d, θ/d, δ^{**}/d, R_θ, and τ_0/q, at every point of interest along the nozzle contour.

Figure 4.13 is a typical portion of the results from a computer printout solution of this problem. Note the transition point between laminar and turbulent boundary layers, as determined by (4.69), which identifies the Tollmien-Schlichting indifference point of Figure 4.2.

Figures 4.14 through 4.18 summarize the boundary layer results for the flow nozzle according to the Walz method.

In Figure 4.14 the results for displacement thickness throughout the nozzle are compared with those of Leutheuser [27], for various throat Reynolds numbers. This figure should be recognized as a generalized plot

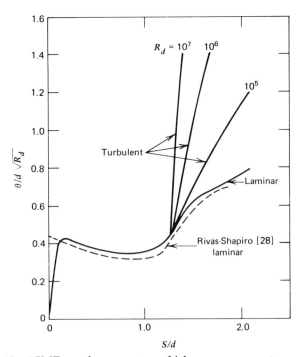

Figure 4.15 ASME nozzle momentum thickness versus contour position.

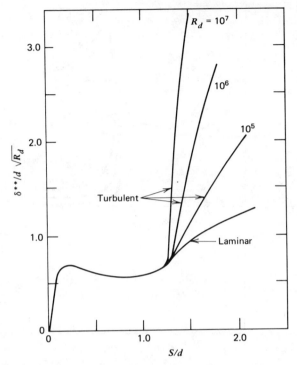

Figure 4.16 ASME nozzle energy thickness versus contour position.

in the laminar boundary layer region in the sense that $(\delta^*/d)\sqrt{R_d}$ is independent of the throat Reynolds number. After transition, however, in the turbulent regime no such simple correlation holds.

In Figure 4.15, a similar generalized plot for momentum thickness in terms of $(\theta/d)\sqrt{R_d}$ is given, and compared with the similar results of Rivas and Shapiro [28].

In Figure 4.16 the characteristics of energy thickness are presented in terms of $(\delta^{**}/d)\sqrt{R_d}$.

Figure 4.17 shows the shape factor (H_{12}) from the Walz analysis, compared with Leutheuser's choice of a constant ($H_{12}=2.554$) in the laminar region, and with the flat plate value of $H_{12}=1.286$ in the turbulent region.

Finally, in Figure 4.18 the loss parameter (τ_0/q), is given in terms of the throat Reynolds number, for both the *completely* laminar boundary layer and the *completely* turbulent boundary layer. Evidently, the nozzle loss is a strong function of the transition criterion used.

In Chapter 14, which deals specifically with flow nozzles, we will apply this boundary layer information toward obtaining nozzle discharge coefficients.

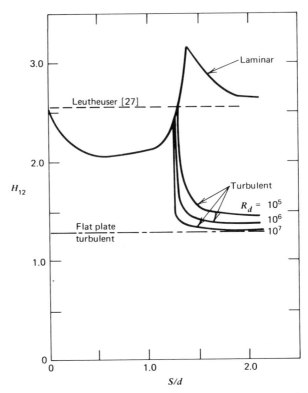

Figure 4.17 ASME nozzle shape factor versus contour position.

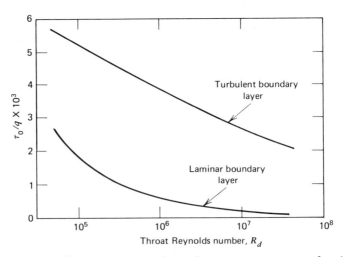

Figure 4.18 Wall shear stress to dynamic pressure parameter for ASME nozzle.

Although these results are believed to be general for the nozzle analyzed, they are given here primarily to illustrate the method and the typical results that can be obtained from a boundary layer solution.

In Reference 16 a similar boundary layer solution is given for a nozzle installed in a pipe, rather than for a plenum inlet. Briefly, if a nonseparating flow nozzle is installed in a pipe with fully developed flow, the boundary layer extends to the pipe centerline. Applying the power law velocity distribution of (5.12) and the boundary layer thickness definitions of (4.14), (4.20), and (4.25), we have, at $R_D = 1.1 \times 10^5$ (where $N=7$),

$$\frac{\delta^*}{R} = \frac{1}{2} - \frac{N^2}{(N+1)(2N+1)} = 0.09167,$$

$$\frac{\theta}{R} = \frac{N^2}{(N+1)(2N+1)} - \frac{N^2}{(N+2)(2N+2)} = 0.06805,$$

$$\frac{\delta^{**}}{R} = \frac{N^2}{(N+1)(2N+1)} - \frac{N^2}{(N+3)(2N+3)} = 0.12010.$$

The required shape factor at nozzle inlet, in this case, is by (4.32)

$$H = \frac{\delta^{**}}{\theta} = \frac{0.12010}{0.06805} = 1.7649.$$

Also, R_θ is determined from the relation

$$R_\theta = R_d \left(\frac{U}{U_{\text{ref}}} \right) \left(\frac{\theta}{d} \right)$$

and leads at once, through (4.62), to the value of Z at the inlet.

REFERENCES

1. D. Bernoulli, *Hydrodynamica*, Strasbourg, 1738. Translated from the Latin by T. Carmody and H. Kobus at the Iowa Institute of Hydraulic Research, Dover Publications, 1968.
2. C. Truesdell, *Essays in the History of Mechanics*, Springer-Verlag, 1968, pp. 106–173, for example.
3. L. M. H. Navier, "Memoire sur les lois du mouvement des fluids," *Mem. Acad. R., Sci.*, Vol. 6, 1837, p. 389. Translated as "Note on the laws of the flow of fluids."
4. H. Rouse and S. Ince, *History of Hydraulics*, Dover Publications, 1963, p. 195.
5. G. G. Stokes, "On the theories of the internal friction of fluids in motion, and of the equilibrium and motion of elastic solids," *Trans. Camb. Phil. Soc.*, Vol. 8, 1845.

174 The Boundary Layer

6. L. Prandtl, "Über Flussigkeitsbewegung bei sehr kleiner Reibung," *Verhandl. III Int. Math. Kongr., Heidelberg,* 1904. Translated as "On fluid motions with very small friction," *NACA TM* 435, 1927.

7. P. R. H. Blasius, "Grenzschichten in Flussigkeiten mit kleiner Reibung," Dissertation, Gottingen, 1907; or *Z. Math. Phys.*, Vol. 56, p. 1, 1908. Translated as "Boundary layers in fluids of small viscosity," *NACA TM* 1256, February 1950.

8. L. Howarth, "On the solution of the laminar boundary layer equations," *Proc. R. Soc. London,* Vol. 164, 1938, p. 547.

9. W. Li and S. Lam, *Principles of Fluid Mechanics*, Addison-Wesley, 1964.

10. L. Prandtl and O. G. Tietjens, *Applied Hydro- and Aeromechanics*, McGraw-Hill, 1934, p. 30. Translated by J. P. Den Hartog.

11. O. Reynolds, "An experimental investigation of the circumstances which determine whether the motion of water will be direct or sinuous, and the laws of resistance in parallel channels," *Phil. Trans. R. Soc. London*, 1883.

12. W. Tollmien, "Über die Entstehung der Turbulenz," *Nachr. Ges. Wiss. Gottingen, Math. Phys. Klasse*, Vol. 21, 1929. Translated as "On the origin of turbulence," *NACA TM* 609, 1931.

13. H. Schlichting, *Boundary Layer Theory*, 1st ed., McGraw-Hill, 1955. Translated by J. Kestin.

14. M. Hansen, "Die Geschwindigkeitsverteilung in der Grenzschicht an der langsangestromten ebenen Platte," *Z. Angew. Math. Mech.*, Vol. 8, 1928, p. 185. Translated as "The velocity distribution in the boundary layer on a flat plate," *NACA TM* 585, 1930.

15. R. P. Benedict and J. S. Wyler, "Analytical and experimental studies of ASME flow nozzles," *Trans. ASME, J. Fluids Eng.*, September 1978, p. 262.

16. R. P. Benedict, "Generalized fluid meter discharge coefficient based solely on boundary layer parameters," *Trans. ASME, J. Eng. Power*, October 1979, p. 572.

17. T. von Karman, "Über laminare und turbulente Reibung," *Z. Angew. Math. Mech.*, Vol. 1, 1921, p. 233. Translated as "On laminar and turbulent friction," *NACA TM* 1092, 1946.

18. E. Gruschwitz, "Die turbulente Reibungschicht in ebener Stromung mit Druckabfall und Druckanstieg," *Ing. Arch.*, Vol. 2, 1931, p. 321. Translated as "The turbulent friction layer in smooth flow with decreasing and increasing pressure."

19. J. K. Vennard and R. L. Street, *Elementary Fluid Mechanics*, 5th ed., John Wiley, 1975.

20. S. B. Au, "Boundary layer development in venturimeters," Ph.D. Thesis, Department of Mechanical Engineering, University of Wales, 1972.

21. S. B. Au, "The prediction of axisymmetrical turbulent boundary layer in conical nozzles," *J. App. Mech., Trans. ASME*, March 1974, p. 20.

22. A. Walz, *Boundary Layers of Flow and Temperature*, M. I. T. Press, 1969. Translated by H. J. Oser.

23. K. Wieghardt, "Über einen Energiesatz zur berechnung laminarer Grenzschichten," *Ing. Arch.*, Vol. 16, 1948, p. 231. Translated as "*The energy principle applied to calculation of a laminar boundary layer.*"
24. H. W. Emmons, "The numerical solution of heat conduction problems," *Trans. ASME*, Vol. 65, August 1943, p. 607; see also G. M. Dusinberre, "Numerical methods for transient heat flow,"*Trans. ASME*, Vol. 67, November 1945, p. 703.
25. R. P. Benedict, "Analog simulation," *Electro-Technol.*, Science and Engineering Series No. 60, December 1963, p. 73.
26. R. P. Benedict and C. A. Meyer, "Electrolytic tank analog for studying fluid flow fields within turbomachinery," *ASME Pap.*, 57-A-120, December 1957.
27. H. J. Leutheuser, "Flow nozzles with zero beta ratio," *Trans. ASME, J. Basic Eng.*, September 1964, p. 538.
28. M. A. Rivas and A. H. Shapiro, "On the theory of discharge coefficients for rounded-entrance flowmeters and venturis," *Trans. ASME*, April 1956, p. 489.

NOMENCLATURE

Roman

A_Z, A_H	boundary layer functions
B_Z, B_H	boundary layer functions
c_f	skin friction coefficient
d	throat diameter
D	pipe diameter
f	friction factor
F_1, F_2, F_3, F_4	boundary layer functions of H
F_x	force in x direction
F_f	force of friction
g	local gravity
g_c	gravitational constant
H	shape factor $= \delta^{**}/\theta$
H_{12}	shape factor $= \delta^*/\theta$
K.E.	kinetic energy flux
L	length
$\dot{m}$	mass flow rate (mass flux)
M	mass
$\dot{M}$	momentum flux

176 The Boundary Layer

- N, NN boundary layer exponents
- p static pressure
- P general parameter
- q dynamic pressure
  $= \rho U^2 / 2 g_c$
- $q \cos \gamma$ velocity vector
- Q volumetric flow rate
- r general radius
- R radius
- R_D Reynolds number
  $= VD/\nu$
- R_d Ud/ν
- R_L UL/ν
- R_x Ux/ν
- R_{δ^*} $U\delta^*/\nu$
- R_θ $U\theta/\nu$
- S distance along contour
- t time
- u x component of velocity
- U potential velocity
- v y component of velocity
- V average velocity
- V^* friction velocity
  $= \sqrt{\tau_0 g_c / \rho}$
- x x direction
- y y direction
- Z transformation function
  $= \theta R_\theta^N$

Greek

- δ boundary layer thickness
- δ^* displacement thickness
- δ^{**} energy thickness
- Δ finite difference
- ϵ unknown function
- λ wavelength

μ dynamic viscosity
ν kinematic viscosity
θ momentum thickness
ρ fluid density
τ shear stress

Mathematical Symbols

d total derivative
∂ partial derivative
∇ Laplacian operator

Subscripts

L laminar
T turbulent
ref   reference

5

Velocity Distributions

> ...many speeds are in each branch, one speed along the walls, another at the center, and every manner of speed in between... —*Gustav Eckstein*

5.1 GENERAL REMARKS

The many factors that affect velocity distribution, such as fluid viscosity, fluid density, mean velocity, pipe size, condition of pipe surface, and turbulence level, are discussed in this chapter. Formulations describing velocity profiles in laminar and turbulent flows, and in the transition region in between, are given in terms of the point velocity, the volumetric mean velocity, the maximum velocity, and the friction velocity. The experimental bases for these formulations are gathered in this chapter. For turbulent flow the power law and the law of the wall are developed, the latter in terms of the laminar sublayer, the buffer zone, and the logarithmic layer. The influence of the pipe wall roughness on the velocity profile and on the pipe factor is shown. A universal form of the law of the wall, applying to both smooth and rough pipes, is discussed. Finally, the effect of a pressure gradient on the validity of the law of the wall is examined.

Because of *viscosity* the various layers of fluid (i.e., the fluid at different radii in the pipe), travel at different velocities (see Figure 5.1). For a fluid of given viscosity and *density* in a pipe of given *size*, the distribution of velocities is further affected by the magnitude of the *mean velocity*. These variables, namely, the fluid viscosity (μ), the fluid density (ρ), the mean velocity (V), and the pipe diameter (D), are collected for economy of thought into a single dimensionless group known as the pipe Reynolds number [1]:

$$R_D = \frac{\rho V D}{\mu} = \frac{VD}{\nu}, \tag{5.1}$$

General Remarks 179

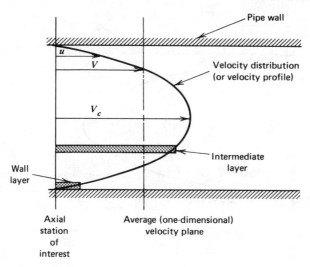

Figure 5.1 Velocity distribution in a pipe (u = point velocity at any radial position, V = volumetric average velocity, V_c = maximum velocity, at centerline).

where $\nu = \mu/\rho$ and is called the *kinematic* viscosity. The Reynolds number is one of the basic parameters used in fluid mechanics. Specifically, in this chapter R_D will serve as one indicator of the velocity distribution.

Of course, there are other factors that also influence the velocity distribution. The *turbulence level* in the flowing stream is one important factor upon which the velocity profile depends. We must distinguish between laminar and turbulent flows, both in the main stream and in the various portions of the boundary layer (see Figure 5.2). Another factor of far-reaching importance is the condition of the bounding wall surface. Whether it is to be judged a *smooth* or a *rough* wall will be determined in part by the absolute roughness (e) of the wall surface (see Figure 5.3). However, as will be shown, e is not the only criterion for judging hydraulic roughness. Still another factor that affects the velocity profile is the proximity of the axial station of interest in the pipe with respect to the pipe inlet. Near the inlet to a circular pipe, the velocity distribution will vary with axial position as the boundary layer develops. Such a condition is described as a distribution in a *developing* stage, wherein the profile varies with position. When the velocity profile no longer changes with distance from the inlet, the boundary layer is said to be *fully developed* (see Figure 5.4).

These various factors that affect the velocity distribution in a pipe are not independent of one another. For example, even though the velocity

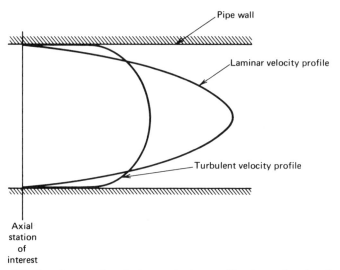

Figure 5.2 Laminar and turbulent velocity profiles in a pipe for the same volumetric average velocity.

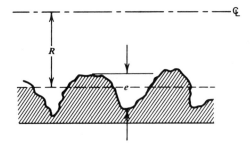

Figure 5.3 Greatly magnified sketch of a rough pipe wall.

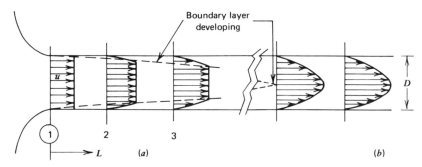

Figure 5.4 Variation of velocity profile with axial position from inlet. (*a*) Developing (varying) velocity profile, from inlet 1 to $L/D \sim 40$. (*b*) Fully developed (identical) velocity profile, independent of axial position.

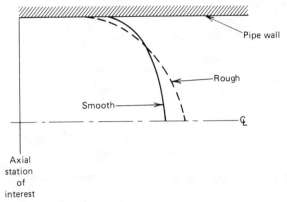

Figure 5.5 Comparison of turbulent velocity profiles in smooth and rough pipes.

profile is fully developed, its form still depends on whether the pipe is smooth or rough (see Figure 5.5).

We consider next, in some detail, the two basic flow regimes, namely, laminar and turbulent flows. Our objective is to be able to describe mathematically velocity profiles such as are shown in Figures 5.2 and 5.5.

5.2 LAMINAR FLOW

When the fluid can be considered to be flowing in layers or laminae (as in Figure 5.1), with the sole linkage between the layers taken to be based on momentum transfer set up as a result of molecular interactions, the flow is called *laminar*. In the fully developed stage, wherein the velocity profile is independent of axial position, it can be shown quite simply that the velocity distribution through any diametral slice (i.e., a plane passing through the pipe that includes the pipe axis) can be described by a parabola.

Briefly, when acceleration forces can be considered negligible, we have, following the notation of Figure 5.6,

$$\Sigma F_x = Ma = 0$$

or

$$p_1 A_p - p_2 A_p - \tau A_s = 0,$$

where A_p represents the area perpendicular to the flow, and A_s represents the *surface* area of the fluid element. Then

$$(p_1 - p_2)(\pi r^2) = \tau(2\pi r L)$$

182 Velocity Distributions

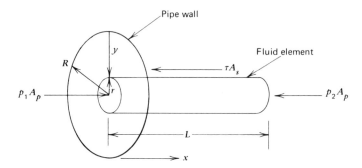

Figure 5.6 Forces acting on fluid element in laminar flow.

or

$$\tau = \frac{\Delta p r}{2L}. \tag{5.2}$$

But τ represents the shear stress, which, for laminar flow in Newtonian fluids, is simply

$$\tau_L = \mu \frac{du}{dr}. \tag{5.3}$$

Combining (5.2) and (5.3) gives

$$\frac{du}{dr} = \left(\frac{\Delta p}{2\mu L}\right) r, \tag{5.4}$$

which represents the velocity gradient at any radial position. When (5.4) is integrated across the complete pipe radius (R), we obtain

$$u = \left(\frac{\Delta p}{2\mu L}\right) \int_0^R r\, dr = \left(\frac{\Delta p}{4\mu L}\right)(R^2 - r^2), \tag{5.5}$$

which is the velocity profile of laminar flow. Note that (5.5) is of the parabolic form, $u = a + br^2$.

The boundary conditions of this profile are as follows:
At $r = 0$, that is, on the pipe centerline,

$$u = V_c = \left(\frac{\Delta p}{4\mu L}\right) R^2. \tag{5.6}$$

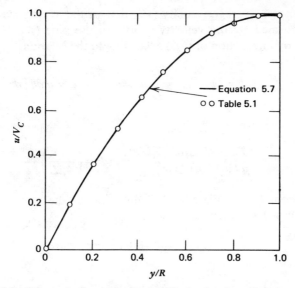

Figure 5.7 Dimensionless velocity profile for laminar flow in a pipe.

At $r = R$, that is, at the walls,

$$u = 0.$$

It follows from (5.5) and (5.6) that the ratio of the point velocity (u) to the maximum velocity (V_c) is

$$\frac{u}{V_c} = 1 - \left(\frac{r}{R}\right)^2. \tag{5.7}$$

For future reference we note that it takes two independent items of information to establish the dimensionless velocity profile. For laminar flow, these items are the geometric factor (r/R), and the laminar exponent, 2, according to (5.7).

The dimensionless, and hence generalized, velocity profile for laminar flow in a pipe is shown in Figure 5.7, based on solutions of (5.7) as given in Table 5.1.

Table 5.1 Generalized Laminar Velocity Profile [based on (5.7)]

y/R:	0	0.1	0.2	0.3	0.4	0.5	0.6	0.7	0.8	0.9	1.0
r/R:	1	0.9	0.8	0.7	0.6	0.5	0.4	0.3	0.2	0.1	0
u/V_c:	0	0.19	0.36	0.51	0.64	0.75	0.84	0.91	0.96	0.99	1.00

184 Velocity Distributions

The *volumetric* average velocity (V), that is, the average velocity of the *paraboloid* formed by the velocity profile in the pipe (see Figure 5.8), is determined by integration of (5.5) according to the relation

$$Q = AV = (\pi R^2)V = \int_A u\, dA = \int_0^R \left[\int_0^{2\pi} ur\, d\theta \right] dr$$

or

$$V = \frac{2}{R^2} \int_0^R \left(\frac{\Delta p}{4\mu L} \right)(R^2 - r^2) r\, dr = \left(\frac{\Delta p}{4\mu L} \right) \frac{R^2}{2}. \tag{5.8}$$

It follows from (5.6) and (5.8) that the ratio of the volumetric average velocity (V) to the maximum velocity (V_c) is

$$\frac{V}{V_c} = \frac{1}{2}, \tag{5.9}$$

which means quite simply that the volumetric average velocity in laminar flow is precisely one half the centerline velocity.

Another average velocity, the *area* average velocity (V_D), that is, the average velocity of the *parabola* formed by the velocity profile in a diametral slice (see Figure 5.9), is determined by integration of the factor $G(x_0)$ in the relation

$$Q = \int_{-R}^{R} \left[\int_{-C}^{C} u(x,y)\, dy \right] dx = \int_{-R}^{R} [G(x)]\, dx,$$

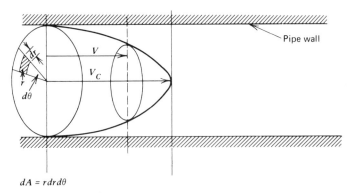

$dA = r\, dr\, d\theta$

Figure 5.8 Volumetric average velocity of parabaloid of laminar flow.

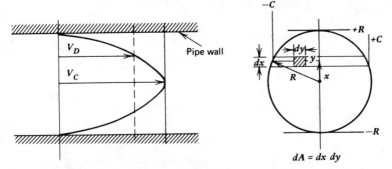

Figure 5.9 Area average velocity of parabola of laminar flow.

where $\pm C$ represents the limits of integration in the y direction and $G(x_0)$ means an evaluation of $G(x)$ at the pipe center [2]. Thus the limits of y when $x=0$ are $-R$ to $+R$, so that

$$2\int_0^R u(r)\,dr = \int_{-R}^R u(x_0, y)\,dy = V_D(2R)$$

or

$$V_D = \frac{1}{R}\int_0^R \left(\frac{\Delta p}{4\mu L}\right)(R^2 - r^2)\,dr = \left(\frac{\Delta p}{4\mu L}\right)\tfrac{2}{3}R^2. \tag{5.10}$$

It follows from (5.6) and (5.10) that the ratio of the area average velocity (V_D) to the maximum velocity (V_c) is

$$\frac{V_D}{V_c} = \frac{2}{3}.$$

5.3 TRANSITION

For reasons that can best be discussed later, and with reference to Figure 5.2, there is evidently one condition when the flow is in the laminar regime and another when it is turbulent. Between these two regimes there is necessarily a transition region. This has been found experimentally to be a function of the pipe Reynolds number, $R_D = VD/\nu$ (see Figure 5.10). Nominally, the value of R_D that marks the inception of turbulence in a pipe is about 2000. Far beyond this critical Reynolds number, the flow is

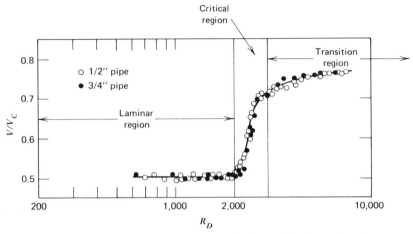

Figure 5.10 Critical Reynolds number, as indicated by the ratio of average to maximum velocity (after Senecal and Rothfus[28]).

bound to be turbulent. Much below the critical Reynolds number, the flow will be laminar. The dramatic increase in V/V_c (see Figure 5.10), that is, in the ratio of average to maximum velocity, above the critical Reynolds number is accompanied by a sudden increase in the pressure drop in the pipe. We will discuss this in greater detail in Chapter 6 on the friction factor. Now, having considered laminar flow, we discuss next the velocity profiles of turbulent flow.

5.4 TURBULENT FLOW IN SMOOTH PIPES

As we have just noted, when the Reynolds number exceeds a critical value, nominally 2000 for pipe flow, the flow deviates considerably from the laminar case. Momentum is now transferred radially as well as axially as a result of the mixing motion that accompanies the onset of turbulence.

Once more, we are speaking of *fully developed* boundary layers, where the velocity profile does not change with axial position. However, unlike the laminar flow case, we must be careful to distinguish between smooth and rough pipes in turbulent flow considerations, admitting that the method of delineating between these two boundary surface conditions is given only in part by e/D, the relative roughness ratio.

The *smooth pipe* condition has received much attention in the literature and will be treated first. Even under the smooth category, the classification must be broken down further. One representation, historically the oldest, of a smooth pipe velocity profile is the so-called *power law*. A slightly newer, much more sophisticated, and the most used of the smooth pipe velocity profiles is the *law of the wall*. The first (the power law) is strictly empirical in background, whereas the second (the law of the wall) is semi empirical, having a theoretical basis for the form of the equations.

5.4.1 Power Law

One of the first relations developed to summarize turbulent velocity profiles in smooth pipes was the power law. Early experimenters plotted their data in terms of u/V_c versus y/R, just as was done so successfully for laminar flow (see Figure 5.7). However, whereas only one curve was required to describe laminar flow, it was found that a series of curves was needed to describe turbulent flow, depending on the pipe Reynolds number (see Figure 5.11). Evidently, use of these conventional coordinates does not produce a generalized plot for turbulent flows.

The data become more ordered when the same information is plotted on a log-log scale of u/V_c and y/R (see Figure 5.12). Since the resulting

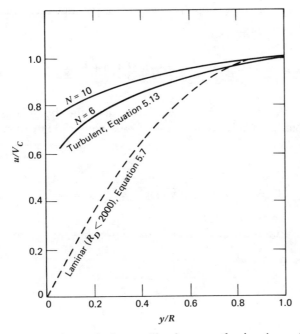

Figure 5.11 Turbulent velocity profiles for smooth pipe (according to the power law).

188   Velocity Distributions

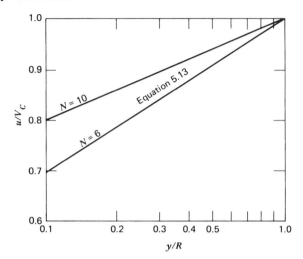

Figure 5.12 Turbulent velocity profiles for smooth pipes (according to the power law).

curves of Figure 5.12 are straight lines, it is clear that the general form of these lines is

$$\log\left(\frac{u}{V_c}\right) = \frac{1}{N}\log\left(\frac{y}{R}\right) + b, \qquad (5.11)$$

which is of the form $y = mx + b$. Note that the boundary condition, $u/V_c = 1$ when $y/R = 1$, requires that $b = 0$. Thus (5.11) can be expressed alternatively as

$$\left(\frac{u}{V_c}\right)^N = \frac{y}{R} \qquad (5.12)$$

or, in the most conventional form of the power law, as

$$\frac{u}{V_c} = \left(\frac{y}{R}\right)^{1/N} = \left(1 - \frac{r}{R}\right)^{1/N}, \qquad (5.13)$$

where N is the reciprocal of the slope of the lines. Dimensionless velocity profiles for smooth pipes in turbulent flow, according to (5.13), have already been shown in Figures 5.11 and 5.12. Numerical solutions to (5.13) are given in Table 5.2.

Table 5.2 Turbulent Velocity Profiles in Smooth Pipes [according to the power law of (5.13)]

y/R \ N	\multicolumn{5}{c}{u/V_c}				
	6	7	8	9	10
0	0	0	0	0	0
0.1	0.681	0.720	0.750	0.774	0.794
0.2	0.765	0.794	0.818	0.836	0.851
0.3	0.818	0.842	0.860	0.875	0.886
0.4	0.858	0.877	0.892	0.903	0.912
0.5	0.891	0.906	0.917	0.926	0.933
0.6	0.918	0.930	0.938	0.945	0.950
0.7	0.942	0.950	0.956	0.961	0.965
0.8	0.963	0.969	0.972	0.976	0.978
0.9	0.982	0.985	0.987	0.988	0.990
1.0	1	1	1	1	1

Note, as with the laminar velocity profile of (5.7), that two items of information are required to establish the dimensionless velocity profiles for smooth pipes in turbulent flow. For the power law, these items are the geometric factor (y/R), and the turbulent exponent $(1/N)$, according to (5.13).

The quantity N evidently is a function of the Reynolds number, as seen from the fact that the data taken at different Reynolds numbers fall on different curves. Although a functional relation between N and R_D is not immediately apparent, an empirical relation between these two variables can be given, as in Table 5.3.

Sometimes, to match the analytical work of Blasius on the friction factor (see Chapter 6), the *general* power called for in (5.13) is replaced by the *specific* power $1/7$, that is $N=7$; and in this guise (5.13) is called the *1/7th power law*.

When the ratio u/V_c of (5.13) is integrated over the cross-sectional area of the pipe, we obtain a dimensionless volumetric average velocity (V/V_c) in terms of the exponent N as follows:

$$Q = AV = (\pi R^2)V = \int_A u\, dA = \int_0^R u(2\pi r\, dr)$$

Table 5.3 Empirical Relation between Pipe Reynolds Number and the Exponent N

R_D:	4×10^3	1.1×10^5	1.1×10^6	3.2×10^6
N:	6	7	8.8	10

**Table 5.4   Ratio of Average to Maximum Velocity
[according to the power law of (5.16)]**

V/V_c:	0.7912	0.8167	0.8366	0.8526	0.8658
N:	6	7	8	9	10

or

$$\frac{V}{V_c} = \frac{2}{R^2} \int_0^R \left(1 - \frac{r}{R}\right)^{1/N} r\, dr. \tag{5.14}$$

The general form and general solution of (5.14) are

$$\int x(ax+b)^M dx = \frac{(ax+b)^{M+2}}{a^2(M+2)} - \frac{b(ax+b)^{M+1}}{a^2(M+1)}, \tag{5.15}$$

so (5.14), integrated in terms of (5.15) and with $x=r$, $a=-1/R$, $b=1$, and $M=1/N$, becomes

$$\frac{V}{V_c} = \frac{2N^2}{(N+1)(2N+1)}. \tag{5.16}$$

Numerical values for the ratio of average to maximum velocity, as given by (5.16), are listed in Table 5.4 for several values of N.

Since the curves shown in Figures 5.11 and 5.12 are seen to apply nearly to the center of the pipe, that is, where y/R approaches 1, it follows that the validity of the power law is not restricted to the wall layer. Note, however, that differentiation of (5.13) indicates that the velocity gradient, du/dy, approaches infinity as y approaches zero. Such a result would require that the shear stress become infinitely large at the wall. Evidently, the power law cannot hold right up to the pipe wall, but must cease to apply in a very thin region adjacent to the wall. Such a thin *laminar sublayer* will be discussed in the next section.

5.4.2   Smooth Law of the Wall

The fact that the exponent in the power law of (5.13) decreases with increasing Reynolds number suggests that there is some unique expression toward which this equation tends at high Reynolds numbers. Mathematically, as the exponent in the power law becomes smaller and smaller, the turbulent velocity profile can be described more and more precisely by a logarithmic function of the independent variable, y.

Turbulent Flow in Smooth Pipes   191

Dimensional analysis provides important clues as to the form of this limiting, high R_D, turbulent velocity profile law for smooth pipes. Starting with the proposition that, in general, the velocity at any point across the pipe is influenced by the radial position (y), the obvious independent variable; the wall shear stress (τ_0), which provides the means of specifying wall and fluid interactions; the fluid density (ρ), upon which the turbulent shear stresses depend; and the fluid viscosity (μ), upon which the laminar shear stresses depend; we have

$$u = \varphi(y, \tau_0, \rho, \mu). \tag{5.17}$$

Thus five variables are felt to be controlling, and they involve three primary dimensions, namely, mass (M), length (L), and time (T). By the Buckingham pi theorem [3] of dimensional analysis, $5-3=2$ dimensionless π terms (or parameters) are required to succinctly represent these variables. Briefly, we have

$$\pi_1 = \left(\rho^a \tau_0^b y^c\right) u$$

and

$$\pi_2 = \left(\rho^A \tau_0^B y^C\right) \mu.$$

When these parameters are expressed in terms of the primary dimensions, there result

$$\pi_1 = (ML^{-3})^a (ML^{-1}T^{-2})^b (L)^c LT^{-1}$$

and

$$\pi_2 = (ML^{-3})^A (ML^{-1}T^{-2})^B (L)^C ML^{-1}T^{-1}.$$

These expressions lead to the following series of simultaneous equations in the exponents (recalling that, to be dimensionless, the net dimensions of M, L, T must be independently zero):

π_1

$$\begin{aligned} M:& \quad 0 = a + b, \\ L:& \quad 0 = -3a - b + c + 1, \\ T:& \quad 0 = -2b - 1, \end{aligned}$$

or

$$a = \tfrac{1}{2}, \quad b = -\tfrac{1}{2}, \quad c = 0.$$

Hence

$$\pi_1 = \frac{u}{(\tau_0/\rho)^{1/2}}.$$

Velocity Distributions

π_2

$$M: \quad 0 = A + B + 1,$$
$$L: \quad 0 = -3A - B + C - 1,$$
$$T: \quad 0 = -2B - 1,$$

or

$$A = -\tfrac{1}{2}, \quad B = -\tfrac{1}{2}, \quad C = -1.$$

Hence

$$\pi_2 = \frac{\mu}{(\rho \tau_0)^{1/2} y}.$$

At this point in the development, it is conventional to introduce the *friction velocity* (V^*), defined as

$$V^* = \sqrt{\frac{\tau_0 g_c}{\rho}} \tag{5.18}$$

in the pound mass-pound force system. Thus we have

$$\pi_1 = \frac{u}{V^*}$$

and

$$\pi_2 = \frac{\mu}{\rho V^* y} = \frac{\nu}{V^* y}.$$

It follows, from the principles of dimensional analysis, that

$$\pi_1 = f(\pi_2)$$

or

$$\frac{u}{V^*} = f\left(\frac{V^* y}{\nu}\right),$$

where we note that any π term can be raised to any power without altering its dimensionality—hence the inversion of π_2.

It is convenient to introduce still another shorthand notation here, namely,

$$u^+ = \frac{u}{V^*}, \tag{5.19}$$

a dimensionless velocity ratio, and

$$y^+ = \frac{V^* y}{\nu}, \tag{5.20}$$

a *new* Reynolds number called the *friction Reynolds number*. Thus the generalized function for the velocity distribution across the *wall layer* for turbulent flow in smooth pipes becomes

$$u^+ = f(y^+). \tag{5.21}$$

At first, this function between u and y was taken to be that given by the power law. However, after Ludwig Prandtl announced his mixing length theory [4] in 1926, the power law interpretation was abandoned in favor of the newer *law of the wall*.

Five important steps can be listed in the development of the law of the wall.

1. The first step was recognition of the *parameters* of primary importance in the wall layer, namely, u^+ and y^+, as obtained from dimensional analysis.
2. A second step forward was the recognition that a major portion of the wall layer could be described by a *logarithmic* function in y, as suggested by the limiting power law, and as indicated functionally by Prandtl's mixing analogy.
3. The third step in the development of the law of the wall was taken when Prandtl suggested the existence of a thin laminar sublayer next to the wall, where a *linear* velocity profile prevailed.
4. A fourth step, not part of the mathematical formulation at all, was the *belief* that the resulting law of the wall remained valid independently of the axial pressure gradient in the pipe.
5. The fifth step was the experimental demonstration that the law of the wall applied to wall layers in *developing* boundary layers, as well as to fully developed turbulent boundary layers.

The various zones covered by the law of the wall, given in general form by (5.21), are discussed next in some detail.

Logarithmic Layer Expecting a logarithmic function of y as the limiting expression for the power law, Prandtl [4] developed a mixing length theory as a means for predicting this logarithmic function analytically. Beginning with the basic idea that shear stress at any radial position is made up, in general, of two parts, he arrived at the expression

$$\tau = \mu \frac{du}{dy} + \rho \left(L \frac{du}{dy} \right)^2. \tag{5.22}$$

194 Velocity Distributions

The first term of (5.22) is the familiar *viscous* contribution to the total shear stress, while the second term is the turbulent portion of the shear stress.

For turbulent flow in general, it was postulated that over most of the flow area the viscous shear stress essentially could be neglected. In such regions (5.22) reduces to

$$\tau_T = \rho\left(L\frac{du}{dy}\right)^2, \tag{5.23}$$

which expresses the turbulent shear stress (τ_T) between any two layers of fluid in terms of Prandtl's mixing length (L), which he claimed was proportional to y only, that is,

$$L = Ky, \tag{5.24}$$

where K is taken as a universal constant.

As a further simplification, Prandtl asserted that the turbulent shear stress (τ_T) at any y in the near vicinity of the wall essentially was given by the *wall* shear stress (τ_0). When (5.23) and (5.24) are combined in light of this assertion, there results

$$\sqrt{\frac{\tau_0}{\rho}} = L\frac{du}{dy} = Ky\frac{du}{dy},$$

which, upon separation of variables, yields

$$\frac{du}{V^*} = \frac{1}{K}\frac{dy}{y}.$$

Integration of this equation leads to the very significant smooth pipe turbulent velocity relation

$$\frac{u}{V^*} = \frac{1}{K}\ln y + C. \tag{5.25}$$

Equation 5.25, which was first given by Prandtl [4] in 1926, can be put in the functional form of (5.21) by introducing the constant $1/K \ln V^*/\nu$ into both terms on the right-hand side of (5.25). There results

$$\frac{u}{V^*} = \frac{1}{K}\ln\frac{V^*y}{\nu} + \left(C - \frac{1}{K}\ln\frac{V^*}{\nu}\right),$$

which, of course, can be reduced to

$$u^+ = A \ln y^+ + B. \tag{5.26}$$

Equation 5.25 also can be obtained from a dimensional analysis, by noting that the velocity gradient in the fully turbulent region is independent of viscosity. In terms of (5.17), this condition can be expressed as

$$\frac{du}{dy} = f(y, \tau_0, \rho).$$

There being four variables and three primary dimensions, a single π term results:

$$\pi = \frac{du}{dy} \frac{y}{\sqrt{\tau_0/\rho}} = \text{constant},$$

which can be integrated to

$$\frac{u}{V^*} = A \ln y + C. \tag{5.25'}$$

Equation 5.26, which follows directly from either (5.25) or (5.25'), involves two very important constants, A and B. A is a dimensionless constant, the reciprocal of which, in the form of K in (5.24) and (5.25), has come to be called von Karman's constant. A (or K) is expected to be very nearly the same for all wall layers in turbulent flow. It specifies the *rate* at which turbulent mixing develops at points progressively further from the wall. B is another dimensionless constant whose value, however, differs for smooth and rough walls.

The first experimental determination of the constants in (5.26) was made by Johann Nikuradse (who was on Prandtl's staff at Gottingen [5] in 1932. He gave for smooth pipes:

$$u^+ = 2.5 \ln y^+ + 5.5 = 5.75 \log y^+ + 5.5. \tag{5.27}$$

Some of the many experimental values for A and B are given in Table 5.5, in terms of *common* (i.e., base 10) logarithms. Several reasons can be given for the variability of A and B, including geometric effects, core flow differences, Reynolds number effects, roughness effects, instrument effects (such as static tap errors), and mathematical difficulties in fitting the data.

In spite of these minor variations, the law of the wall may be considered, broadly speaking, to be established experimentally, and may be applied to

Velocity Distributions

Table 5.5 Summary of Constants Used in Smooth Pipe Log Profile $(u^+ = A \log y^+ + B)$

Source	Date	A	B
Nikuradse [5]	1932	5.75	5.50
Ludwieg and Tillman [21]	1949	5.75	5.20
Preston [22]	1954	5.50	5.80
Clauser [23]	1954	5.60	4.90
Coles [24]	1956	5.75	5.10
Smith and Walker [25]	1958	5.00	7.15
Pao [26]	1961	5.50	5.75
Patel [14]	1965	5.50	5.45
Rainbird [27]	1967	5.32	6.22
Lindley [15]	1970	5.60	4.90
Au [16]	1972	5.70	5.30

any turbulent boundary layers over smooth pipe walls, when the pressure gradients are not too severe (see Section 5.4.5).

Laminar Sublayer. Again considering turbulent flow in general, Prandtl postulated that there must still remain a thin region next to the pipe wall where the turbulent mode of transfer of momentum gave way to a molecular mode of momentum transfer such as characterizes laminar flow. In other words, very close to the wall, turbulent mixing stresses must be damped out or suppressed by the presence of the wall, and viscous stresses alone must predominate. But this would require a thin layer of fluid in laminar motion near the wall, even for the most turbulent flow in the pipe. Since the fluid velocity at the wall must be zero, and since the laminar sublayer must be very thin, it is generally accepted that the velocity profile in this region is *linear*.

In this laminar sublayer the controlling variables are u, y, τ_0, and μ only, that is, the density is omitted from (5.17) because the turbulent stresses are negligible in this region compared to the viscous stresses. Then, by dimensional analysis, we obtain the single π term

$$\pi = \frac{u\mu}{\tau_0 y} = \text{constant};$$

or, by a slight rearrangement and by introducing the friction velocity, we have

$$\frac{u}{V^*} = \frac{V^* y}{\nu}, \qquad (5.28)$$

which, in the usual shorthand notation, becomes

$$u^+ = y^+. \tag{5.29}$$

Actually, the assumption of a linear velocity profile *alone* leads to (5.28) and (5.29), as indicated by the following steps:

1. Viscosity is defined in the laminar region by

$$\tau_0 = \mu \frac{du}{dy}.$$

2. When the velocity gradient is linear, this becomes

$$\frac{\tau_0}{\mu} = \frac{u}{y}.$$

3. Upon rearrangement we have

$$\frac{u}{\tau_0/\rho} = \frac{y\rho}{\mu},$$

which, of course, is equivalent to (5.28) and (5.29).

Some writers, notably B. Miller [6], have seriously questioned* the laminar sublayer hypothesis of turbulent flow. But most authorities accept, as basic to both mass and heat transfer processes, the existence of a laminar sublayer, and we proceed as if it were firmly established experimentally. Indeed, the successful application of the law of the wall is *almost* a proof that Prandtl was correct.

It is interesting to note just how the laminar sublayer thickness depends on the pipe Reynolds number, and to note the order of magnitude of this thickness. We can rewrite (5.28) and (5.29), in terms of (5.1) and (5.18), as

$$u^+ = y^+ = \left(\frac{VD}{\nu}\right)\left(\frac{V^*}{V}\right)\left(\frac{y}{D}\right).$$

This can be further simplified, in terms of (5.40), as

$$u^+ = y^+ = R_D \sqrt{\frac{f}{8}} \left(\frac{y}{D}\right). \tag{5.30}$$

*On p. 357 of Reference 6, Miller states, "The thickness of the laminar film must be a statistical thing, a kind of average of values which differ from place to place over the surface of the pipe, and from instant to instant at each place, these values ranging from zero upward."

Table 5.6 Thickness of Laminar Sublayer as a Function of the Pipe Reynolds Number

R_D	f_s (6.12)	$\delta_L/D \times 10^3$ (5.31)
10^4	0.03089	8.0465
10^5	0.01799	1.0544
10^6	0.01165	0.1310
10^7	0.00810	0.0157

When the extent of the laminar sublayer is taken as $y^+ = 5$, we can, setting $y = \delta_L$, where δ_L signifies the thickness of the laminar sublayer, rearrange (5.30) to

$$\frac{\delta_L}{D} = \frac{5}{R_D}\sqrt{\frac{8}{f}}, \qquad (5.31)$$

which expresses the approximate thickness of the laminar sublayer for turbulent flow in pipes, as a function of the pipe Reynolds number. Solutions to (5.31) are presented numerically in Table 5.6, and graphically in Figure 5.13, in terms of Prandtl's equation (6.12) for the friction factor. These show that δ_L is less than 1% of D at $R_D = 10^4$, and further indicate that δ_L decreases rapidly as R_D increases. The large fluctuations in velocity and the thinness of the laminar sublayer combine to make it difficult to obtain velocity profile information in this region.

We will have reason to discuss again the basis of the laminar sublayer when we consider the overall law of the wall.

Buffer Zone. After formulating the two-part law of the wall, namely, the *purely laminar* sublayer, in which the velocity varied linearly with y, and the *fully turbulent* layer, in which the velocity varied logarithmically with y, it was recognized that the intersection of these profiles introduced a discontinuity in the velocity and in the velocity gradient that was not realistic. It was felt [7] that there must be a transition region[*] in which velocities and velocity gradients changed gradually and continuously from the linear to the logarithmic profile. Thus a *buffer zone* was postulated[†] in which viscous and turbulent stresses are of the same magnitude. This

---

[*] As far back as 1910, Prandtl said, "There must be an *intermediate* zone between laminar and turbulent flow, whose thickness certainly could not be zero."
[†] First by A. Eagle and R. M. Ferguson [8] in 1930, and then by T. von Karman [9] in 1934.

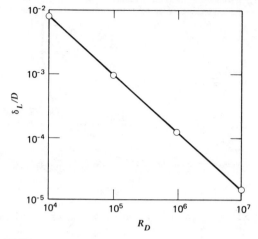

Figure 5.13 Thickness of laminar sublayer (δ_L) as a function of Reynolds number (R_D).

means that the *general* analysis, as given by (5.21), is required to describe the velocity profile in the buffer zone, where the functional form of (5.21) depends on knowledge of both the laminar sublayer and the turbulent logarithmic layer. Theodor von Karman, in 1934, gave an equation for the velocity profile in this transition region as

$$u^+ = 5\ln y^+ - 3.05 = 11.5 \log y^+ - 3.05, \qquad (5.32)$$

which was said to apply for $5 < y^+ < 30$. Others gave more complex equations and said the buffer zone extended from a y^+ of 5 to a y^+ of about 60 or 70.

Summary of Law of the Wall

1. There is a wall layer in turbulent pipe flow where most of the velocity variation occurs.
2. The wall layer is made up of three segments: a laminar sublayer, a buffer zone, and a logarithmic turbulent layer (see Figure 5.14).
3. A law of the wall has been formulated to describe the velocity profile in the wall layer, whose general relation is $u^+ = f(y^+)$. Typical functional relations, one for each zone of the wall layer, are given in Table 5.7. A schematic plot of u^+ versus $\log y^+$ is given in Figure 5.15.
4. The thickness of the laminar sublayer and of the buffer zone is reduced as the pipe Reynolds number increases (see Figure 5.13).

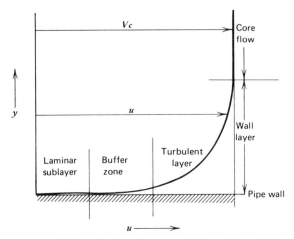

Figure 5.14 Schematic drawing of wall layer in turbulent flow.

Table 5.7 Various Relations for Turbulent Velocity Profiles for Smooth Pipes

Region	General Functional Relation	Range	Typical Specific Functional Relation
Laminar sublayer	$u^+ = y^+$	$0 < y^+ < 5$	$u^+ = y^+$
Buffer zone	$u^+ = f(y^+)$	$5 < y^+ < 30$	$u^+ = 5 \ln y^+ - 3.05$
Turbulent layer	$u^+ = A \ln y^+ + B$	$y^+ > 30$	$u^+ = 2.5 \ln y^+ + 5.5$

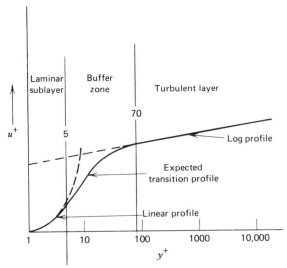

Figure 5.15 Semilog plot of u^+ and y^+ for turbulent velocity profile in smooth pipe.

5. The law of the wall is universal* in the sense that it applies equally well to *developing* turbulent boundary layers and to *fully developed* turbulent pipe flows.
6. A really satisfactory formulation for the velocity distribution over the entire pipe flow area must include and match together the various wall layer components and the core flow in such a manner that the velocity varies smoothly and continuously.

5.4.3 Further Discussion of Law of the Wall

Nikuradse Shift. It was first pointed out by Miller [6] that Nikuradse's published work showed a shift in the original data of seven units on the y^+ scale. On pursuing this question, it appears that the original data, falling above the hypothesized laminar sublayer curve, were shifted in the belief that Prandtl's laminar film hypothesis was more correct than were the data (see Figure 5.16). Of course, this shift was entirely arbitrary since the expected point of tangency with the suggested laminar curve was unknown.

One important consequence of this shift is the von Karman representation of the velocity profile in the buffer zone (see Figure 5.17). This was based Nikuradse's adjusted values and has the added shortcomings that it introduces a discontinuous match with the velocity profile in the fully turbulent layer at $y^+ = 30$.

In spite of these rather dubious curve fitting techniques, the Nikuradse shift affects the velocity profile only for y^+ values less than 50, and the law of the wall remains well accepted and generally useful over most of the pipe flow area.

Average Velocity for Turbulent Flow. The average velocity across a pipe with turbulent flow can be estimated if the pipe Reynolds number is high (so that the laminar sublayer and the buffer zone are very thin), and if the flow is fully developed (so that velocity changes in the core are small), by assuming that the velocity distribution in the fully turbulent wall layer (as given by the logarithmic profile) describes the velocity across the whole pipe. From (5.25), evaluated at the pipe center as well, and Nikuradse's experimental determination of von Karman's constant, there results the general velocity expression

$$u = V_c - 2.5 V^* \ln \frac{R}{y}. \tag{5.33}$$

*This was first demonstrated in 1962 by M. R. Head and I. Rechenberg in a very significant experiment as reported in Reference 10.

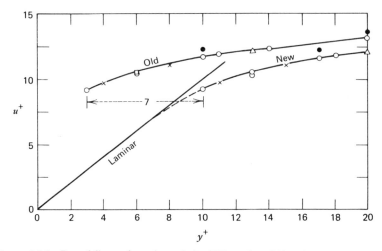

Figure 5.16 Prandtl's explanation of the Nikuradse shift (after Miller [6]).

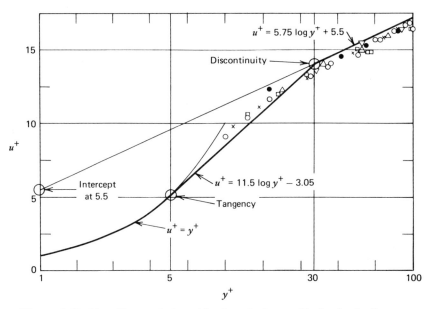

Figure 5.17 Von Karman's logarithmic velocity profile in the buffer zone (after Miller [6], using Nikuradse's [5] points).

The volumetric average velocity is obtained by combining the general volumetric flow rate equation with the turbulent velocity relation of (5.33) as follows:

$$Q = AV = \int_A u\,dA = 2\pi \int_0^R \left(V_c - 2.5V^* \ln\frac{R}{y}\right) r\,dr.$$

With $r = R - y$ and $dr = -dy$ there results

$$V = \frac{2}{R^2}\int_0^R (V_c r\,dr - 2.5V^* \ln R r\,dr + 2.5V^* \ln y r\,dr)$$

or

$$V = V_c - 2.5V^* \ln R + \frac{5V^*}{R^2}\left(-R\int_R^0 \ln y\,dy + \int_R^0 y\ln y\,dy\right). \quad (5.34)$$

The general form and general solution of the logarithmic integrals of (5.34) are

$$\int y^N (\ln y)^M dy = \frac{y^{N+1}(\ln y)^M}{N+1} - \frac{M}{N+1}\int y^N (\ln y)^{M-1} dy, \quad (5.35)$$

so that (5.34), integrated in terms of (5.35), becomes

$$V = V_c - \tfrac{3}{2}AV^* = V_c - 3.75V^*. \quad (5.36)$$

But (5.36) is bound to define an average velocity that is *greater* than actual, because the log law of (5.33) does not yield a zero velocity gradient at the pipe center (see Figure 5.18). Nikuradse's experiment suggests, instead of (5.36), the more empirical relation

$$V = V_c - 4.07V^*. \quad (5.37)$$

Comparison of Power Law and Logarithmic Profile. To draw comparisons between the power law of (5.13) and the log law of (5.27), it is necessary to express the latter in terms of information similar to that used in the former. Specifically, we need to express the log law in terms of the geometric factor (y/R) and the pipe Reynolds number (R_D). This can be done as follows. From a slightly modified form of (5.33) we have

$$\frac{u}{V_c} = 1 + 2.5\left(\frac{V^*}{V_c}\right)\ln\frac{y}{R}. \quad (5.38)$$

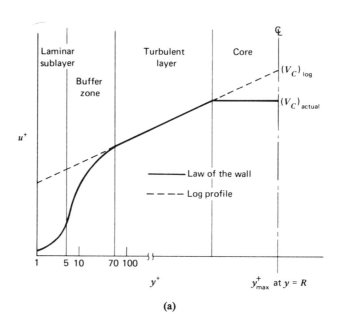

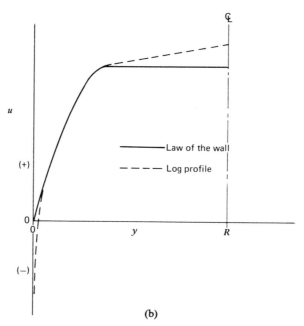

Figure 5.18 Comparison of fully turbulent logarithmic profile and the law of the wall profile.

To get $(u/V_c)_{\log}$ in terms of y/R and R_D, it is clear that we must express V^*/V_c as a function of R_D. The simplest entry to R_D is through the friction factor (f) to be discussed in greater detail in Chapter 6. Briefly, when the Darcy-Weisbach equation, namely,

$$\Delta p = \left(f\frac{L}{D}\right)\frac{\rho V^2}{2g_c} \tag{5.39}$$

in the pound mass-pound force system, is combined with (5.2) and (5.18), there results

$$\frac{V^*}{V_c} = \sqrt{\frac{f}{8}}\left(\frac{V}{V_c}\right). \tag{5.40}$$

In terms of (5.36) this can be reduced further to

$$\frac{V^*}{V_c} = \frac{1}{\sqrt{8/f} + 3.75}. \tag{5.41}$$

It remains only to insert suitable values for $f = f(R_D)$ to express $(u/V_c)_{\log}$ in terms of the required parameters, y/R and R_D. For Reynolds numbers that correspond to the power law parameters, $N = 6$, 7, and 10, values for the turbulent velocity profile in smooth pipes, according to the log law of (5.38) and (5.41), are given in Table 5.8. These same power law and log law velocity profiles are compared in Figure 5.19. The required friction factors are those based on Prandtl's equation, (6.12), as given in the next chapter.

Table 5.8 Turbulent Velocity Profiles in Smooth Pipes [according to the log law of (5.38) and (5.41)]

R_D		u/V_c	
y/R	4×10^3	1.1×10^5	3.2×10^6
0	∞	∞	∞
0.1	0.67852	0.76957	0.82328
0.2	0.77530	0.83894	0.87648
0.3	0.83191	0.87951	0.90760
0.4	0.87207	0.90830	0.92967
0.5	0.90322	0.93063	0.94680
0.6	0.92868	0.94888	0.96079
0.7	0.95020	0.96431	0.97262
0.8	0.96884	0.97767	0.98287
0.9	0.98529	0.98946	0.99191
1.0	1	1	1
f_s	0.03992	0.01775	0.00963
V^*/V_c	0.05585	0.04003	0.03070

206 Velocity Distributions

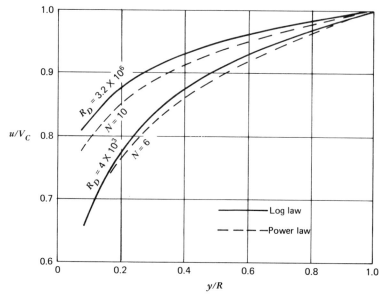

Figure 5.19 Comparison of several turbulent velocity profiles for smooth pipe according to the power and log laws (see Tables 5.2 and 5.8).

5.4.4 Other Forms of Law of the Wall

Reichardt's Expression. In 1951, H. Reichardt [11], at the Max Planck Institute, gave a single expression to describe the velocity profile in the complete wall layer and the core, in a manner that was compatible with the experimental evidence. For $y^+ > 30$ this equation is in agreement with the log law of (5.27). Moreover, it defines an essentially linear velocity gradient in the laminar sublayer region, with the velocity approaching zero at the wall, as does the laminar sublayer relation of (5.29), but in direct opposition to the log law, which approaches $-\infty$ very near the wall (see Figure 5.18b). Reichardt's equation is

$$u^+ = 2.5\ln(1+0.4y^+) + 7.8\left(1 - e^{-y^+/11} - \frac{y^+}{11}e^{-0.33y^+}\right). \quad (5.42)$$

Some of Nikuradse's and Reichardt's experimental points, determined for turbulent flows in smooth pipes, are compared in Figure 5.20 with Reichardt's equation, (5.42).

Deissler's Equations. Also in 1951, R. G. Deissler [12], at the NACA Lewis Laboratories, presented many experimental data for fully developed

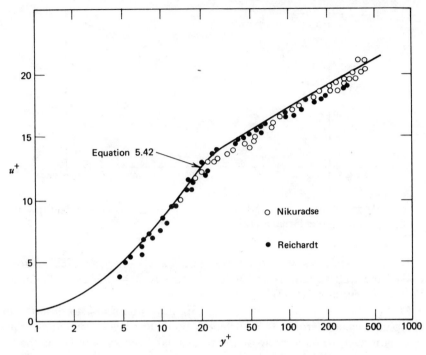

Figure 5.20 Turbulent velocity profile (according to Reichardt [11]) in smooth pipe.

turbulent flow in smooth pipes for Reynolds numbers ranging from 10,000 to 200,000. He represented his data by two empirical expressions:

$$y^+ = \frac{1}{N} \frac{(1/\sqrt{2\pi})\int_0^{Nu^+} e^{-(Nu^+)^2/2} d(Nu^+)}{(1/\sqrt{2\pi})e^{-(Nu^+)^2/2}}, \qquad (5.43)$$

where N was determined experimentally to be 0.109 for the inner layers, with $y^+ < 26$; and

$$u^+ = 2.78 \ln y^+ + 3.8 \qquad (5.44)$$

for the logarithmic layer, where $y^+ > 26$. It should be mentioned that (5.43) is best evaluated by noting that the denominator is the normal error function of Nu^+, while the numerator is the integral of this function. Both of these quantities are available in standard mathematical and statistical tables.

208 Velocity Distributions

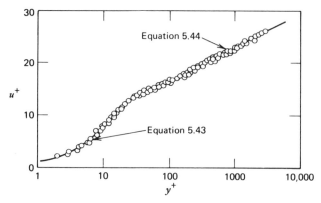

Figure 5.21 Experimental and empirical representations of turbulent flow in smooth pipes (after Deissler [12]).

Deissler's work is summarized in Figure 5.21. Numerical solutions to the law of the wall equations of Table 5.7, Reichardt's equation, (5.42), and Deissler's equations, (5.43) and (5.44), are given in Table 5.9.

Many others see, e.g., R. B. Dean [13]), have attempted to describe the complete velocity profile in a turbulent boundary layer by a single equation. However, to date, none of the results has been as popular as the law of the wall and Reichardt's expression.

Table 5.9 Several Velocity Profiles for Turbulent Flow in a Smooth Pipe Compared with the Law of the Wall

Zone	y^+	Law of the Wall [per Table 5.7]	Reichardt's Profile [per (5.42)]	Deissler's Equations [(5.43) and (5.44)]
Laminar sublayer	1	1.0000	1.0092	0.9962
	2	2.0000	2.0332	1.9678
	5	5.0000	4.9147	4.5950
Buffer zone	10	8.4629	8.4195	7.8100
	20	11.9287	12.0077	11.5190
	50	15.2800	15.3285	14.6754
Log layer	100	17.0129	17.0830	16.6024
	200	18.7458	18.7861	18.5293
	500	21.0365	21.0583	21.0766
	1000	22.7694	22.7849	23.0036

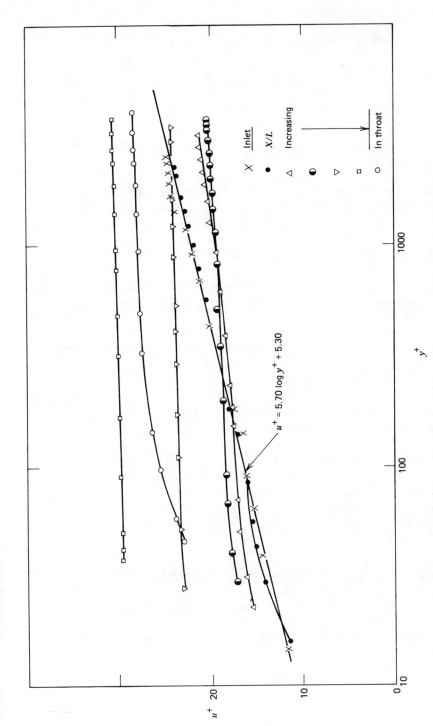

Figure 5.22 Development of velocity profile through a venturimeter (after Au [16]).

5.4.5 Pressure Gradient Effect on the Law of the Wall

Although, as initially conceived, the law of the wall was believed to hold for *all* turbulent flows, independently of the axial pressure gradient, it now appears from the results of several experimenters that this is not the case.

V. C. Patel [14], in 1965, noted that in sufficiently strong favorable pressure gradients (where *favorable* indicates that the pressure is *dropping* in the flow direction), the law of the wall breaks down. He further indicated that this breakdown is to be associated with a reversion (i.e., a falling back) of the turbulent boundary layer toward laminar flow. D. Lindley [15], in 1969, followed the same reasoning and supported it with further experimental data obtained in venturi tests.

S. B. Au [16], in 1972, provided detailed measurements, again obtained in venturi tests, where the favorable pressure gradients were those set up by the converging inlet of the venturi, which document the departure of the velocity profile from the log law in the wall region as the flow is sampled further and further downstream of the venturi inlet. A typical plot, based on Au's work, is shown in Figure 5.22, where it can be seen that after the inlet region, where the profile does follow the law of the wall, the velocity profile moves upward near the wall region (smaller y^+) and downward in the outer range (larger y^+). As the flow is sampled by pressure tapping, further and further into the converging section, where the flow is now strongly accelerated, the velocity profile departs further and further from the law of the wall. However, as the pressure gradient decreases in magnitude, somewhere in the throat section of the venturi, a redevelopment toward the fully developed turbulent state is seen to take place.

It was Patel who concluded that, although the pressure gradients that exist in fully developed *pipe* flow have a negligible effect on the velocity distribution in the wall region (i.e., that the law of the wall generally applies in *pipe* flow), for flows at certain severe gradients (such as are encountered in nozzles and venturis), the law of the wall breaks down completely.

5.5 TURBULENT FLOW IN ROUGH PIPES

In considering laminar flows, we overlooked the roughness of the pipe wall when determining the velocity distribution, as given by (5.7). However, we indicated that in turbulent flow the condition of the pipe wall (i.e., smooth or rough) influences the velocity profile. Equations were developed to describe the velocity distribution for turbulent flow cases where the pipe was judged to be *smooth*, notably the power law and the law of the wall. We now consider the effect of wall roughness on the velocity profile, and at the same time find out when and why roughness plays its part.

5.5.1 Wall Roughness

Wall roughness is a term loosely used to describe the complex size, shape, and spacing of the protrusions found on the inner wall of a pipe (see Figure 5.3). Although it is normal practice to assign a single number, the relative roughness (e/D), to signify the condition of the pipe wall, it should be recognized that the absolute roughness (e) is of necessity some sort of statistical attempt to indicate only the size of the protrusions.

It has been found experimentally that a pipe with a given relative roughness will behave sometimes as a smooth pipe, and at other times as a rough pipe, depending on the pipe Reynolds number. The velocity profiles in these two situations encountered in the same pipe are entirely different. The explanation of this dual-role phenomenon constitutes one of the successes of the boundary layer theory in general, and of the laminar film hypothesis in particular.

Hydraulically, the roughness of a pipe is ultimately determined by the size of the absolute roughness with respect to the thickness of the laminar boundary layer. Thus, for laminar flow, since the protrusions are all contained within a laminar boundary layer, at all Reynolds numbers the flow resistance is caused entirely by viscous shear stresses, and the pipe roughness *does not* influence the velocity distribution at all. However, for turbulent flow, two regimes are encountered. When the *roughness* Reynolds number,

$$e^+ = \frac{V^* e}{\nu} \qquad (5.45)$$

[patterned after the y^+ of (5.20)], is less than 5, the protrusions are necessarily contained wholly within the laminar sublayer, so that again viscous shear alone determines the flow resistance, and roughness has *no effect* on the flow. When e^+ is greater than 70, however, the protrusions extend into the turbulent wall layer. Now, it is the form drag of the protrusions that determines the flow resistance, rather than the viscous effects, and hence roughness has a *profound effect* on the velocity profile. For e^+ between 5 and 70, there is a transition region where flow resistance is the result of both viscous shear and form drag.

5.5.2 Rough Law of the Wall

Accordingly, Prandtl modified his smooth pipe equation, (5.25), to the form

$$u^+ = A \ln \frac{y}{e} + C_1 \qquad (5.46)$$

to describe turbulent flows in rough pipes, that is, he replaced the distance from the wall (y) in (5.25) with a relative roughness (y/e) in (5.46). Once again, it was Nikuradse [17] who supplied the first experimental verification of (5.46) and provided the first numerical evaluation of C_1. His data, plotted in Figure 5.23, yielded for rough pipes in the fully turbulent region

$$u^+ = 2.5\ln\frac{y}{e_s} + 8.5, \tag{5.47}$$

where it is significant to note that von Karman's constant, $K=1/A$, was found to apply also for turbulent flow in rough pipes, just as it had for smooth pipes.

A generalized plot that compares, smooth, rough, and in-between pipes can be given in terms of the roughness Reynolds number of (5.45) and the factor

$$F = u^+ - 2.5\ln\frac{y}{e}. \tag{5.48}$$

For smooth pipes we have, from (5.27) and (5.48),

$$F_{\text{smooth}} = 2.5\ln e^+ + 5.5. \tag{5.49}$$

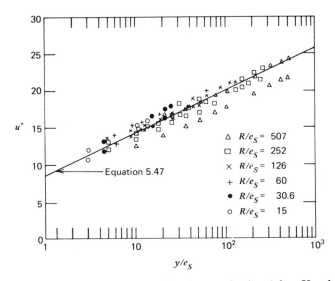

Figure 5.23 Turbulent velocity profile for rough pipe (after Knudsen and Katz [29], using Nikuradse's [17] points).

Turbulent Flow in Rough Pipes

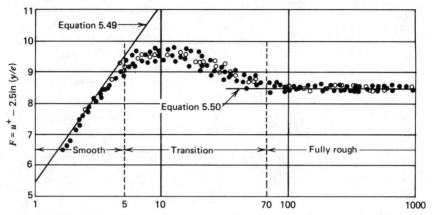

Figure 5.24 Comparison of law of wall for smooth and rough pipes (after Schlichting [20], using Nikuradse's [5, 17] points).

For rough pipes we have, from (5.47) and (5.48),

$$F_{\text{rough}} = 8.5. \qquad (5.50)$$

The straight lines of (5.49) and (5.50), along with the experimental points of Nikuradse, are presented on the semilogarithmic plot of Figure 5.24.

5.5.3 Pipe Factor

The average velocity in the pipe is greatly affected by the pipe roughness, which, in turn, is a function of the pipe Reynolds number. A *pipe factor* is defined as

$$\sigma = \frac{V}{V_c}. \qquad (5.51)$$

A plot of σ versus R_D is shown, just as Nikuradse first determined it, in Figure 5.25. The upper line, labeled $R/e_s = 507$, can be taken as representing an essentially smooth pipe. Parameters of increasing roughness are shown peeling off below the smooth curve. These curves indicate a flattening of the velocity profile with increased roughness. The pipe factor is quite useful in computing flow rate from centerline velocity measurements.

Example 5.1. In Example 2.3 we obtained the centerline velocity in a smooth 12 in. pipe as 27.25 ft/sec when the fluid density was 62.4 lbm/ft³.

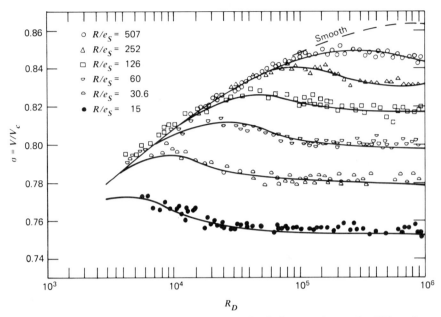

Figure 5.25 Pipe factor (σ) as a function of relative roughness (e_S/R) and pipe Reynolds number (R_D) (after Knudsen and Katz [29], using Nikuradse's ([17] points).

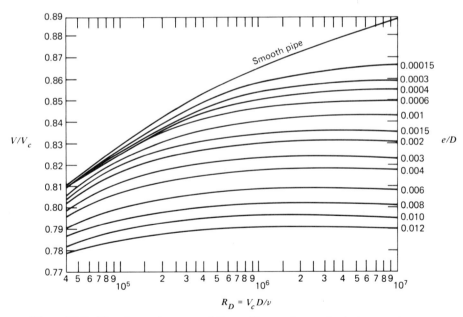

Figure 5.26 Pipe factor in terms of Reynolds number and relative roughness (after Binder [30]).

214

If the Reynolds number, based on this velocity, was 2.5×10^6, what would be the volumetric average velocity and the mass flow rate in this pipe?

Solution: Referring to Figure 5.26, which gives the pipe factor in terms of a Reynolds number based on the maximum velocity at the pipe centerline, we find for a smooth pipe at $R_D = 2 \times 10^6$ the pipe factor, $V/V_c = 0.875$. Hence

$$V = 0.875 \times 27.25 = 23.84 \text{ ft/sec}$$

and

$$\dot{m} = \rho A V = 62.4 \times \frac{\pi (1)^2}{4} \times 23.84 = 1168.4 \text{ lbm/sec.}$$

5.6 UNIVERSAL LAW OF THE WALL

When the rough pipe equation, (5.47), is evaluated at the pipe centerline, and the difference between the point velocity and the maximum velocity is taken, the result is (5.33), the smooth pipe equation. Of course, integration of (5.33) yields the average velocity of (5.36) for the rough pipe, as for the smooth. Thus, although *point* and *average* velocities are quite different numerically in the smooth and rough pipes, when the average velocity in a pipe is *subtracted* from the point velocity, for both smooth and rough pipes, we get the *same* equation. In other words, for smooth or rough pipes, from (5.33) and (5.36) we obtain

$$u^+ = 2.5 \ln \frac{y}{R} + 3.75 + \frac{V}{V^*}. \tag{5.52}$$

This result, which we could not have anticipated, is taken as an indication that turbulence is independent of the condition of the pipe wall, and suggests that a further generalization exists. It suggests also that an equation can be developed to describe the turbulent velocity distribution in a pipe, *independent* of its roughness.

When the equations already given for smooth and rough pipes are written in the form*

$$\frac{V_c - u}{V^*} = f\left(\frac{y}{R}\right), \tag{5.53}$$

*The form of (5.53) was first given, in 1914, by T. E. Stanton and J. R. Pannell [18].

216 Velocity Distributions

the velocity distribution curves versus the geometric factor (y/R) collapse to a single curve for all values of the pipe Reynolds number and for all degrees of roughness. Thus, for smooth pipes, from (5.33), written in the form of (5.53), we have

$$\frac{V_c - u}{V^*} = 2.5 \ln \frac{R}{y}, \qquad (5.54)$$

while for rough pipes, from (5.47), evaluated at the pipe centerline as well, and written in the form of (5.53), we also arrive at (5.54).

It is interesting to note that in 1858 Henry Philibert Gaspard Darcy [19] published experimental work on pipe flow, which Nikuradse [17] numerically arranged in the form of (5.53) as

$$\left(\frac{V_c - u}{V^*}\right)_{\text{Darcy}} = 5.08\left(1 - \frac{y}{R}\right)^{3/2}. \qquad (5.55)$$

The curve for both smooth and rough pipes, as given by (5.54), is compared with Darcy's work, as given by (5.55), in Figure 5.27. It can be seen that Darcy's results agree quite well with the more modern representation for $y/R > 0.25$.

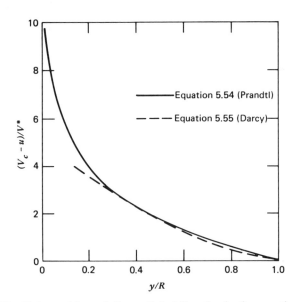

Figure 5.27 Universal law of the wall, holding for both smooth and rough pipes.

5.7 KINETIC ENERGY COEFFICIENT

The velocity profiles that we have discussed and described mathematically in this chapter are of practical use in solving real pipe flow problems. Not only do the various velocity profiles serve as the basis of the rational friction factor formulations to be dealt with in Chapter 6, but also they bear directly on the energy and momentum equations of real pipe flow. The kinetic energy coefficient, to be defined in this section, is just one example of how a knowledge of the velocity profile improves our ability to make more meaningful energy balances for pipe flow.

From the basic concept that mass flow rate ($\dot{m}$) generally can be given in terms of the average velocity (V) of the profile as

$$\dot{m} = \rho V A, \tag{5.56}$$

and in terms of the point velocity (u) as

$$\dot{m} = \int_A \rho u \, dA, \tag{5.57}$$

we can define, for constant density flows, the average velocity as

$$V = \frac{1}{A} \int_A u \, dA. \tag{5.58}$$

Similarly, the *true* kinetic energy flow rate ($\dot{K}.E.$), known also as the kinetic energy flux, generally is given as

$$\dot{K}.E. = K.E._{\text{true}} \dot{m} = \int K.E._{\text{point}} \, d\dot{m}, \tag{5.59}$$

where the fluid kinetic energy at a *point* is defined as

$$K.E._{\text{point}} = \frac{u^2}{2g_c} \tag{5.60}$$

on a per pound mass basis. From (5.59) and (5.60) we have

$$K.E._{\text{true}} = \frac{1}{\dot{m}} \int \left(\frac{u^2}{2g_c} \right) d\dot{m}. \tag{5.61}$$

218 Velocity Distributions

Another kinetic energy expression can be defined, as the counterpart of (5.60), based on the *average* velocity of (5.58), as

$$\text{K.E.}_{\text{ave}} = \frac{V^2}{2g_c}. \tag{5.62}$$

Whenever this average kinetic energy expression is used, as in an energy balance, a *correction* term is necessary to account for the difference between (5.61) and (5.62) so that

$$\text{K.E.}_{\text{true}} = \alpha \, \text{K.E.}_{\text{ave}}. \tag{5.63}$$

From (5.61), (5.62), and (5.63), it follows that α can be written as

$$\alpha = \frac{(1/\dot{m}) \int (u^2/2g_c) \, d\dot{m}}{(V^2/2g_c)}. \tag{5.64}$$

For pipe flow, where $A = \pi R^2$, $dA = 2\pi r \, dr$, and $d\dot{m} = \rho u (2\pi r \, dr)$, (5.64) simplifies to

$$\alpha = \frac{2}{R^2} \int_0^R \left(\frac{u}{V}\right)^3 r \, dr. \tag{5.65}$$

The α of (5.65) is known as the kinetic energy coefficient. It remains to evaluate α for the various velocity profiles.

5.7.1 Parabolic Profile of Laminar Flow

From u/V_c defined by (5.7) and V/V_c defined by (5.9) it follows that

$$\left(\frac{u}{V}\right)_L = 2\left[1 - \left(\frac{r}{R}\right)^2\right]. \tag{5.66}$$

Inserting (5.66) into the general definition of α, as given by (5.65), yields

$$\alpha_L = \frac{16}{R^2} \int_0^R \left[1 - \left(\frac{r}{R}\right)^2\right]^3 r \, dr = 2. \tag{5.67}$$

This result indicates that just *twice* as much kinetic energy is carried by the parabolic velocity distribution of laminar flow as there is in a one-dimensional uniform velocity profile of the *same* average velocity.

5.7.2 1/7 Power Law Profile of Turbulent Flow

From (5.58), with u/V_c defined by (5.13), and with $r = R - y$, we have

$$V = \frac{2V_c}{R^2} \int_0^R \left(\frac{y}{R}\right)^{1/7} (R-y)\,dy = \frac{98}{120} V_c. \tag{5.68}$$

Thus

$$\left(\frac{u}{V}\right)_{\text{power law}} = \left(\frac{120}{98}\right)\left(\frac{y}{R}\right)^{1/7}. \tag{5.69}$$

Inserting (5.69) into (5.65) yields

$$(\alpha_T)_{\text{power law}} = \frac{2}{R^2}\left(\frac{120}{98}\right)^3 \int_0^R \left(\frac{y}{R}\right)^{3/7}(R-y)\,dy = 1.05838. \tag{5.70}$$

This result indicates how much flatter is the turbulent velocity profile (as given by the power law) than the laminar profile. It further suggests that we can all but neglect the kinetic energy coefficient for turbulent flows, except when working toward the most precise results.

5.7.3 Logarithmic Profile of Turbulent Flow

From u/V defined by (5.33), it follows that

$$\left(\frac{u}{V}\right)_{\text{log law}} = \left(\frac{V_c}{V}\right) + 2.5\left(\frac{V^*}{V}\right) \ln \frac{y}{R}, \tag{5.71}$$

which can be written in the simplified form

$$X = W + 2.5z \ln Y. \tag{5.71a}$$

Thus

$$(\alpha_T)_{\text{log law}} = \frac{2}{R^2} \int_0^R X^3 r\, dr,$$

where the cube of x is

$$X^3 = W^3 + 7.5z W^2 \ln Y + 18.75 Wz^2 (\ln Y)^2 + 15.625 z^3 (\ln Y)^3.$$

220 Velocity Distributions

Making the proper substitutions, we have

$$(\alpha_T)_{\text{log law}} = \frac{2}{R^2}\left[W^3\int_0^R r\,dr + 7.5zW^2\int_0^R (\ln Y)r\,dr \right.$$
$$\left. + 18.75Wz^2\int_0^R (\ln Y)^2 r\,dr + 15.625z^3\int_0^R (\ln Y)^3 r\,dr\right]. \tag{5.72}$$

The terms involving $\ln Y$ in (5.72) can be integrated by means of (5.35). For example, the term

$$\frac{2}{R^2}(7.5zW^2)\int_0^R (\ln Y)r\,dr$$

can be written as

$$\frac{15zW^2}{R^2}\int_1^0 (\ln Y)[-R(Y-1)](-R\,dY)$$

since $r = (-RY + R)$ and $dr = -R\,dY$. Simplifying, we have

$$15zW^2\left[\int_1^0 Y\ln Y\,dY - \int_1^0 \ln Y\,dY\right],$$

which by (5.35) integrates to

$$15zW^2(\tfrac{1}{4} - 1) = -11.25zW^2.$$

After similar treatment of each term in (5.72) there results

$$(\alpha_T)_{\text{log law}} = W^3 - 11.25W^2 z + 65.625 Wz^2 - 175.78125 z^3$$

or, translated back to velocity ratio terms,

$$(\alpha_T)_{\text{log law}} = \left(\frac{V_c}{V}\right)^3 - 11.25\left(\frac{V_c}{V}\right)^2\left(\frac{V^*}{V}\right)$$
$$+ 65.625\left(\frac{V_c}{V}\right)\left(\frac{V^*}{V}\right) - 175.78125\left(\frac{V^*}{V}\right)^3. \tag{5.73}$$

But V_c/V has been defined by (5.36), and V^*/V by (5.40); hence it follows that (5.73) can be written as

$$(\alpha_T)_{\text{log law}} = \left(1+3.75\sqrt{\frac{f}{8}}\right)^3 - 11.25\left(1+3.75\sqrt{\frac{f}{8}}\right)^2\sqrt{\frac{f}{8}}$$

$$+ 65.625\left(1+3.75\sqrt{\frac{f}{8}}\right)\left(\frac{f}{8}\right) - 175.78125\left(\frac{f}{8}\right)\sqrt{\frac{f}{8}}.$$

When like terms are gathered this finally reduces to

$$(\alpha_T)_{\text{log law}} = 1 + 2.9296875f - 1.5537014f^{3/2}. \quad (5.74)$$

The formulation of f that is consistent with the log law is that given by Prandtl as (6.12). When (5.74) is evaluated at a Reynolds number consistent with the 1/7 power law (namely, at $R_D = 10^5$; see Table 5.3), where $f_s = 0.01799$ (see Table 6.1), we obtain

$$(\alpha_T)_{\text{log law}} = 1.04896. \quad (5.75)$$

Equation 5.74 also can be approximated by the brief form

$$(\alpha_T)_{\text{log law}} \simeq 1 + 2.7f \quad (5.76)$$

which yields, at $R_D = 10^5$, the result

$$(\alpha_t)_{\text{log law}} \simeq 1.04857.$$

These solutions show how similar are the results obtained from the power law and the log law velocity profiles, and again indicate how flat are turbulent velocity profiles compared with laminar profiles.

5.8 MOMENTUM CORRECTION FACTOR

Not only must the *average* kinetic energy expression be modified by the kinetic energy coefficient just described, but also by similar reasoning, so must the *average* momentum flux be modified by the momentum correction factor, which is defined next.

The *true* momentum flux ($\dot{M}_{\text{true}}$) generally is given as

$$\dot{M}_{\text{true}} = \int u \, d\dot{m}. \quad (5.77)$$

222 Velocity Distributions

Another momentum expression can be defined, as the counterpart of (5.77), based on the *average* velocity of (5.58), as

$$\dot{M}_{ave} = V\dot{m}. \tag{5.78}$$

Whenever this average momentum flux expression is used, as in a momentum balance, a *correction* term is necessary to account for the difference between true and average values so that

$$\dot{M}_{true} = \beta \dot{M}_{ave}. \tag{5.79}$$

From (5.77), (5.78), and (5.79), it follows that β can be written as

$$\beta = \frac{\int u\, d\dot{m}}{V\dot{m}}. \tag{5.80}$$

For pipe flow through quantities previously defined, (5.80) simplifies to

$$\beta = \frac{2}{R^2} \int_0^R \left(\frac{u}{V}\right)^2 r\, dr. \tag{5.81}$$

The β of (5.81) is known as the momentum correction factor. It remains to evaluate β for the various velocity profiles.

5.8.1 Parabolic Profile of Laminar Flow

Inserting (5.66) into the general definition of β, as given by (5.81), yields

$$\beta_L = \frac{8}{R^2} \int_0^R \left[1 - \left(\frac{r}{R}\right)^2\right]^2 r\, dr = \frac{4}{3}. \tag{5.82}$$

5.8.2 1/7 Power Law Profile of Turbulent Flow

Inserting (5.69) into (5.81) gives

$$(\beta_T)_{\text{power law}} = \frac{2}{R^2}\left(\frac{120}{98}\right)^2 \int_0^R \left(\frac{y}{R}\right)^{2/7}(R-y)\, dy = \frac{50}{49}. \tag{5.83}$$

5.8.3 Logarithmic Profile of Turbulent Flow

From (5.71), written in the simplified form of (5.71a), we have

$$(\beta_T)_{\text{log law}} = \frac{2}{R^2} \int_0^R X^2 r \, dr,$$

where the square of x is

$$X^2 = W^2 + 5Wz \ln Y + 6.25 z^2 (\ln Y)^2.$$

Thus we have

$$(\beta_T)_{\text{log law}} = \frac{2}{R^2} \left[W^2 \int_0^R r \, dr + 5Wz \int_0^R (\ln Y) r \, dr + 6.25 z^2 \int_0^R (\ln Y)^2 r \, dr \right], \quad (5.84)$$

where once again (5.35) is used to evaluate the $\ln Y$ integrals above, yielding

$$(\beta_T)_{\text{log law}} = W^2 - 7.5 Wz + 12.5(1.75) z^2, \quad (5.85)$$

which reduces, in terms of (5.36) and (5.40), to

$$(\beta_T)_{\text{log law}} = 1 + 0.9765625 f. \quad (5.86)$$

When (5.86) is evaluated at $R_D = 10^5$, where $f_s = 0.01799$, there results

$$(\beta_T)_{\text{log law}} = 1.01757. \quad (5.87)$$

A summary of the kinetic energy coefficients and the momentum correction factors for the various velocity profiles is given in Table 5.10.

Table 5.10 Comparison of Kinetic Energy and Momentum Correction Factors for Various Velocity Profiles

Velocity Profile	Kinetic Energy Coefficient, α	Momentum Correction Factor, β
Parabolic—laminar	2.000	1.333
Power law—turbulent	1.058	1.020
Log law—turbulent	1.049	1.018

REFERENCES

1. O. Reynolds, "An experimental investigation of the circumstances which determine whether the motion of water will be direct or sinuous, and the laws of resistance in parallel channels," *Phil. Trans. R. Soc. London*, 1883.
2. R. P. Benedict and J. S. Wyler, "Determining flow rate from velocity measurements," *Instrum. Control Syst.*, February 1974, p. 47.
3. E. Buckingham, "Model experiments and forms of empirical equations," *ASME Trans.*, Vol. 35, 1915, p. 263.
4. L. Prandtl, "Über die Ausgebildete Turbulenz," *Proc. II Int. Congr. Appl. Mech.*, Zurich, 1926, p. 62. Translated as "On the development of turbulence," *NACA TM* 425, 1927.
5. J. Nikuradse, "Gesetzmassigkeiten der turbulenten Stromung in glatten Rohren," *Forsch.-Arb. Ing.-Wesen*, No. 356, 1932. Translated as "Laws of turbulent flow in smooth pipes."
6. B. Miller, "The laminar-film hypothesis," *Trans. ASME*, May 1949, p. 357.
7. L. Prandtl, "Eine Beziehung zwischen Warmeaustausch und Stromungswiderstand der Flussigkeiten," *Phys. Z.*, Vol. 11, 1910, p. 1072. Translated as "A relation between heat convection and flow resistance in liquids."
8. A. Eagle and R. M. Ferguson, "The coefficients of heat transfer from tube to water," *Proc. Inst. Mech. Eng.*, Vol. 2, 1930, p. 985.
9. T. von Karman, "Aspects of turbulence problems," *Proc. IV Int. Congr. Appl. Mech.*, Cambridge, England, 1934.
10. M. R. Head and I. Rechenberg, "The Preston tube as a means of measuring skin friction," *J. Fluid Mech.*, Vol. 14, 1962, p. 1.
11. H. Reichardt, "Vollstandige Darstellung der turbulenten Geschwindigkeitsverteilung in glatten Leitungen," *Z. Angew. Math. Mech.*, Vol. 31, 1951, p. 208. Translated as "Complete description of the turbulent velocity distribution in smooth pipes."
12. R. G. Deissler, "Investigation of turbulent flow and heat transfer in smooth tubes, including the effects of variable fluid properties," *Trans. ASME*, February 1951, p. 101.
13. R. B. Dean, "A single formula for the complete velocity profile in a turbulent boundary layer," *Trans. ASME, J. Fluids Eng.*, December 1976, p. 723.
14. V. C. Patel, "Calibration of the Preston tube and limitations on its use in pressure gradients," *J. Fluid Mech.*, Vol. 23, Part 1, 1965, p. 185.
15. D. Lindley, "An experimental investigation of the flow in a classical venturimeter," *Proc. Inst. Mech. Eng.*, Vol. 184, Part 1, No. 8, 1969–70, p. 133.
16. S. B. Au, "Boundary layer development in venturimeters," PH.D Thesis, Department of Mechanical Engineering, University of Wales, 1972.
17. J. Nikuradse, "Stromungsgesetze in rauhen Rohren," *Forsch. Arb. Ing.-Wesen*, No. 361, 1933. Translated as "Laws of flow in rough pipes," *NACA TM* 1292, November, 1950.

18. T. E. Stanton and J. R. Pannell, "Similarity of motion in relation to the surface friction of fluids," *Phil. Trans. R. Soc. London*, Vol. 214, 1914, p. 199.
19. H. P. G. Darcy, "Recherches experimentales relatives aux mouvement de l'eau dans les tuyaux," *Mem. Acad. Inst. Imp. Fr.*, Vol. 15, 1858, p. 141. Translated as "Experimental research on the flow of water in pipes."
20. H. Schlichting, *Boundary Layer Theory*, 1st ed., McGraw-Hill, 1955. Translated by J. Kestin.
21. H. Ludwieg and W. Tillman, "Untersuchungen über die Wandschubspannung in turbulenten Reibungssehichten," *Ing. Arch.*, Vol. 17, 1949, p. 288. Translated as "Investigations of the wall-shearing stress in turbulent boundary layers," *NACA TM* 1285, 1950.
22. J. H. Preston, "The determination of turbulent skin friction by means of Pitot tubes," *J. R. Aeronaut. Soc.*, Vol. 58, February 1954, p. 109.
23. F. H. Clauser, "Turbulent boundary layers in adverse pressure gradients," *J. Aeronaut. Sci.*, February 1954, p. 91.
24. D. Coles, "The law of the wake in the turbulent boundary layer," *J. Fluid Mech.*, Vol. 1, 1956, p. 191.
25. D. S. Smith and J. H. Walker, "Skin friction measurements in incompressible flow," *NACA TN* 4231, 1958; see also NASA TR 26, 1959.
26. R. H. F. Pao, *Fluid Mechanics*, John Wiley, 1961.
27. W. J. Rainbird, "Errors in measurement of mean static pressure of a moving fluid due to pressure holes," *Natl. Res. Counc. Can. Rep.*, DME/NAE, Vol. 3, 1967, p. 55.
28. V. E. Senecal and R. R. Rothfus, "Transition flow of fluids in smooth tubes," *Chem. Eng. Prog.*, Vol. 49, No. 10, 1953, p. 533.
29. J. G. Knudsen and D. L. Katz, *Fluid Dynamics and Heat Transfer*, McGraw-Hill, 1958, Chapter 7.
30. R. C. Binder, *Fluid Mechanics*, 4th ed., Prentice-Hall, 1962, p. 117.

NOMENCLATURE

Roman

- a acceleration
- A area
- C integration constant
- d total derivative
- D pipe diameter
- e absolute roughness
- e^+ roughness Reynolds number $= V^* e / \nu$

226 Velocity Distributions

- f friction factor
- F factor
- F_x force in x direction
- g_c gravitational constant
- G function
- K von Karman's constant
- K.E.   kinetic energy
- $\dot{\text{K.E.}}$ kinetic energy flux
- L length
- m slope
- $\dot{m}$ mass flow rate
- M mass
- $\dot{M}$ momentum flux
- N exponent
- p static pressure
- Q volumetric flow rate
- r general radius
- R pipe radius
- R_D Reynolds number $= VD/\nu$
- u x component of velocity
- u^+ $= u/V^*$
- V volumetric average velocity
- V_c maximum centerline velocity
- V_D area average velocity
- V^* friction velocity $= \sqrt{\tau_0 g_c/\rho}$
- x, y coordinates
- y^+ friction Reynolds number $= V^* y/\nu$

Greek

- α kinetic energy coefficient
- β momentum correction factor
- δ_L laminar sublayer thickness
- Δ finite difference

θ circumferential coordinate
μ dynamic viscosity
ν kinematic viscosity
ρ fluid density
σ pipe factor
      $= V/V_c$
τ shear stress
φ function

Subscripts

L laminar
T turbulent
s smooth
ave   average

6
The Friction Factor

> In every creature high enough to have a circulation there is friction between vessels and blood, also between blood and blood, so there must be pressure to keep the stream going... —*Gustav Eckstein*

6.1 GENERAL REMARKS

All of the standard formulations for the friction factor in the various flow regimes, for *fully developed* boundary layers, are collected in this chapter. The experimental and theoretical bases for these formulations are given where available. Graphical and tabular comparisons are made between these formulations, the experimental data as reported in the literature, and previously published curves. The amazing success of the universal law of the wall in predicting the forms of the law of friction is discussed. Reasons for the divergence between Nikuradse's sand-roughened pipe data and commercially rough pipe data in the transition region are discussed. Suggestions are made for determining the effective relative roughness for a given pipe of interest. The tie-in between the friction factor and the skin friction coefficient is examined, along with the Clauser plot for exploiting the skin friction coefficient. Some relations between the various friction coefficients and the boundary layer parameters are given. Formulations other than the standard ones are presented in the interest of completeness. The effect of chemical additives on the friction factor of the solvent is considered, and, finally, the apparent friction factor for use in *developing* boundary layers is also discussed.

The Darcy-Weisbach equation

$$\frac{\Delta p}{\rho} = f\left(\frac{L}{D}\right)\left(\frac{V^2}{2g_c}\right) \tag{6.1}$$

was given [1, 2] in about 1850 to express the pressure loss in a piping system. The *number* of velocity heads ($V^2/2g_c$) lost, for a given pressure drop (Δp), is expressed by the product of the friction factor (f) and the geometric factor (L/D), that is, the length/diameter ratio. The velocity head is to be evaluated using the volumetric average velocity (V) of continuity. The friction factor under discussion here is that corresponding to fully developed velocity profiles, both laminar and turbulent, which are encountered only after 25 or more diameters downstream of a pipe inlet [3, 4]. Furthermore, f in (6.1) can signify the friction factor in both smooth and rough pipes.

6.2 DIMENSIONAL ANALYSIS

A simple dimensional analysis indicates the parameters that influence f. We begin by listing all the variables believed to affect the pressure drop in a pipe:

$$\Delta p = \varphi(V, \rho, \mu, D, L, e),$$

where ρ is fluid density, μ is fluid viscosity, and e is the absolute roughness of the pipe. Since the number of variables is seven, and the number of primary dimensions involved is three, it follows that four dimensionless π terms will be required to express the functional relationship between Δp and the remaining variables. Following the Buckingham π theorem, we can write one convenient set of π's as

$$\pi_1 = (\rho^{a_1} V^{b_1} D^{c_1}) \Delta p,$$
$$\pi_2 = (\rho^{a_2} V^{b_2} D^{c_2}) \mu,$$
$$\pi_3 = (\rho^{a_3} V^{b_3} D^{c_3}) e,$$
$$\pi_4 = (\rho^{a_4} V^{b_4} D^{c_4}) L,$$

from which the exponents, a_i, b_i, and c_i, are to be determined by solving the simultaneous equations derived by setting the sum of the exponents equal to zero. The resulting π's are, by inspection,

$$\pi_1 = \frac{\Delta p}{\rho V^2},$$
$$\pi_2 = \frac{\mu}{\rho V D},$$
$$\pi_3 = \frac{e}{D},$$
$$\pi_4 = \frac{L}{D}.$$

230 The Friction Factor

These combine, according to

$$\pi_1 = \varphi(\pi_2, \pi_3, \pi_4),$$

as

$$\frac{\Delta p}{\rho} = \varphi\left(R_D, \frac{e}{D}, \frac{L}{D}\right) V^2. \tag{6.2}$$

But, on comparing (6.2) with (6.1), it follows that

$$f = f\left(R_D, \frac{e}{D}\right), \tag{6.3}$$

that is, the friction factor is, in general, a function of the pipe Reynolds number and the relative roughness.

6.3 LAMINAR FLOW

In 1839, G. H. L. Hagen, a German hydraulic engineer, published [5] the results of his tests on the flow of water through three *brass* capillaries, of diameters 2.55, 4.02, and 5.91 mm, and of lengths 47.4, 109, and 105 cm, respectively. Almost immediately thereafter, in 1840, 1841, and 1846, Jean Louis Marie Poiseuille [6], the Parisian physician and physicist, presented his results on the viscosity of water and the pressure drop caused by water flowing in glass capillaries. His object was to apply his results to the flow of blood in capillary veins. Since the Reynolds numbers were below 2000 in both Hagen's and Poiseuille's experiments, laminar flow was assured, and the flow was stabilized under the influence of viscous shear forces alone. Actually, as O. Reynolds [7] later showed, even if there were initial flow disturbances at the inlet to these capillaries, such disturbances would be damped out for Reynolds numbers below 2000, whereas at larger Reynolds numbers the initial disturbances would be amplified to develop into the irregular motions typical of turbulent flow.

The equation that Hagen and Poiseuille independently gave to express the pressure loss in laminar flow, and that represents the exact solution to the general differential equations of L. M. J. Navier and G. G. Stokes [8, 9] for the boundary conditions of laminar flow, can be written as

$$\Delta p = \frac{128 \mu Q L}{\pi D^4 g_c}. \tag{6.4}$$

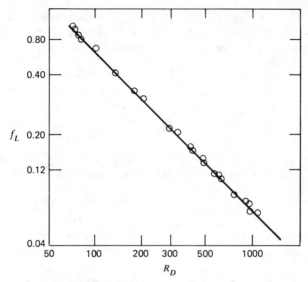

Figure 6.1 Laminar friction factor as a function of pipe Reynolds number (Hagen's [5] test points).

By equating the Darcy-Weisbach general equation, (6.1), for pressure drop, which involves f, with the Hagen-Poiseuille equation, (6.4), for laminar flow, we obtain an expression for the laminar friction factor:

$$f_L = \frac{64\mu}{\rho VD} = \frac{64}{R_D}. \tag{6.5}$$

Note that f_L is a function of the pipe Reynolds number, as predicted by (6.3), but is independent of the pipe roughness, as required by (6.3), in this flow regime since any turbulence caused by protrusions on the pipe wall is necessarily damped out in solely viscous flow.

Over the years, much experimental work has borne out the validity of (6.5) and hence the usefulness of (6.1) and (6.4). The logarithm of f_L plots as a straight line versus the logarithm of R_D in the laminar region, according to (6.5), with a slope of -1 and an intercept (where $R_D = 1$ and the log of $R_D = 0$) of the logarithm of 64 ($= 1.80618$), as shown in Figure 6.1, where the experimental points given are those of Hagen.

6.4 CRITICAL ZONE

For pipe Reynolds numbers between 2000 and 4000, the friction factor can have large uncertainties and is essentially indeterminate, and hence this region is called the *critical zone*. However, for $R_D > 4000$ the friction factor

232 The Friction Factor

again becomes reasonably determinate, but there are now three regions to consider: the smooth pipe region, the fully rough pipe region, and a transition region in between. These will be discussed next, with a separate treatment for each.

6.5 TURBULENT FLOW IN SMOOTH PIPES

Although no exact solutions are available from the general differential equations for turbulent flow, it is a fact that turbulent flow occurs more frequently in most commercial applications—hence the great interest in this flow regime.

In 1911, P. R. H. Blasius, a student of Prandtl at the Kaiser Wilhelm Institute at Gottingen, showed analytically [10] that $f_{smooth} = f(R_D)$ only. Blasius examined the available experimental data on losses in pipes, in light of his dimensional analysis [see, e.g., (6.2) and (6.3)]. In 1913, he published, under the same title as that of his 1911 paper, an article that compared his predicted functional relationship between f and R_D graphically with the Cornell University data of V. Saph and E. W. Schoder [11], as shown in Figure 6.2. Thus Blasius was the first to plot f versus R_D. He obtained the empirical relation

$$f_{\text{Blasius}} = 0.3164 R_D^{-1/4} \tag{6.6}$$

as best representing the available data for smooth circular pipes at pipe Reynolds numbers up to 10^5.

It was T. E. Stanton (a student and colleague of O. Reynolds at Owens College, and later superintendent of the Engineering Department at the National Physical Laboratory) and J. R. Pannell [12] who first indicated, in 1914, by air and water tests, that the $f - R_D$ relationship is independent of the fluid. And W. Frossel [13] first demonstrated in 1936 that the $f - R_D$ relationship is independent of Mach number effects (i.e., is independent of compressibility).

When $N=7$, the Blasius equation of (6.6) is entirely consistent with the power law of (5.13). This was first shown by Prandtl [14] in 1925, and can be seen from the following brief development. From (5.18) and (5.40) we have

$$\tau_0 = \frac{f}{8} \frac{\rho V^2}{g_c}.$$

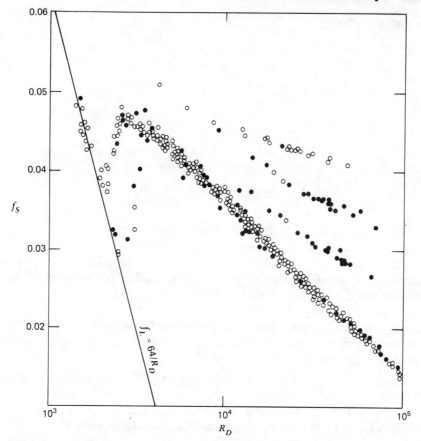

Figure 6.2 Turbulent friction factor for smooth pipe (after Blasius' original analysis, using data from Saph and Schoder [11]).

Combined with (6.6), this gives

$$\tau_0 = 0.03326 \frac{\rho}{g_c} V^{7/4} \nu^{1/4} R^{-(1/4)}.$$

Equating this with (5.18) yields

$$\frac{V}{V^*} = 30.066^{4/7} \left(\frac{V^* R}{\nu} \right)^{1/7}.$$

234 The Friction Factor

From the identity

$$\frac{V_c}{V^*} = \left(\frac{V}{V^*}\right)\left(\frac{V_c}{V}\right),$$

and with $V/V_c \simeq 0.8$, from (5.16) or Table 5.4, there result

$$\frac{V_c}{V^*} = 8.74\left(\frac{V^*R}{\nu}\right)^{1/7}$$

and

$$\frac{u}{V^*} = 8.74\left(\frac{V^*y}{\nu}\right)^{1/7}.$$

The power law of (5.13) follows from the ratio of u to V_c.

Equation 6.6 was the first of many correlations given to express the variation of friction factor with Reynolds number for turbulent flow in *smooth pipes*, where, as in laminar flow, $f = f(R_D)$ only.

In 1933, Prandtl [15], using his boundary layer theory, his mixing length hypothesis, and the law of the wall, developed a theoretical *law of friction* for smooth pipes in turbulent flow. Briefly, the volumetric average velocity of (5.36), namely,

$$V = V_c - 3.75 V^*, \qquad (6.7)$$

is based on the velocity profile in the log layer and applies to both smooth and rough pipes. Moreover, the maximum velocity, namely,

$$V_c = V^*\left[2.5\ln\left(\frac{V^*R}{\nu}\right) + 5.5\right], \qquad (6.8)$$

is based on the log layer point velocity of (5.27) for smooth pipes, evaluated at the pipe centerline. The maximum friction Reynolds number of (6.8) can be written in the useful forms

$$\frac{V^*R}{\nu} = y^+_{max} = \left(\frac{V^*}{V}\right)\left(\frac{VD}{\nu}\right)\left(\frac{1}{2}\right) = \frac{R_D\sqrt{f}}{4\sqrt{2}}. \qquad (6.9)$$

From (5.40) and (6.7) it follows that f can be expressed as

$$f = 8\left(\frac{V^*}{V}\right)^2 = \frac{8}{[(V_c/V^*) - 3.75]^2}, \qquad (6.10)$$

which, on combining with (6.8) and (6.9), becomes

$$f_S = \frac{8}{\left[2.5\ln\left(R_D\sqrt{f}/4\sqrt{2}\right)+1.75\right]^2}$$

or finally, in terms of the base 10 logarithm (i.e., $\log x = 0.43429448 \ln x$),

$$\frac{1}{\sqrt{f_S}} = 2.0352 \log\left(R_D\sqrt{f_S}\right) - 0.9129. \tag{6.11}$$

In (6.11), f_S is the friction factor for smooth pipes, based on the law of the wall theory and Nikuradse's experimental determination of von Karman's constant ($K = 1/2.5$) and the smooth pipe intercept ($B = 5.5$). However, some slight modification in the numerical values in (6.11) is required, primarily because the law of the wall concerns *wall layer* phenomena, whereas f describes loss in terms of the *full* cross-sectional area pipe flow. The straight line that best fits the data of Nikuradse [16], in *rounded form*, is known as Prandtl's equation for the friction factor in a smooth pipe in turbulent flow, namely,

$$\frac{1}{\sqrt{f_S}} = 2\log\left(R_D\sqrt{f_S}\right) - 0.8. \tag{6.12}$$

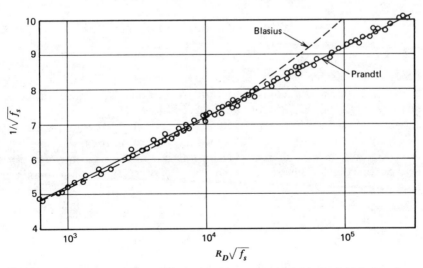

Figure 6.3 Comparison between Prandtl's (6.12) and Blasius' (6.6) for the turbulent friction factor for smooth pipe (after Schlichting [41]).

Prandtl's equation and the Blasius equation, (6.12) and (6.6), are shown plotted in Figure 6.3, along with Nikuradse's data, which substantiates the straight line character of (6.12).

6.6 TURBULENT FLOW IN FULLY ROUGH PIPES

Whereas the friction factor is independent of wall roughness in laminar flow, roughness is in general of fundamental importance in turbulent pipe flow.

In 1858, Darcy [1] made comprehensive tests on 22 pipes of cast iron, lead, wrought iron, asphalted iron, and glass. Each of these pipes was 100 m (about 328 ft) long and 1.2 to 50 cm (about $\frac{1}{2}$ to 20 in.) in diameter; they were of various ages.

Nikuradse analyzed the Darcy data in present terms and found for turbulent flow in rough pipes that f varied only slightly with R_D, for a given roughness; f decreased with R_D, the rate of this decrease being slower for greater relative roughness values; and, above a certain large value of R_D, f became independent of R_D.

Adding greatly to Darcy's work, Nikuradse [17], in 1932, did his own famous experiment with artificially roughened pipes. He coated the inside of the pipes with carefully graded Gottingen sand grains, glued in place. He used the mesh size of the grading screens to help establish his scale of relative roughness, that is,

$$\frac{e_S}{R} = \frac{\text{diameter of Gottingen sand, uncoated}}{\text{radius of pipe, uncoated}}.$$

Because of this historical precedence, pipe roughness is still most often given in terms of an equivalent sand roughness. Nikuradse varied the relative roughness, e_S/R, from about $1/500$ to $1/15$.

Nikuradse's data were analyzed in the fully rough region, where $f \neq f(R_D)$, by von Karman, who developed an expression for f_{rough}, employing the same data as were used to develop the law of the wall in rough pipes. Briefly, for the volumetric average velocity of (6.7), and the maximum velocity, namely,

$$V_c = V^*\left(2.5 \ln \frac{R}{\nu} + 8.5\right), \tag{6.13}$$

which is based on the log layer point velocity of (5.47) for rough pipes, evaluated at the pipe centerline, we can write (6.10) as

$$f_R = \frac{8}{[2.5\ln(R/e)+4.75]^2}$$

or finally, in terms of the base 10 logarithm,

$$\frac{1}{\sqrt{f_R}} = 2.0352\log\left(\frac{R}{e}\right) + 1.6794. \tag{6.14}$$

In (6.14), f_R is the friction factor for rough pipes, based on the law of the wall theory and Nikuradse's experimental determination of von Karman's constant ($K=1/2.5$) and the rough pipe intercept ($C=8.5$). However, some slight modification of the numerical values in (6.14) is required, again primarily because the wall layer phenomenon does not exactly describe the loss in terms of the full pipe flow. The straight line that best fits the data of Nikuradse, *in rounded form*, is known as von Karman's equation for the friction in a fully rough pipe in turbulent flow, namely,

$$\frac{1}{\sqrt{f_R}} = 2\log\left(\frac{R}{e_S}\right) + 1.74. \tag{6.15}$$

6.7 TRANSITION BETWEEN SMOOTH AND ROUGH PIPES

If solutions to the equations for friction factor in smooth and rough pipes are plotted on a single graph (as in Figure 6.4), a noticeable *gap* between these two regions is apparent. The basis of the coordinates used in Figure 6.4 is as follows. It was found that the family of friction factor curves versus Reynolds number for the various e/D's could be coalesced to a single curve by subtracting the reciprocal of $\sqrt{f_{actual}}$ from both sides of (6.15). This amounts to defining for the *dependent variable* a function φ such that

$$\varphi = \frac{1}{\sqrt{f_R}} - \frac{1}{\sqrt{f_{actual}}}.$$

For fully rough pipes, $\varphi = 0$. For smooth pipes,

$$\varphi = \left[1.74 - 2\log\left(\frac{2e_S}{D}\right)\right] - \left[2\log\left(R_D\sqrt{f_S}\right) - 0.8\right],$$

238 The Friction Factor

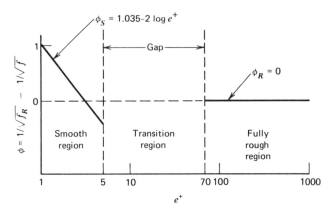

Figure 6.4 Friction factors for smooth and rough pipes.

or

$$\varphi_S = 2.54 - 2\log\left[R_D\sqrt{f_S}\left(\frac{2e_S}{D}\right)\right]. \tag{6.16}$$

But the *independent variable* of φ in (6.16), that is, the bracketed term, can be expressed in terms of the *roughness Reynolds number*, e^+ [see (5.45)], as follows:

$$e^+ = \frac{V^* e_S}{\nu} = \left(\frac{V^*}{V}\right)\left(\frac{VD}{\nu}\right)\left(\frac{e_S}{D}\right) = \sqrt{\frac{f}{8}}\, R_D\left(\frac{e_S}{D}\right), \tag{6.17}$$

where e_S is the equivalent Nikuradse *uncoated* sand diameter. The use of the roughness Reynolds number is suggested by viewing the effective roughness as the ratio of the absolute roughness (i.e., the projection size, e_S) to the laminar boundary layer thickness (δ_L), that is,

$$\frac{e_S}{\delta_L} \propto \frac{e_S}{\nu/V^*} = e^+.$$

Hence φ_{smooth} can be written in terms of (6.16) and (6.17) as

$$\varphi_S = 2.54 - 2\log e^+ - 2\log(2\sqrt{8})$$

or, finally,

$$\varphi_S = 1.035 - 2\log e^+. \tag{6.18}$$

Transition between Smooth and Rough Pipes

This is a straight line of slope -2 and of intercept 1.035 on the φ versus $\log e^+$ plot. Note the similarity between Figure 6.4 for friction factor and Figure 5.24 for velocity profiles.

Two very different answers are available in the literature to bridge the gap between the smooth pipe and the rough pipe characteristics:

1. Nikuradse, in 1933, offered much *experimental evidence* on the transition region, based on his artificially roughened pipe tests. Although he gave no formulation to correlate the friction factor data in the transition region, his experimental points are presented here in Figure 6.5 in terms of the Colebrook coordinates, developed above for use in Figure 6.4, and again in Figure 6.6 in terms of the Blasius coordinates.

2. C. F. Colebrook [18], in collaboration with C. M. White, developed, in 1939, a *mathematical function* which he claimed gave a transition curve between smooth and rough pipes that more closely agreed with *actual measurements* on most forms of naturally rough commercial pipes. Briefly, Colebrook took Prandtl's smooth pipe law of friction, namely, (6.12), and expressed it in the equivalent forms

$$\frac{1}{\sqrt{f_S}} = 1.74 - 2\log\left(\frac{18.7}{R_D\sqrt{f_S}}\right)$$

and

$$\frac{1}{\sqrt{f_S}} = -2\log\left(\frac{2.51}{R_D\sqrt{f_S}}\right).$$

Then he took von Karman's fully rough pipe law of friction, namely, (6.15), and expressed it in the form

$$\frac{1}{\sqrt{f_R}} = -2\log\left(\frac{e/D}{3.76}\right).$$

Colebrook then simply combined the expressions for the friction factor for smooth and rough pipes into a single transition equation of the equivalent forms

$$\frac{1}{\sqrt{f_T}} = -2\log\left(\frac{e/D}{3.76} + \frac{2.51}{R_D\sqrt{f_T}}\right)$$

240 The Friction Factor

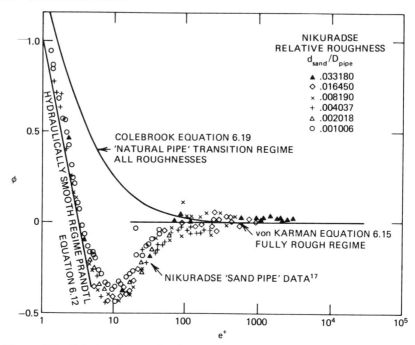

Figure 6.5 Friction factors in the transition region between smooth and rough pipes, in terms of the Colebrook coordinates [42].

and (6.19)

$$\frac{1}{\sqrt{f_T}} = 1.74 - 2\log\left(2\frac{e}{D} + \frac{18.7}{R_D\sqrt{f_T}}\right),$$

where f_T signifies the transition friction factor.

Note that Colebrook's expressions for the friction factor in the transition region reduce to Prandtl's smooth pipe equations when the relative roughness approaches zero, and reduce to von Karman's fully rough pipe equations at very high pipe Reynolds numbers.

It is one of those amazingly fortuitous happenings in engineering, and a tribute to Colebrook's insight, that this *simple addition* of the two limiting equations for the friction factor, namely, that for smooth and that for fully rough pipes, *also* predicts a transition region friction factor (f_T) which is more realistic concerning the performance of commercial pipe than are Nikuradse's extensive experimental data which led to the two accepted limiting equations. Colebrook's entirely *empirical* arrangement of Prandtl's and von Karman's equations has only one thing in its favor: it works. The Colebrook function, (6.19), for the friction factor in

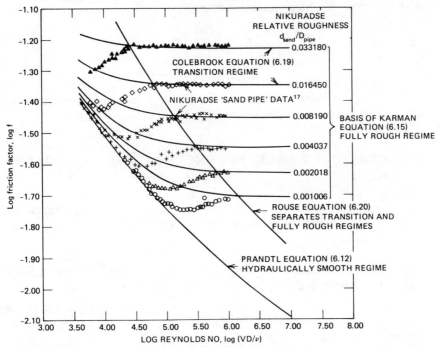

Figure 6.6 Friction factors in the transition region between smooth and rough pipes, in terms of the Blasius coordinates [42].

the transition region is shown plotted in Figure 6.5 and 6.6, for comparison with Nikuradse's experimental data. It should be noted here that Colebrook did not stop with the analytical solution of (6.19), but proceeded to compare the results of many groups of tests on various types of commercial pipe surfaces. He found that each class of commercial pipe gave a curve of the same form, and that the curves agreed closely with each other and with (6.19).

H. Rouse, long-time director of the Iowa Institute of Hydraulic Research [19], also showed that a large number of experimental points, obtained from commercial pipes of natural roughness, agreed far better with the Colebrook formulation in the transition region than with the Nikuradse sand pipe data. The points used by Rouse to further substantiate the Colebrook function are also shown in Figure 6 of Reference 20.

Finally, concerning this transition region, Rouse [19], in 1943, gave a limiting equation that delineated between the transition region, where $f = f(R_D, e/D)$, and the fully rough regime, where $f = f(e/D)$ only. This

242 The Friction Factor

Rouse limit line can be given in the form

$$R_D^* = \frac{200}{\sqrt{f}\,(e/D)}. \tag{6.20}$$

Beyond the critical Reynolds number of transition, as given by (6.20), the friction factor becomes essentially independent of the pipe Reynolds number. The Rouse limit line is plotted in Figure 6.6.

6.8 REASONS FOR DIFFERENCES IN THE TRANSITION REGION

Nikuradse's data, shown in Figures 6.5 and 6.6, drop far below Colebrook's formulation in the transition region. Some reasons for this difference are as follows:

1. The sand grains used by Nikuradse formed protrusions that were *more uniform in size* than the irregular protrusions found in commercial pipes.
2. The closely packed sand grain protrusions were also *spaced more uniformly* than the more randomly spaced protrusions in commercial pipes.
3. Thus the larger of the mixed-sized commercial protrusions would project beyond the laminar sublayer and break it up, whereas the more uniform sand grains would remain hidden within the laminar sublayer for a longer time. This would cause the Nikuradse pipe to appear *hydraulically smooth*, and thus Nikuradse's experimental points would follow the smooth pipe characteristic to *higher Reynolds numbers* than would the commercial pipe points.
4. Above a certain Reynolds number, given approximately by (6.20), which decreases with increasing roughness, the laminar sublayer becomes so thin that the sand pipe appears fully rough.
5. As to which of these two transition characteristics is more realistic, it is clear that the artificial nature of Nikuradse's sand pipes weighs against the use of his values in the transition region, and hence the Colebrook function, (6.19), is recommended to represent f_T in this region.

Of some concern is Nikuradse's sand-fastening technique, since his relative roughness scale was based on the *uncoated* size of the sand grains. Nikuradse went from sand diameters of 0.1 mm (about 0.004 in.) to 1.6 mm (about 0.063 in.) in building up his sand pipes, and yet used the same lacquer and techniques to fasten the sand to the pipes. Hence we would have expected differences in the data obtained with pipes made up of widely different grain sizes, albeit of the *same* relative roughness. Such differences are not apparent in his data (see Figures 6.5 and 6.6), however, so we must suppose that the gluing did not significantly affect the relative roughness.

6.9 ENGINEERING CHARTS FOR DETERMINING FRICTION FACTOR

6.9.1 Moody Plot

The plots shown in Figures 6.3, 6.4, and 6.5 are believed to be less convenient to use than a composite plot of all regions of interest on the coordinate system used in Figures 6.1, 6.2, and 6.6. On such a coordinate system (first devised by Blasius), Louis F. Moody [20], in 1944, first presented such a composite plot including the straight line laminar friction factor curve (f_L) of (6.5); the smooth pipe turbulent friction factor curve (f_S) of (6.12); the various fully rough turbulent friction factor curves (f_R) of (6.15); and the transition friction factors (f_T) of (6.19).

This information is organized in Figure 6.7 and is known with good reason as the *Moody plot*. Also shown on this plot is the Rouse limit line of (6.20), which separates the transition region from the fully rough region. In the critical zone, between the laminar and turbulent regimes, one can join the laminar line with the proper roughness curve by a straight line, with as much reason as not.

Because the von Karman and Colebrook equations are based on Nikuradse's sand roughness scale, we are bound to express any pipe roughness on the Nikuradse scale as long as we continue to use either of these two equations for rough pipes. Moody [20] first gave the effective roughness of commercial pipes on the Nikuradse scale. His results for *new* pipes are given in Figure 6.8.

In connection with the Nikuradse sand scale, we note that, although other experimenters (see, e.g., J. M. Robertson et al. [21]) have also used sand-roughened pipes to obtain pipe friction factor data, none has been able to duplicate Nikuradse's results except Hermann Schlichting, who used sand from the same source as did Nikuradse. When it is realized that longitudinal spacing, lateral spacing, size distribution, and sharpness all affect the transition character, this lack of agreement between the various experimenters is not surprising.

As mentioned, these charts of Figures 6.7 and 6.8 apply to *clean, new* pipes only. It is a fact that pipe surface deteriorates rapidly with age. This

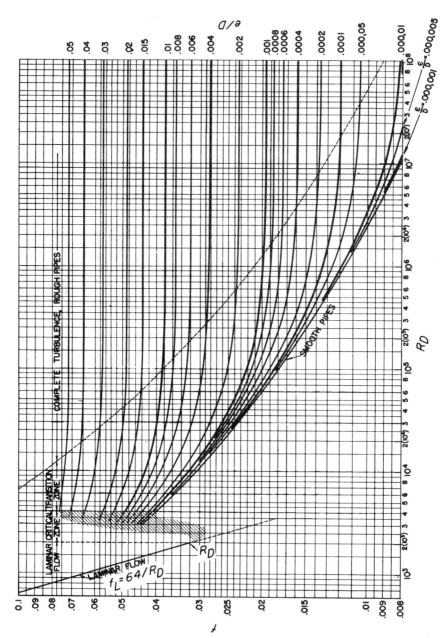

Figure 6.7 Complete friction factor map (after Moody [20]).

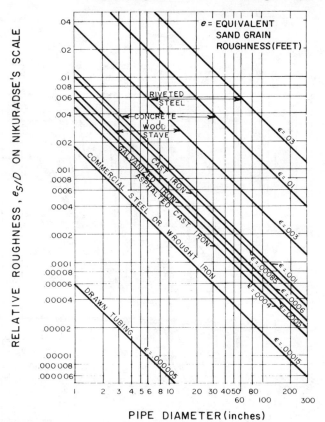

Figure 6.8 Commercial pipe roughness in terms of Nikuradse's sand grain scale (after Moody [20]).

has two effects: the pipe roughness changes with time, and the pipe inner diameter usually decreases with age. For example, one writer, A. T. Ippen [22], indicated, in 1944, that a 4 in. galvanized steel pipe with an $e_{new} = 0.00045$ ft changed to $e_{used} = 0.00090$ ft in 3 years, as observed at the Hydraulic Laboratory of Lehigh University.

The best estimate of the *effective roughness* of a given pipe on the Nikuradse sand grain scale can be obtained by testing hydraulically a representative section of the pipe. The test should be run at a pipe Reynolds number that exceeds the Rouse limit line. Then, in the fully rough regime, the relative roughness (e/D) that makes the von Karman equation, (6.15), yield the experimentally determined f_R (obtained from pressure drop data) is the required effective relative roughness on the Nikuradse scale. Of course, it is this *effective roughness* that is to be used with the Moody plot.

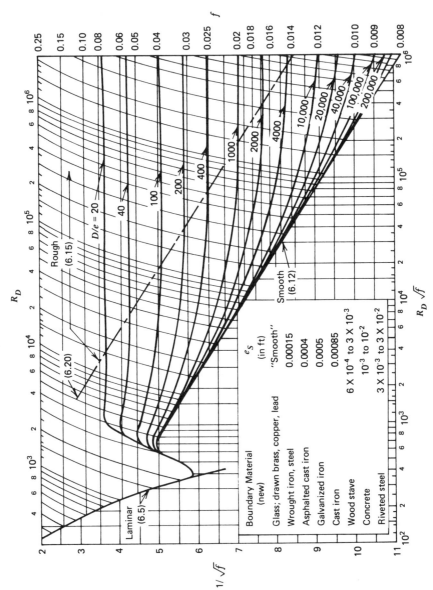

Figure 6.9 Complete friction factor map (after Rouse [19]).

Because of the lack of a satisfactory technique for measuring pipe roughness, our knowledge of the friction factor for rough pipes, in the transition or the fully rough region, is believed to be no better than ±10%.

6.9.2 Rouse Plot

Lest one get the impression that the only coordinates of practical use for presentation of the friction factor are those used by Blasius, Stanton, and Moody, as in Figures 6.1, 6.2, 6.6, and 6.7, we present also the *Rouse plot* [19] of Figure 6.9. Here the coordinates used are $1/\sqrt{f}$ and $R_D\sqrt{f}$. These are not really a surprising choice, since they are the very variables that appear in the Prandtl, von Karman, and Colebrook equations. Furthermore, as pointed out by Rouse himself [23], the friction factor is *not* inextricably embodied in the independent variable since we can write the identity

$$R_D\sqrt{f} = \sqrt{\frac{2g_c \Delta p\, D^3}{\rho L \nu^2}}.$$

Of course, on the Rouse plot, the laminar characteristic no longer appears as a straight line, but this is compensated for by the fact that the smooth pipe characteristic in turbulent flow does appear as a straight line. Furthermore, all the transition curves between the smooth and fully rough regions will be geometrically similar.

Only time will tell which of these coordinate systems proves more convenient for the determination of the friction factor. Thus far, the Moody plot seems to be the more popular choice.

6.10 THE SKIN FRICTION COEFFICIENT

As one component of the total drag coefficient (along with profile drag and secondary flow drag), there has been defined, notably by von Karman [24], a *mean* skin friction coefficient as

$$C_{fM} = \frac{\tau_0}{\rho V^2/2g_c}, \qquad (6.21)$$

based on the wall shear stress and the volumetric average velocity.

When the Darcy friction factor* of (6.1) is combined with the wall shear stress of (5.2), there results

$$f = 4\left(\frac{\tau_0}{\rho V^2/2g_c}\right). \qquad (6.22)$$

---

*Some authors, mostly in heat transfer work, define the Fanning friction factor (f') to be identical with the mean skin friction coefficient (C_{fM}) of (6.21).

248 The Friction Factor

It follows from (6.21) and (6.22) that

$$f = 4C_{fM}. \tag{6.23}$$

Thus all of the equations given that involve f can also be expressed in terms of C_{fM}. For example, the Prandtl equation, (6.12), for the friction factor in a smooth pipe in turbulent flow can also be written as

$$\frac{1}{2\sqrt{C_{fM}}} = 2\log\left(2R_D\sqrt{C_{fM}}\right) - 0.8$$

or

$$\frac{1}{\sqrt{C_{fM}}} = 4\log\left(R_D\sqrt{C_{fM}}\right) - 0.4. \tag{6.24}$$

A *local* skin friction coefficient has been defined, patterned after (6.21), as

$$C_f = \frac{\tau_0}{\rho V_c^2 / 2g_c}. \tag{6.25}$$

It follows from (6.21) and (6.25) that

$$C_f = \sigma^2 C_{fM}, \tag{6.26}$$

where σ is the pipe factor, V/V_c, of (5.51).

By (5.40), in terms of (6.23) and (6.26), there results

$$V^* = \sqrt{\frac{f}{8}}\ V = \sqrt{\frac{C_f}{2}}\ V_c. \tag{6.27}$$

Thus, for example, Prandtl's *log layer* portion of the law of the wall, that is, (5.27), can be written in terms of the local skin friction coefficient as

$$\frac{u}{V_c} = \sqrt{\frac{C_f}{2}} \left[5.75 \log\left(\frac{V_c y}{\nu}\sqrt{\frac{C_f}{2}}\right) + 5.5 \right]. \tag{6.28}$$

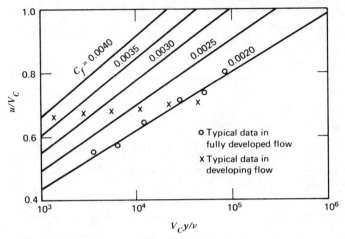

Figure 6.10 Clauser plot for extracting friction information from velocity profile measurements [based on (6.28)].

6.10.1 Clauser Plot

F. H. Clauser [25], in 1954, suggested that information about the frictional effects could be extracted, via (6.28), from data that did not present measurements of Δp explicitly, but did give information on the velocity profile. For parameters of fixed C_f, and for given maximum pipe Reynolds numbers (i.e., for given $V_c y/\nu$), a plot of u/V_c of (6.28) can be obtained, as shown in Figure 6.10. Such a plot, called a *Clauser plot*, provides a useful coordinate system on which to examine frictional effects in turbulent flow. When the data follow a fixed C_f line, we can conclude that the flow is fully developed. However, when the data cross fixed C_f lines, we can conclude that the pressure gradient is not zero and that the flow is in a developing mode (see Figure 6.10).

6.11 OTHER EXPRESSIONS FOR THE TURBULENT FRICTION FACTOR

Several expressions for the friction factor can be given other than those already presented in Sections 6.3, 6.5, 6.6, 6.7, and 6.10. Although the expressions given in this section are more empirical, with less traceability to a theoretical basis than the others, they have one advantage: all of the friction factors given here are in *explicit* form, meaning simply that the friction factor can be obtained directly in terms of R_D and/or e_S/D, or R_θ and H.

6.11.1 For Smooth Pipes

In terms of R_D only, we have:

Date	Source	Use	Expression	
1930	Nikuradse (as quoted by Au [26])	Any R_D	$f_S = 0.00332 + 0.221 R_D^{-0.237}$	(6.29)
1932	Drew et al. [27]	$R_D < 3 \times 10^6$	$f_S = 0.0056 + 0.5 R_D^{-0.32}$	(6.30)
1958	Knudsen and Katz [28]	$10^5 < R_D < 10^6$	$f_S = 0.184 R_D^{-0.2}$	(6.31)
1965	Techo et al. [29]	Any R_D	$f_S = \left[0.86859 \ln\left(\dfrac{R_D}{1.964 \ln R_D - 3.8215} \right) \right]^{-2}$	(6.32)

6.11.2 For Transition

Between smooth and fully rough pipes. We have:

Data	Source	Use	Expression	
1947	Moody [30]	Above critical R_D	$f_T = 0.0055 \left[1 + \left(20{,}000 \dfrac{e_S}{D} + \dfrac{10^6}{R_D} \right)^{1/3} \right]$	(6.33)
1976	Swamee and Jain [31]	Above critical R_D	$f_T = \dfrac{0.25}{\left[\log\left(\dfrac{e/D}{3.7} + \dfrac{5.74}{R_D^{0.9}} \right) \right]^2}$	(6.34)

As will be noted in the tables that follow, (6.33) deviates seriously from the Colebrook formulation, whereas (6.34) well represents the Colebrook formulation.

6.11.3 For Fully Rough Pipes

In terms of e_S/D only, we have:

Date	Source	Use	Expression	
1975	Reynolds [32]	$0.0005 < \dfrac{e_S}{D} < 0.05$	$f_R = 0.16 \left(\dfrac{e_S}{D} \right)^{0.31}$	(6.35)

Other Expressions for the Turbulent Friction Factor 251

All of these formulations for friction factor are compared numerically in Tables 6.1, 6.2, and 6.3 in terms of smooth, transition, and rough pipes.

In Table 6.1 we compare the results of Nikuradse's (6.29), Drew's (6.30), Knudsen-Katz's (6.31), and Techo et al.'s (6.32) with those of Blasius' (6.6) and Prandtl's (6.12). Prandtl's equation is judged to be the most acceptable for smooth pipes, while Techo et al.'s is seen to be the best explicit equation for f_S.

In Table 6.2 we compare the results of Moody's (6.33) and Swamee–Jain's (6.34) with those of Colebrook's (6.19); the latter is considered the most acceptable implicit formulation for the transition region while Swamee–Jain's is seen to be the best explicit equation for f_T.

In Table 6.3 we compare the results of Reynold's (6.35) with those of von Karman's (6.15); the latter is believed to be the most acceptable formulation for rough pipes.

Table 6.1 Comparison of Formulations for Friction Factor for Smooth Pipes in Turbulent Flow

R_D	Prandtl (6.12)	Techo et al. (6.32)	Nikuradse (6.29)	Drew et al. (6.30)	Knudsen-Katz (6.31)	Blasius (6.6)
4×10^3	0.03992	0.03980	0.03427	0.04078	—	0.03978
5×10^3	0.03740	0.03732	0.03268	0.03836	—	0.03763
1×10^4	0.03089	0.03087	0.02823	0.03184	—	0.03164
2×10^4	0.02589	0.02590	0.02446	0.02662	—	0.02660
3×10^4	0.02349	0.02350	0.02252	0.02406	—	0.02404
4×10^4	0.02197	0.02199	0.02126	0.02244	—	0.02237
5×10^4	0.02090	0.02091	0.02033	0.02128	—	0.02116
1×10^5	0.01799	0.01801	0.01775	0.01816	0.01840	0.01779
2×10^5	0.01564	0.01565	0.01557	0.01566	0.01602	—
3×10^5	0.01447	0.01447	0.01444	0.01444	0.01477	—
4×10^5	0.01371	0.01372	0.01371	0.01366	0.01394	—
5×10^5	0.01316	0.01317	0.01318	0.01310	0.01334	—
1×10^6	0.01165	0.01165	0.01168	0.01161	0.01161	—
2×10^6	0.01037	0.01038	0.01042	0.01042	—	—
3×10^6	0.00972	0.00972	0.00977	0.00983	—	—
4×10^6	0.00929	0.00929	0.00934	—	—	—
5×10^6	0.00898	0.00898	0.00903	—	—	—
1×10^7	0.00810	0.00810	0.00817	—	—	—

Table 6.2 Comparison of Formulations for Friction Factor for Pipes in Transition Region in Turbulent Flow

e/D R_D	1×10^{-5}			1×10^{-4}			1×10^{-3}			1×10^{-2}		
	Colebrook*	Moody†	Swamee‡	Colebrook	Moody	Swamee	Colebrook	Moody	Swamee	Colebrook	Moody	Swamee
5×10^{3}	.03746	.03767	.03786	.03756	.03777	.03797	.03854	.03870	.03910	.04728	.04602	.04860
1×10^{4}	.03094	.03105	.03099	.03108	.03120	.03115	.03242	.03263	.03267	.04313	.04232	.04404
5×10^{4}	.02095	.02048	.02080	.02127	.02091	.02116	.02403	.02431	.02418	.03907	.03870	.03946
1×10^{5}	.01806	.01743	.01792	.01853	.01809	.01845	.02218	.02259	.02234	.03848	.03819	.03875
5×10^{5}	.01331	.01265	.01325	.01444	.01423	.01447	.02023	.02091	.02035	.03800	.03777	.03810
1×10^{6}	.01188	.01134	.01185	.01344	.01343	.01351	.01994	.02067	.02003	.03794	.03772	.03801
5×10^{6}	.00956	.00955	.00964	.01234	.01265	.01240	.01969	.02048	.01973	.03789	.03767	.03793
1×10^{7}	.00900	.00918	.00906	.01216	.01254	.01222	.01966	.02045	.01969	.03789	.03767	.03792

*Colebrook, (6.19)
†Moody, (6.33)
‡Swamee–Jain, (6.34)

Table 6.3 Comparison of Formulations for Friction Factor for Fully Rough Pipes in Turbulent Flow

e/D	R/e	von Karman (6.15)	Reynolds (6.35)
1×10^{-5}	50000	.00806	—
5×10^{-5}	10000	.01054	—
1×10^{-4}	5000	.01198	—
2×10^{-4}	2500	.01372	—
4×10^{-4}	1250	.01589	.01415
6×10^{-4}	833	.01740	.01604
1×10^{-3}	500	.01963	.01880
2×10^{-3}	250	.02341	.02330
4×10^{-3}	125	.02840	.02889
6×10^{-3}	83	.03210	.03276
1×10^{-2}	50	.03788	.03838
3×10^{-2}	17	.05713	.05395
5×10^{-2}	10	.07149	.06321

6.11.4 For Skin Friction Coefficient

In addition to these expressions, there remain two well-known formulations for the skin friction coefficient for smooth pipes, in terms of the boundary layer parameters, R_θ and H. These are as follows:

Date	Source	Use	Expression
1938	Squire and Young [33]	Any R_D	$C_{fM} = \dfrac{0.0576}{[\log(4.075 R_\theta)]^2}$

$$= \frac{1}{[1.81 \ln R_\theta + 2.54]^2} \quad (6.36)$$

1949 Ludwieg and Tillmann [34] Any R_D $C_{fM} = 0.246 \times 10^{-0.678 H_{12}} \times R_\theta^{-0.268}$

$$(6.37)$$

Equation 6.37 is the most famous and most often used of all the explicit empirical equations that define the frictional characteristics of smooth pipes in terms of boundary layer parameters. Results from this expression are in good agreement with the Clauser plot, and hence are in agreement with the law of the wall.

6.12 EFFECT OF ADDITIVES ON FLUID FRICTION

It has been found experimentally that very small concentrations, on the order of a few parts per million by weight, of a dissolved high polymer substance can reduce the frictional resistance in turbulent flow to as low as one-fourth that of the pure solvent. J. W. Hoyt of the Naval Undersea Research and Development Center at Pasadena has reported on this phenomenon at some length [35]. He indicates that the viscosities of such high polymer solutions, as measured by conventional viscometers, are actually somewhat higher than those of the pure solvent. Thus it comes as a distinct surprise to the hydrodynamicist that the turbulent friction is reduced. The conclusion is that any macromolecular substance of sufficiently high molecular weight (say 50,000 or more), with a generally linear structure, will lower the friction characteristic of any fluid which is a solvent.

P. S. Virk and his co-workers [36] have indicated that in the limit the maximum drag reduction relation for polymer addition can be given by

$$\frac{1}{\sqrt{C_{fM}}} = 19\log\left(R_D\sqrt{C_{fM}}\right) - 32.4. \tag{6.38}$$

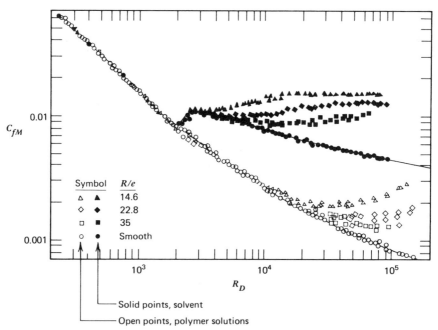

Figure 6.11 Comparison of frictional characteristics of a pure solvent with those of a polymer solution (after Virk [37]).

This is to be compared with (6.24), which represents the normal Newtonian turbulent friction law for pure fluids in smooth pipes. Such a comparison is perhaps best appreciated by referring to Figure 6.11, which contrasts the friction characteristics of a pure solvent with those of a polymer solution.

6.13 FRICTION IN DEVELOPING BOUNDARY LAYERS

The apparent friction factor in flows where the boundary layer and hence the velocity profile are developing (i.e., changing) is quite different from that in fully developed flows, where the velocity profile is fixed. In developing boundary layers there is a pressure gradient over and above that associated with the frictional effect because of accelerations in the core flow as the boundary layer thickness changes.

The friction factor, according to the Darcy-Weisbach equation, (6.1), is based on the measured pressure drop caused by friction alone in a fully developed region. The measured pressure drop in the developing case, however, includes both the friction pressure drop and a pressure drop caused by changes in momentum flux associated with changes in the velocity profile. Evidently, the friction factor based on a measured pressure drop in a developing boundary layer region is not the usual one. For this reason the friction factor of developing flow is called the *apparent* friction factor. Whenever the measured pressure drop exceeds that of fully developed pipe flow, f_{app} exceeds f. For example, it has been estimated [38] that for the inlet region of a pipe, where the core flow is accelerating as the laminar boundary layer grows, only 39% of the measured pressure drop is caused by wall friction, whereas 61% is the result of momentum changes.

The best available experimentally determined *local* apparent friction factor data for developing laminar boundary layers at the entry region of a pipe are those given by Stephen Kline and Ascher Shapiro [39], in 1952–1953. Their results were summarized by them in the empirical equation

$$f_{app} = \frac{6.87}{[R_D(x/D)]^{1/2}}. \qquad (6.39)$$

This is the apparent friction factor *at a point*. What is usually desired is the integrated (or *mean*) apparent friction factor ($\overline{f_{app}}$), where the integration (or averaging) is carried out over the dimensionless distance, (L/D), that is,

$$\overline{f_{app}}\left(\frac{L}{D}\right) = \int_0^{L/D} f_{app} d\left(\frac{x}{D}\right). \qquad (6.40)$$

256 The Friction Factor

When (6.39) is integrated according to (6.40), there results the useful expression

$$\overline{f_{app}} = \frac{13.74}{[R_D(L/D)]^{1/2}}. \tag{6.41}$$

Note that the mean value is just two times the local value. The $\overline{f_{app}}$ of (6.41) has been shown to be valid for $L/(DR_D) < 10^{-3}$. For greater values of $L/(DR_D)$, the theoretical work of H. L. Langhaar [3] is recommended.

The empirical work of Kline and Shapiro and the theoretical work of Langhaar, both applying to developing laminar boundary layers, are compared with the fully developed laminar flow work of Hagen and Poiseuille, as given by (6.5), in Figure 6.12.

For developing turbulent boundary layers, R. G. Deissler [40] has presented graphical friction factors, assuming the complete absence of an initial laminar boundary layer. However, the mean apparent friction factor of flat plate theory, that is,

$$\overline{f_{app}} = \frac{0.296}{[R_D(L/D)]^{0.2}}, \tag{6.42}$$

as given by H. Schlichting [41], for example, is believed to be as good as any at this time.

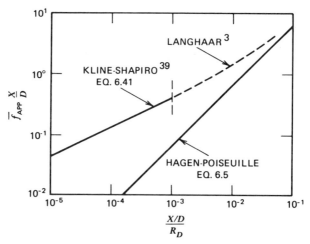

Figure 6.12 Comparison of friction factors for laminar flow.

REFERENCES

1. H. P. G. Darcy, "Experimental research on the flow of water in pipes" (in French), *Mem. Acad. Sci. Inst. Imp. Fr.*, Vol. 15, 1858, p. 141.
2. J. Weisbach, *Mechanics of Engineering* (in German), Van Nostrand, 1872. Translated by E. B. Coxe.
3. H. L. Langhaar, "Steady flow in the transition length of a straight tube," *J. Appl. Mech.*, Vol. 9, and *Trans. ASME*, Vol. 64, June 1942, p. A-56.
4. A. H. Shapiro and R. D. Smith, "Friction coefficients in the inlet length of smooth, round tubes," *NACA TN* 1785, November 1948.
5. G. H. L. Hagen, "Über die Bewegung des Wassers in angen cylindrischen Rohren," *Poggendorfs Ann. Phys. Chem.*, Vol. 46, 1839, p. 423. Translated as "On the flow of water in narrow cylindrical pipes."
6. J. L. M. Poiseuille, "Recherches experimentales sur le mouvement des liquides dans les tubes des tres petits diametres," *C. R. Acad. Sci.*, 1841. Translated as "Experimental research on the flow of liquids through pipes of very small diameters."
7. O. Reynolds, "An experimental investigation of the circumstances which determine whether the motion of water will be direct or sinuous, and the laws of resistance in parallel channels," *Phil. Trans. R. Soc. London*, 1883.
8. L. M. H. Navier, "Memoire sur les lois du mouvement des fluids," *Mem. Acad. R. Sci.*, Vol. 6, 1827, p. 389. Translated as "Note on the laws of the flow of fluids."
9. G. G. Stokes, "On the theories of the internal friction of fluids in motion, and of the equilibrium and motion of elastic solids," *Trans. Camb. Phil. Soc.*, Vol. 8, 1845.
10. P. R. H. Blasius, "Das Achnlichkeitsgesetz bei Reibungsvorgangen in Flussigkeiten," *Phys. Z.*, Vol. 12, 1911, p. 1175. Translated as "The law of similarity applied to friction phenomena."
11. V. Saph and E. W. Schoder, "An experimental study of the resistance to the flow of water in pipes," *Trans. ASCE*, Vol. 51, 1903, p. 944.
12. T. E. Stanton and J. R. Pannell, "Similarity of motion in relation to the surface friction of fluids," *Phil. Trans. R. Soc. London*, Vol. 214, 1914, p. 199.
13. W. Frossel, "Stromung in glatten, geraden Rohren mit Uber- und Unterschallgeschwindigkeit," *Forsch.-Arb. Ing.-Wesen*, Vol. 7, March-April 1936, p. 75. Translated as "Flow in smooth straight pipes at velocities above and below sound velocity," *NACA TM* 844, 1938.
14. L. Prandtl, "Ergeb. Aerodyn. Versuchanst. Gottingen," Series 3, 1927. Translated as "Reports of the Aerodynamic Research Institute at Gottingen."
15. L. Prandtl, "Neuere ergebnisse der turbulenzforschung," *Z. VDI*, Vol. 77, 1933, p. 105. Translated as "Recent results of turbulence research."
16. J. Nikuradse, "Laws of turbulent flow in smooth pipes" (in German), *Forsch.-Arb. Ing.-Wesen*, No. 356, 1932.

258 The Friction Factor

17. J. Nikuradse, "Laws of flow in rough pipes" (in German), *Forsch.-Arb. Ing.-Wesen*, No. 361, 1933; see also *NACA TM* 1292, 1950.
18. C. F. Colebrook, "Turbulent flow in pipes, with particular reference to the transition region between the smooth and rough pipe laws," *J. Inst. Civil Eng.*, London, Vol. 11, 1938–1939, p. 133.
19. H. Rouse, "Evaluation of boundary roughness," Proc. II Hydraulic Conf., *Univ. Iowa Bull.* 27, 1943.
20. L. F. Moody, "Friction factors for pipe flow," *Trans. ASME*, November 1944, p. 676.
21. J. M. Robertson, J. D. Martin, and T. H. Burkhart, "Turbulent flow in rough pipes," *Ind. Eng. Chem. Fundam.*, Vol. 7, No. 2, May 1968, p. 253.
22. A. T. Ippen, In discussion of Reference 20, p. 678.
23. H. Rouse, In discussion of Reference 20, p. 680.
24. T. von Karman, "Über laminare und turbulente Reibung," *Z. Angew. Math. Mech.*, Vol. 1, 1921, p. 233. Translated as "On laminar and turbulent friction," *NACA TM* 1092, 1946.
25. F. H. Clauser, "Turbulent boundary layers in adverse pressure gradients," *J. Aeronaut. Sci.*, February 1954, p. 91.
26. S. B. Au, "Boundary Layer Development in Venturimeters," Ph.D. Thesis, Department of Mechanical Engineering, University of Wales, 1972, p. 20.
27. T. B. Drew, E. C. Koo, and W. H. McAdams, "The friction factor for clean round pipes," *Trans. AIChE*, Vol. 28, 1932, p. 56.
28. J. G. Knudsen and D. L. Katz, *Fluid Dynamics and Heat Transfer*, McGraw-Hill, 1958, p. 173.
29. R. Techo, R. R. Tickner, and R. E. James, "An accurate equation for the computation of the friction factor for smooth pipes from the Reynolds number," *J. Appl. Mech.*, June 1965, p. 443.
30. L. F. Moody, "An approximate formula for pipe friction factors," *Mech. Eng.*, Vol. 69, December 1947, p. 1005.
31. P. K. Swamee and A. K. Jain, "Explicit equations for pipe flow problems," *Proc. ASCE, J. Hyd. Div.*, Vol. 102, Hy 5, May 1976, p. 657.
32. A. J. Reynolds, *Turbulent Flow in Engineering*, Wiley-Interscience, 1974, p. 202.
33. H. B. Squire and A. D. Young, "The calculation of the profile drag of aerofoils," Aeronautical Research Council Report and Memorandum, 1838, 1938.
34. H. Ludwieg and W. Tillmann, "Untersuchungenüber die Wandschubspennung in turbulenten Reibungsschichten," *Ing. Arch.*, Vol. 17, p. 288. Translated as "Investigations of the wall-shearing stress in turbulent boundary layers," *NACA TM* 1285, 1950.
35. J. W. Hoyt, "The effect of additives on fluid friction," *Trans. ASME, J. Basic Eng.*, June 1972, p. 258.
36. P. S. Virk, E. W. Merrill, H. S. Mickley, K. A. Smith, and E. L. Mollo-Christensen, "The Toms phenomenon: turbulent pipe flow of dilute polymer solutions," *J. Fluid Mech.*, Vol. 30, 1967, p. 305.

37. P. S. Virk, "Drag reduction in rough pipes," *J. Fluid Mech.*, Vol. 45, 1971, p. 225.
38. A. H. Shapiro, R. Siegel, and S. J. Kline, "Friction factor in the laminar entry region of a smooth tube," *Proc. 2nd U.S. Natl. Congr. Appl. Mech.*, June 1954, p. 733.
39. S. J. Kline and A. H. Shapiro, "Experimental investigation of the effects of cooling on friction and boundary layer transition for low-speed gas flow at the entry of a tube," *NACA TN* 3048, November 1953.
40. R. G. Deissler, "Analysis of turbulent heat transfer and flow in the entrance regions of smooth passages," *NACA TN* 3016, October 1953.
41. H. Schlichting, *Boundary Layer Theory*, 1st ed., 1955. Translated by J. Kestin, p. 433.
42. R. P. Benedict, "Friction in pipe flow," *Instrum. Control Syst.*, December 1969, p. 91.

NOMENCLATURE

Roman

C_f skin friction coefficient
D pipe diameter
e absolute pipe roughness
e^+ roughness Reynolds number $= V^*e/\nu$
f friction factor
g_c gravitational constant
H_{12} shape factor $= \delta^*/\theta$
L length
R pipe radius
R_D Reynolds number $= VD/\nu$
R_θ $= U\theta/\nu$
p static pressure
Q volumetric flow rate
u x component of velocity
V volumetric average velocity
V_c maximum centerline velocity
V^* friction velocity $= \sqrt{\tau_0 g_c/\rho}$
x x coordinate
y y direction
y^+ friction Reynolds number $= V^*y/\nu$

Greek

- Δ   finite difference
- μ dynamic viscosity
- ν kinematic viscosity
- ρ fluid density
- φ function

Subscripts

- app   apparent
- L laminar
- M mean
- R rough
- S smooth, sand
- T transition

III

FLOW OF REAL FLUIDS IN PIPES

In this part we consider in some detail the real problem of fluid flow in pipes, that is, flow with losses arising because of fluid viscosity, unrestrained expansions, mixing and turbulence effects, and so on. Wherever possible, the boundary layer and the friction factor are used to explain the loss mechanism. In general, however, the description of losses remains highly empirical in nature, so that many experimental equations, curves, and tables are necessary to characterize losses through the various piping components. Generalizations are drawn whenever recognized for liquid, gaseous, and two-phase flow in pipes.

7
Flow of Real Liquids in Pipes

> ...the aorta has been studied for its place in a geometry of pipes, for the way a viscous fluid leaping into it would stretch it... —*Gustav Eckstein*

7.1 GENERAL REMARKS

In this chapter we admit that real liquids flow in pipes with *losses* introduced by the viscosity of the fluid, by eddies carried by the fluid, by sudden area changes in the pipe, by secondary flow losses around bends, and so on. The main conservation equations used will be those of continuity and of energy. Whereas continuity remains unchanged in the real pipe flow case, the energy equation will have an additional term to account for losses. An application of real liquid flow involving the conservation of momentum also will be given to illustrate its use. The friction factor of Chapter 6 will come into its own in this chapter as it is applied in equation, table, and chart form toward solutions of real liquid flow in pipes. Finally, a generalized pipe flow solution is given in terms of a constant density flow function and a constant density pipe flow chart.

7.2 THE ONE-DIMENSIONAL ENERGY EQUATION WITH LOSSES

The Bernoulli equation deals with flow in the absence of losses. When we combine the steady flow general energy equation of (1.30) and (3.19), namely,

$$\delta Q + \delta W = du + p\,dv + \alpha \frac{V\,dV}{g_c} + \frac{g}{g_c}\,dZ \qquad (7.1)$$

264 Flow of Real Liquids in Pipes

with the first law of thermodynamics, given by (3.13) as

$$\delta Q + \delta F = du + p\, dv, \qquad (7.2)$$

we obtain

$$\delta W = \delta F + \frac{dp}{\rho} + \alpha \frac{V\, dV}{g_c} + \frac{g}{g_c}\, dZ, \qquad (7.3)$$

where all three equations are on a foot-pound force per pound mass basis. Equation 7.3 applies to real fluid flow and indicates that any external work transferred across system boundaries will be used to overcome losses (δF), change the pressure (dp/ρ), accelerate the fluid ($V\, dV/g_c$), and/or change the elevation of the fluid ($g/g_c \times Z$). It is the introduction and application of the general loss term (δF) in (7.3) that primarily concerns us in this chapter.

First, however, a word is in order about the kinetic energy coefficient (α) of (7.3), which was previously defined in Section 5.7. The coefficient α was introduced in real fluid flow analyses to account for velocity profile effects on the kinetic energy terms of (7.3). Although the use of α represents a refinement in pipe flow calculations, it is quite common to ignore the α effect by setting α's equal to unity in most pipe flow analyses. Some of the reasons that can be given for this simplification are as follows [1]:

1. In turbulent flow, $\alpha_T \simeq 1$ anyway, according to (5.70) for the power law velocity distribution, and (5.75) for the log law distribution.
2. In laminar flow, although $\alpha_L \simeq 2$, according to (5.67), the $V^2/2g_c$ term is usually small compared to the other terms of (7.3), so that use of $\alpha > 1$ hardly matters. In fact, *in general*, the kinetic energy term is smaller than the other terms of (7.3).
3. The effect of α tends to cancel in (7.3) because it appears on both sides of the integrated form of the equation.
4. Finally (and not to be overlooked), any values chosen for the loss terms in (7.3) are generally the results of approximations, so that to engineering accuracy there is no need to use refined values of α in (7.3).

In view of these practical reasons for neglecting the effect, we usually write (7.3) as

$$\delta W = \delta F + \frac{dp}{\rho} + \frac{V\, dV}{g_c} + \frac{g}{g_c}\, dZ \qquad (7.4)$$

and call this the *one-dimensional* energy equation with losses, applicable in this differential form to both real liquid and real gas pipe flows.

7.3 THE HEAD LOSS EQUATION

The general thermodynamic loss term (δF) of (7.4) is usually called the head loss (h_{loss}) in pipe flow work, and is defined from (7.4) for workless, constant density flow as

$$\int_1^2 \delta F = (h_{\text{loss}})_{1,2} = \frac{p_1 - p_2}{\rho} + \frac{V_1^2 - V_2^2}{2g_c} + \frac{g}{g_c}(Z_1 - Z_2). \qquad (7.5)$$

Here we recall that the Darcy-Weisbach equation has been given to empirically express the pressure drop in a pipe, according to (4.8) and (6.1), as

$$\frac{\Delta p}{\rho} = f \frac{L}{D} \frac{V^2}{2g_c} \qquad (7.6)$$

in terms of the dimensionless friction factor (f) of Chapter 6. On comparison, we note that the Darcy-Weisbach equation of (7.6) also expresses the head loss of (7.5) whenever the density is constant (for then ρ is independent of pressure), when the pipe diameter is constant (for then $V_1 = V_2$), and when the pipe is horizontal (for then $Z_1 = Z_2$). In view of these restrictions on (7.6), an empirical form for the head loss of (7.5) that is more general than the Darcy-Weisbach equation of (7.6) is given next.

7.4 THE LOSS COEFFICIENT

7.4.1 Definition

A loss coefficient can be defined in general by the relation

$$(h_{\text{loss}})_{1,2} = (K_{1,2})_x \frac{V_x^2}{2g_c}, \qquad (7.7)$$

where $K_{1,2}$ represents the loss coefficient from 1 to 2, and the subscript x indicates that the numerical value of K is intimately connected with the velocity used in the head loss formulation. It should be noted here that the loss coefficient (K) always represents the number of velocity heads ($V^2/2g_c$) lost.

266 Flow of Real Liquids in Pipes

For a *viscous pipe*, that is, for a constant diameter horizontal pipe with loss, the loss coefficient of (7.7) is simply

$$K_{1,2} = f\frac{L}{D}, \tag{7.8}$$

as can be seen by comparing (7.6) and (7.7).

For piping elements other than the constant diameter pipe—an elbow, a valve, a pipe inlet, a pipe exit, or a nozzle, for example—the loss coefficient is given as a number which characterizes the piping element. In Chapter 10, discussions, developments, and realistic numbers will be given by graph and table for the losses of the most common piping elements. Here it is sufficient to note that such values exist.

7.4.2 Combining Loss Coefficients

At this point in the discussion we raise a logical question: What is the overall loss of a number of piping elements connected in series for constant density flow? This is easily answered by the following brief development [2]:

1. By (7.7) and reason:

$$(h_{\text{loss}})_{\text{overall}} = K_1 \frac{V_1^2}{2g_c} + K_2 \frac{V_2^2}{2g_c} + \cdots + K_N \frac{V_N^2}{2g_c}.$$

2. But by continuity:

$$A_1 V_1 = A_2 V_2 = A_N V_N = A_R V_R,$$

where the subscript N signifies the Nth loss element and the subscript R signifies a reference location that may or may not be an actual geometric location in the piping complex.

3. When each loss term is expressed in terms of a common reference area and a common reference velocity, there results

$$(h_{\text{loss}})_{\text{overall}} = \left[K_1 \left(\frac{A_R}{A_1}\right)^2 + K_2 \left(\frac{A_R}{A_2}\right)^2 + \cdots + K_N \left(\frac{A_R}{A_N}\right)^2 \right] \frac{V_R^2}{2g_c}. \tag{7.9}$$

Equation 7.9 is the required equation for the overall head loss in a series piping network for constant density flow. It tells us that the loss coefficients are additive, provided that they are all based on the same reference area.

7.4.3 Deriving a Loss Coefficient

Very few loss coefficients can be derived analytically. Instead, most loss coefficients must be determined experimentally. One of the oldest examples of a theoretically derived loss coefficient is that of an abrupt enlargement. On the basis of the work of Carnot and Borda (1800–1850), we can apply the momentum, continuity, and energy equations for a constant density fluid to obtain a very realistic value for the loss coefficient of a sudden enlargement [3, 4].

Briefly, from (7.5) we have, for the horizontal abrupt enlargement of Figure 7.1,

$$(h_{loss})_{1,2} = \frac{p_1 - p_2}{\rho} + \frac{V_1^2 - V_2^2}{2g_c}. \tag{7.10}$$

A momentum balance can be written in differential form, for the fluid element between stations 1 and 2 in Figure 7.1, as

$$[p - (p + dp)]A_2 = \frac{(\rho A_2 V_2)}{g_c} dV, \tag{7.11}$$

based on the assumption that the pressure at 1 acts over the area A_2, and

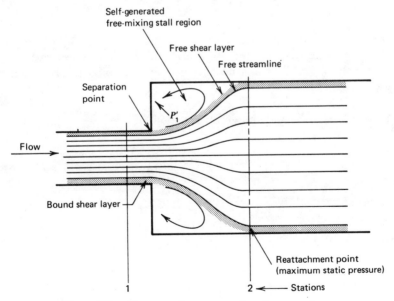

Figure 7.1 Abrupt enlargement, showing notation.

268 Flow of Real Liquids in Pipes

that shearing stresses along the free streamline between 1 and 2 are negligible. Equation 7.11 is integrated to give

$$\frac{p_1 - p_2}{\rho} = \frac{V_2(V_2 - V_1)}{g_c},$$

which, combined with (7.10), yields the Carnot-Borda equation:

$$(h_{\text{loss}})_{1,2} = \frac{(V_1 - V_2)^2}{2g_c}. \tag{7.12}$$

When (7.12) is equated to the head loss equation of (7.7), there results

$$(K_{1,2})_1 = \left(1 - \frac{V_2}{V_1}\right)^2 = \left(1 - \frac{A_1}{A_2}\right)^2 = (1 - \beta^2)^2, \tag{7.13}$$

where $(K_{1,2})_1$ means the loss coefficient from 1 to 2, that is, across the abrupt enlargement, based on the velocity at 1.

Also of significance in discussing flow with losses is the total pressure ratio. A general expression can be given as

$$\frac{p_{t2}}{p_{t1}} = 1 - (1 - R_1)K_1 = \frac{1}{1 + (1 - R_2)K_2}, \tag{7.14}$$

where R_1 is the inlet pressure ratio (p_1/p_{t1}), K_1 is the loss coefficient in terms of the inlet velocity, and R_2 and K_2 refer to similar quantities at the exit state. Equation 7.14 can be obtained by the following brief development:

1. From (7.5), (7.7), and the definition of total pressure of (2.5), we have

$$p_{t1} - p_{t2} = K_1 \left(\frac{\rho V_1^2}{2g_c}\right). \tag{7.15}$$

2. Dividing through by p_{t1} and noting that

$$1 - R_1 = \frac{\rho V_1^2 / 2g_c}{p_{t1}}, \tag{7.16}$$

we obtain (7.14). For the abrupt enlargement, (7.14) can be expressed

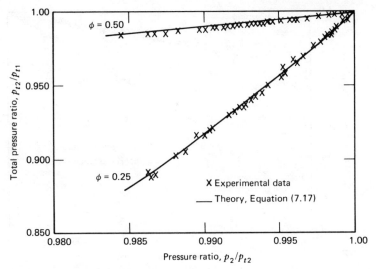

Figure 7.2 Experimental results for abrupt enlargements in pipes, obtained in water for area ratios (ϕ) of 0.25 and 0.5.

explicitly as

$$\left(\frac{p_{t2}}{p_{t1}}\right)_{\substack{\text{abrupt}\\\text{enlargement}}} = 1 - (1 - R_1)(1 - 2\beta^2 + \beta^4). \tag{7.17}$$

The static pressure recovery is another parameter of importance when dealing with the abrupt enlargement. From (7.7), (7.10), and (7.13), we can write

$$\frac{p_2 - p_1}{\rho} = \frac{V_1^2 - V_2^2}{2g_c} - (1 - \beta^2)^2 \frac{V_1^2}{2g_c}$$

or

$$\left(\frac{p_2 - p_1}{\rho}\right)_{\substack{\text{abrupt}\\\text{enlargement}}} = 2(\beta^2 - \beta^4)\frac{V_1^2}{2g_c} \tag{7.18}$$

for the static pressure recovery across an abrupt enlargement. Equation 7.18 also can be given in terms of the static pressure ratio (p_2/p_1) by

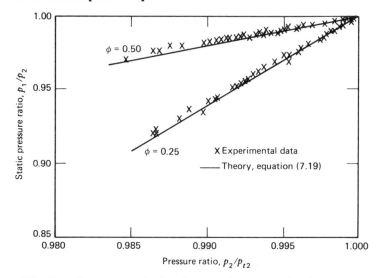

Figure 7.3 Experimental results for abrupt enlargements in pipes, obtained in water for area ratios (ϕ) of 0.25 and 0.5.

combining it with (7.16) to obtain

$$\frac{p_2}{p_1} = 1 + \left(\frac{1-R_1}{R_1}\right)(2\beta^2 - 2\beta^4). \qquad (7.19)$$

These theoretical equations for the total and static pressure ratios across an abrupt enlargement have been confirmed experimentally over the years and are quite acceptable from the engineering viewpoint. Recent results [4] are given in Figures 7.2 and 7.3 for area ratios, $\phi = A_1/A_2$, of 0.25 and 0.5.

7.5 SOLUTION OF REAL LIQUID FLOW IN PIPES

7.5.1 Solution Steps

Certain general steps can be outlined that, when followed, lead to solutions to pipe flow problems involving losses.

1. We determine the pipe Reynolds number according to (5.1), that is,

$$R_D = \frac{\rho V D}{\mu} = \frac{VD}{\nu}. \qquad (7.20)$$

Solution of Real Liquid Flow in Pipes 271

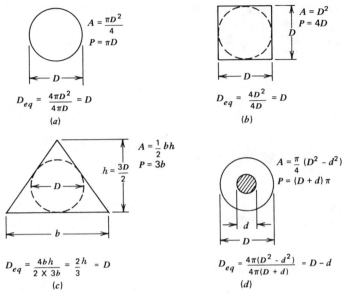

Figure 7.4 Equivalent diameter is an arbitrary definition of a value calculated so that the ratio of pressure forces acting over the flow area to the frictional forces acting along the wetted perimeter of the duct is the same for circular and noncircular pipes: $Deq = 4A/P$.

2. If $R_D < 2000$, we conclude that the flow is laminar and use the friction factor of (6.5), namely,

$$f_L = \frac{64}{R_D}. \qquad (7.21)$$

3. If $R_D > 3000$, we conclude that the flow is turbulent, and determine first the relative roughness (e/D) via Figure 6.8.
4. Given $R_D > 3000$ and e/D, we determine next the turbulent friction factor (f_T) via the Moody plot of Figure 6.7, or by Table 6.2, or by the Colebrook equation of (6.19).
5. We next obtain the head loss by the Darcy-Weisbach equation of (4.8), (6.1), and (7.6), in terms of the pipe friction factor, the pipe length, and the pipe diameter.
6. Finally, from the energy equation of (7.5), we can get either Δp, $\Delta K.E.$, or $\Delta P.E.$, depending on the unknown in the problem. Note that, only if the pipe is horizontal and constant in area, does the head loss equal $\Delta p/\rho$ directly. Note further that the velocity head of (7.5) is to be based

on the volumetric average velocity of the pipe, as defined by the continuity equation. Note finally that, if the pipe cross section is not circular, an equivalent diameter is defined such that

$$D_{eq} = \frac{4(\text{area})}{\text{perimeter}}. \tag{7.22}$$

Several applications of (7.22) are given in Figure 7.4.

7.5.2 Numerical Examples

Example 7.1. Consider points A and B, which are 4000 ft apart in a new steel pipe of 6 in. I.D., with B 100 ft higher than A, $p_A = 100$ psig, $p_B = 50$ psig, and $g = g_c$ (see Figure 7.5). How much medium fuel oil will flow at 70°F?

Solution: Following the solution steps of Section 7.5.1, we have:

1. The Reynolds number of (7.20) cannot be determined yet since no velocity is available. Hence we will assume a typical value for turbulent pipe flow of $R_D = 10^5$.
2. The relative roughness for the 6 in. steel pipe is, from Figure 6.8, $e/D = 0.0003$.
3. From the Moody plot of Figure 6.7, at the intersection of $R_D = 10^5$ and $e/D = 0.0003$, we find $f = 0.0195$. Note that this value is subject to checking when V and R_D are determined.

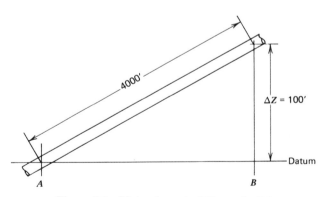

Figure 7.5 Piping layout of Example 7.1.

Solution of Real Liquid Flow in Pipes

4. From (7.5) the head loss is given as

$$h_{\text{loss}} = \frac{p_A - p_B}{\rho_{\text{oil}}} + \frac{g}{g_c}(Z_A - Z_B)$$

$$= \frac{50 \times 144}{0.854 \times 62.3} - 1(100)$$

$$(h_{\text{loss}})_{A,B} = 135.3 - 100 = 35.3 \text{ ft-lbf/lbm},$$

where the specific gravity of medium fuel oil is, from Table 7.1, 0.854, and density is given in general by

$$\rho_x = SG_x \times \rho_{H_2O}. \tag{7.23}$$

5. But by (7.7) and (7.8):

$$h_{\text{loss}} = f \frac{L_{A,B}}{D} \frac{V^2}{2g_c}$$

$$= \left(\frac{0.0195 \times 4000}{\frac{1}{2}}\right) \frac{V^2}{2g_c}$$

$$(h_{\text{loss}})_{A,B} = 156 \frac{V^2}{2g_c}.$$

6. Equating steps 4 and 5 yields

$$\frac{V^2}{2g_c} = \frac{35.3}{156} = 0.2263 \text{ft-lbf/lbm},$$

$$V = \sqrt{2 \times 32.174 \times 0.2263} = 3.816 \text{ ft/sec},$$

and the volumetric flow rate,

$$Q = AV = \frac{\pi \left(\frac{1}{2}\right)^2}{4} \times 3.816 = 0.7493 \text{ ft}^3/\text{sec}.$$

7. The assumption of $R_D = 10^5$ can now be checked. For the calculated velocity we have

$$R_D = \frac{VD}{\nu} = \frac{3.816 \times \frac{1}{2}}{4.12 \times 10^{-5}} = 0.463 \times 10^5,$$

where the viscosity of medium fuel oil at 70°F was obtained from Table 7.1. For this Reynolds number a better friction factor is 0.0225. This

274 Flow of Real Liquids in Pipes

Table 7.1 Some Properties of Liquids

T (°F)	ρ (lbm/ft^3)	Saturated Water μ (lbm/ft sec × 10^4)	ν (ft^2/sec × 10^5)	Medium Fuel Oil S.G., pure	ν (ft^2/sec × 10^5)
32	62.420	11.780	1.887	—	—
40	62.424	10.331	1.655	0.865	6.55
50	62.415	8.732	1.399	0.861	5.55
60	62.365	7.577	1.215	0.858	4.75
70	62.300	6.573	1.055	0.854	4.12
80	62.220	5.849	0.940	0.851	3.65
90	62.115	5.187	0.835	0.847	3.19
100	62.000	4.569	0.737	0.843	2.78
120	61.700	3.856	0.626		
140	61.380	3.155	0.514		
212	59.810	1.875	0.313		

leads to a head loss of $180 V^2/2g_c$, a velocity of 3.552 ft/sec, and a Q of 0.6974 ft^3/sec.

8. A second check of R_D, namely,

$$R_D = \frac{3.552 \times \frac{1}{2}}{4.12 \times 10^{-5}} = 0.431 \times 10^5,$$

shows that we have essentially converged on an acceptable answer, and we take $Q = 0.7$ ft^3/sec.

Example 7.2. An 8 in. I.D. cast iron pipe has been in service for 10 years. At a water flow rate of 1000 gal/min at 70°F, find the pressure drop per 100 ft of horizontal pipe.

Solution:

1. The velocity can be determined from continuity as

$$V = \frac{Q}{A}$$

$$= \frac{(1000 \text{ gal/min} \times 1 \text{ min}/60 \text{ sec} \times 231 \text{ in.}^3/\text{gal} \times 1 \text{ ft}^3/1728 \text{ in.}^3)}{(\pi/4)\left(\frac{8}{12}\right)^2}$$

$$= \frac{2.228 \text{ ft}^3/\text{sec}}{0.349 \text{ ft}^2} = 6.383 \text{ ft/sec}.$$

Hence the Reynolds number is

$$R_D = \frac{6.383 \times 8/12}{1.055 \times 10^{-5}} = 4.0 \times 10^5,$$

where the viscosity of water at 70°F was obtained from Table 7.1.

2. An equation that relates change in pipe roughness with time has been given [5] as

$$\left(\frac{e}{D}\right)_{used} = \left(\frac{e}{D}\right)_{new}\left(1 + \frac{t}{5}\right), \tag{7.24}$$

where t is the service time in years. From Figure 6.8 we find that $(e/D)_{new} = 0.0013$ for 8 in. cast iron pipe. Hence by (7.24)

$$\left(\frac{e}{D}\right)_{used} = 0.0013(1+2) = 0.0039.$$

3. From Figure 6.7, at $R_D = 4 \times 10^5$ and $e/D = 0.0039$, we find that $f = 0.028$.

4. From (7.5), (7.7), and (7.8):

$$h_{\text{loss per 100 ft}} = \frac{\Delta p}{\rho_{H_2O,\, 70°}} = \left(f \frac{100}{\frac{8}{12}}\right)\frac{V^2}{2g_c}$$

or

$$\Delta p = (-)\left(\frac{0.028 \times 100}{\frac{8}{12}}\right)\left(\frac{62.3 \times 6.383^2}{2g_c}\right) = -165.673 \text{ psf},$$

where the density of water at 70°F was obtained from Table 7.1. Thus the decrease in pressure per 100 ft of pipe length is

$$\Delta p = -\frac{165.673}{144} = -1.15 \text{ psi}.$$

Example 7.3. In Example 2.1, a problem was given based on Figure 2.2. We estimated the pressure heads at A and B in terms of the *ideal* flow of water at $\rho = 62.4$ lbm/ft^3, $Q = 0.47856$ ft^3/sec, and $g = g_c$. Now we are in a position to redo this same problem with the following reasonable estimates of the losses assumed for the various piping elements involved:

$K_{3 \text{ in. inlet}} = 0.5$ $\qquad$ $K_{6 \text{ in. elbow}} = 0.6$
$K_{3 \text{ in. elbows}} = 0.4$ each $\qquad$ $K_{6 \text{ in. valve}} = 3$
$K_{3 \text{ to } 6 \text{ in. enlargement}} = 0.5625$ $\qquad$ $K_{6 \text{ in. exit}} = 1$
$f_{3 \text{ in. pipe}} = 0.02$ $\qquad$ $f_{6 \text{ in. pipe}} = 0.03$

For the sake of illustration, we will use 180 ft of 3 in. pipe and 100 ft of 6 in. pipe.

Solution: Applying (7.5) between 1 and 2, we have

$$\frac{p_1}{\rho} + \frac{V_1^2}{2g_c} + Z_1 - (h_{\text{loss}})_{1,2} = \frac{p_2}{\rho} + \frac{V_2^2}{2g_c} + Z_2,$$

$$0 + 0 + 25 - (h_{\text{loss}})_{1,2} = 0 + 0 + 0.$$

It is clear that the overall head loss is 25 ft, as discussed in Example 2.1. As a check we can apply (7.9), choosing the velocity in the 6 in. pipe as a reference, to obtain

$$(h_{\text{loss}})_{\text{overall}} = \left[K_{3 \text{ in. inlet}} \left(\frac{6 \text{ in.}}{3 \text{ in.}}\right)^4 + K_{3 \text{ in. elbows}} \left(\frac{6}{3}\right)^4 + K_{\text{enlarge}} \left(\frac{6}{3}\right)^4 \right.$$
$$\left. + K_{3 \text{ in. pipe}} \left(\frac{6}{3}\right)^4 + K_{6 \text{ in. elbow}} \left(\frac{6 \text{ in.}}{6 \text{ in.}}\right)^4 + K_{6 \text{ in. valve}} + K_{6 \text{ in. pipe}} + K_{6 \text{ in. exit}} \right] \frac{V_6^2}{2g_c}.$$

Factoring out the $(6/3)^4$ term, which is 16, there results

$$(h_{\text{loss}})_{\text{overall}} = [16(0.5 + 2 \times 0.4 + 0.5625 + 14.4) + 0.6 + 3 + 6 + 1] \frac{V_6^2}{2g_c}$$

$$= (260.2 + 10.6) \frac{V_6^2}{2g_c} = 270.8 \frac{V_6^2}{2g_c}.$$

Since

$$V_6 = \frac{Q}{A_6} = \frac{0.47856}{\pi \left(\frac{1}{2}\right)^2 / 4} = 2.437286 \text{ ft/sec}$$

and

$$\frac{V_6^2}{2g_c} = 0.09231621 \text{ ft-lbf/lbm},$$

Solution of Real Liquid Flow in Pipes 277

we have

$$(h_{\text{loss}})_{\text{overall}} = 270.8 \times 0.09231621 = 25 \text{ ft-lbf/lbm}.$$

Applying (7.5) between 1 and A will yield p_A/ρ for comparison with the ideal solution:

$$\frac{p_1}{\rho} + \frac{V_1^2}{2g_c} + Z_1 - (h_{\text{loss}})_{1,A} = \frac{p_A}{\rho} + \frac{V_A^2}{2g_c} + Z_A.$$

From Example 2.1,

$$\frac{V_A^2}{2g_c} = 1.47706 \text{ ft-lbf/lbm}.$$

Thus we have

$$0 + 0 + 25 - K_{3 \text{ in. inlet}} \frac{V_3^2}{2g_c} = \frac{p_A}{\rho} + 1.47706 + 20$$

or

$$\left(\frac{p_A}{\rho}\right)_{\text{actual}} = 25 - 20 - 1.47706 - 0.5 \times 1.47706 = 2.78441 \text{ ft-lbf/lbm},$$

compared with the ideal solution of $(p_A/\rho)_{\text{ideal}} = 3.523$. Of course, the difference is due to the inlet loss. Applying (7.5) between 1 and B yields the required p_B/ρ, where the head loss from 1 to B differs from the overall head loss by the exit loss of one 6 in. velocity head. There results

$$0 + 0 + 25 - 269.8 \frac{V_6^2}{2g_c} = \frac{p_B}{\rho} + \frac{V_6^2}{2g_c} + 0$$

or

$$\left(\frac{p_B}{\rho}\right)_{\text{actual}} = 25 - (269.8 + 1)\frac{V_6^2}{2g_c} = 25 - 270.8\frac{V_6^2}{2g_c} = 0.$$

We conclude, as we should, that the pressure head at B is actually zero. This again stresses the fact that the potential energy available from the elevation difference between 1 and 2 is just used up by the flow losses between 1 and 2. Note that between B and 2 the energy equation reveals that the velocity head at B is just lost across the exit of the pipe. Note

278 Flow of Real Liquids in Pipes

further that the principle of manometry again checks our results, for p_B must equal p_2 (since they are at the same elevation), and that $p_2 = 0$ psig.

Example 7.4 In Example 2.2 a problem was given based on Figure 2.3. We estimated the elevation of water that could be maintained in reservoir 2 by a pump delivering 70 horsepower to the system at a flow rate of $Q = 8$ ft^3/sec. We further estimated the pressures at A and B on either side of the pump. Now we are in a position to redo this same problem with the following reasonable estimates of the losses assumed for the various piping elements involved:

$$K_{18 \text{ in. inlet}} = 0.5 \qquad K_{12 \text{ in. valve}} = 5$$
$$K_{18 \text{ in. pipe}} = 0.4 \qquad K_{12 \text{ in. exit}} = 1$$
$$K_{12 \text{ in. pipe}} = 8$$

Solution: Applying (7.4) between 1 and 2, we have

$$\frac{p_1 - p_2}{\rho} + \frac{V_1^2 - V_2^2}{2g_c} + \frac{g}{g_c}(Z_1 - Z_2) - (h_{\text{loss}})_{1,2} + (W_{\text{on}})_{1,2} = 0.$$

But $p_1 = p_2$, $V_1 = V_2 = 0$, and $g/g_c = 1$. There results

$$Z_2 = Z_1 - (h_{\text{loss}})_{1,2} + (W_{\text{on}})_{1,2},$$

where h_{loss} is by (7.9)

$$(h_{\text{loss}})_{\text{overall}} = \left[0.5 \left(\frac{12}{18}\right)^4 + 0.4 \left(\frac{12}{18}\right)^4 + 8 + 5 + 1 \right] \frac{V_{12}^2}{2g_c}.$$

From Example 2.2, $V_{12}^2/2g_c = 1.6124$ ft-lbf/lbm, and $W_{\text{on}} = 77.12$ ft-lbf/lbm; hence

$$(h_{\text{loss}})_{\text{overall}} = (0.17778 + 14)1.6124 = 22.86025 \text{ ft-lbf/lbm}$$

and

$$Z_2 = 20 - 22.86025 + 77.12 = 74.26 \text{ ft}.$$

This is to be compared with the *ideal* elevation at 2 of 97.12 ft. Of course, this difference is entirely accounted for by the system losses.

The pressures at A and B are estimated from the energy equations. The pressure at A is obtained from (7.5) as

$$\frac{p_1}{\rho} + \frac{V_1^2}{2g_c} + Z_1 - (h_{\text{loss}})_{1,A} = \frac{p_A}{\rho} + \frac{V_A^2}{2g_c} + Z_A,$$

Solution of Real Liquid Flow in Pipes 279

where $(h_{\text{loss}})_{1,A}$ includes both the inlet loss and the 18 in. viscous pipe loss. There results

$$0+0+20-(0.5+0.4)\frac{V_{18}^2}{2g_c} = \frac{p_A}{\rho} + \frac{V_{18}^2}{2g_c} + 10$$

or

$$\frac{p_A}{\rho} = 20 - 10 - 1.9\frac{V_{18}^2}{2g_c}.$$

From Example 2.2, $V_{18}^2/2g_c = 0.318493$ ft-lbf/lbm; hence

$$\frac{p_A}{\rho} = 10 - 1.9(0.318493) = 9.39486 \text{ ft-lbf/lbm}$$

and

$$(p_A)_{\text{actual}} = \rho(9.39486) = 586.23947 \text{ psfg} = 4.07111 \text{ psig}.$$

This compares favorably with the ideal value of 4.195 psig because there is very little loss from inlet to A.

The pressure at B is obtained by applying (7.4) between A and B as

$$\frac{p_A}{\rho} + \frac{V_A^2}{2g_c} + Z_A - (h_{\text{loss}})_{A,B} + W_{A,B} = \frac{p_B}{\rho} + \frac{V_B^2}{2g_c} + Z_B.$$

Here, as in Example 2.2, we will assume that there is no loss across the pump, and that A and B are at the same elevation. There results

$$9.39846 + 0.318493 + 77.12 = \frac{p_B}{\rho} + 1.612372$$

or

$$\frac{p_B}{\rho} = 85.220981 \text{ ft-lbf/lbm}$$

and

$$(p_B)_{\text{actual}} = \frac{62.4 \times 85.220981}{144} = 36.9291 \text{ psig}.$$

This again compares favorably with the ideal solution of 37.0533 psig.

Example 7.5. In Section 2.6.3 we considered the ideal incompressible pressure recovery across an abrupt enlargement. Now we are in a position to compare the ideal results with the more realistic results of Section 7.4.3, which include losses. For a step in the flow of $\beta = 0.5$, the actual pressure recovery is, by (7.18),

$$\left(\frac{p_2 - p_1}{\rho}\right)_{\text{actual}} = 2(0.5^2 - 0.5^4)\frac{V_1^2}{2g_c} = \frac{6}{16}\frac{V_1^2}{2g_c}.$$

The ideal solution indicated for this geometry a recovery of 15/16 of the inlet velocity head. Thus the ideal solution yields a pressure rise that is just 2.5 times the actual rise. Of course, this difference is caused by losses in the free mixing stall region shown in Figure 7.1.

The total pressure ratio across an abrupt enlargement is unity in the ideal loss-free case. However, from the real analysis, via (7.17) we find

$$\left(\frac{p_{t2}}{p_{t1}}\right)_{\text{actual}} = 1 - (1 - 0.5^2)^2(1 - R_1),$$

which, for $R_1 = 0.9$, yields a total pressure ratio of 0.94375. This indicates there is about a 6% irrecoverable drop in total pressure across the abrupt enlargement for this case.

7.6 GENERALIZED CONSTANT DENSITY PIPE FLOW

7.6.1 Generalized Flow Function

A generalized constant density equation can be given [6] in terms of continuity and energy by combining the relation $\dot{m} = \rho A V = $ constant with the relation $V = [2g_c(p_t - p)/\rho]^{1/2}$ to yield

$$A_1^2 p_{t1}\left(1 - \frac{p_1}{p_{t1}}\right) = A_2^2 p_{t2}\left(1 - \frac{p_2}{p_{t2}}\right)$$

or, more simply,

$$\left(\frac{p_{t1}}{p_{t2}}\right)\left(\frac{A_1}{A_2}\right)^2 \Gamma_{i,1} = \Gamma_{i,2}, \qquad (7.25)$$

where $\Gamma_i = 1 - p/p_t = 1 - R$, and represents a generalized incompressible flow function.

Since Γ_i varies between 0 and 1 as does R, it follows that the envelope of all possible values of R and Γ_i is a straight line, as shown in Figure 7.6. Because of its generality, any constant density process can be shown on this envelope, as from state 1 to 2 in Figure 7.6.

A brief review of some of the simplified thermodynamic processes is in order at this point.

Isentropic Process. In an isentropic process, there being no loss and no heat transfer, the thermodynamic state of a constant density fluid can change only because of a change in area and/or a change in the elevation of the flow passage.

The Bernoulli equation, which holds in the absence of losses, can be given in terms of the total pressure as

$$\frac{dp_t}{\rho} + \frac{g}{g_c} dZ = 0$$

or, in integrated forms,

$$p_{t2} = p_{t1} + \rho \frac{g}{g_c}(Z_1 - Z_2) \tag{7.26}$$

and

$$\frac{p_{t1}}{p_{t2}} = \frac{1}{1+Z_t}, \tag{7.27}$$

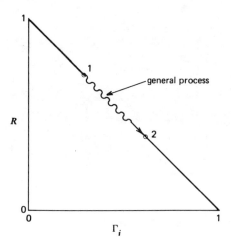

Figure 7.6 Generalized constant density flow envelope.

282 Flow of Real Liquids in Pipes

where

$$Z_t = \rho \frac{g}{g_c} \frac{(Z_1 - Z_2)}{p_{t1}}. \tag{7.28}$$

Equations 7.26 and 7.27 confirm the interesting point that, even in an isentropic process of a constant density fluid, the total pressure can change whenever the pipe elevation changes.

The entropy produced in any constant density process can be given in terms of the basic identity of (3.27) as

$$\int_1^2 dS = \int_1^2 \frac{du + p\,dv}{T},$$

which, in the constant density case, is simply

$$S_2 - S_1 = C \ln \frac{T_2}{T_1} \tag{7.29}$$

where C is the specific heat capacity of the liquid. Equation 7.29 points up the fact that in an isentropic process the temperature must remain constant.

On an h-S diagram the isentropic process for a constant density fluid plots as shown in Figure 7.7. On the generalized R-Γ_i plot the isentropic process, of course, plots as shown in Figure 7.6.

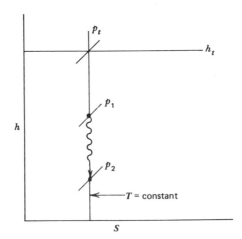

Figure 7.7 Isentropic process—constant density.

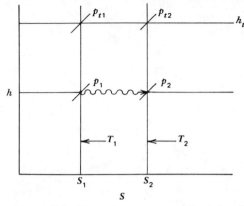

Figure 7.8 Fanno process—constant density.

Adiabatic Process with Loss. In an adiabatic process with loss, there is no heat transfer, and the thermodynamic state of a constant density fluid can change with area change and/or elevation change, as well as with losses. *Fanno* flow is a special case of adiabatic flow with loss in which neither the area nor the elevation changes. In Fanno flow, while the total enthalpy necessarily is constant, the total pressure must decrease. Such a process is shown on the h-S diagram of Figure 7.8. On the generalized R-Γ_i diagram the adiabatic process with loss plots as shown in Figure 7.6.

Diabatic Process without Loss. In a diabatic process without loss, there is heat transfer, and the thermodynamic state of a constant density fluid can change with area change and/or elevation change, as well as with heat transfer. *Rayleigh* flow is a special case of diabatic loss-free flow in which neither the area nor the elevation changes. In Rayleigh flow, while the total enthalpy necessarily changes, the total pressure must remain constant. Such a process is shown on the h-S diagram of Figure 7.9. As with all processes, the diabatic loss-free process plots on the R-Γ_i diagram as shown in Figure 7.6.

Diabatic Process with Loss. An *isothermal* process is a special case of diabatic flow in which neither the area nor the elevation changes, and the heat transfer just counterbalances the loss in such a manner that the temperature remains constant. Of course, according to (7.29), if $T = C$, entropy is constant. This tells us that the heat transfer is *out* of the system (i.e., cooling) since losses are always such as to produce an entropy increase. Cooling means a decrease in total enthalpy, so by the energy equation the total pressure must decrease as well. The isothermal process is

284 Flow of Real Liquids in Pipes

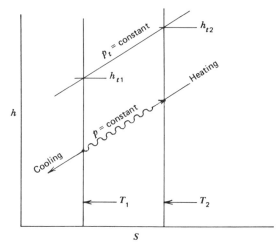

Figure 7.9 Rayleigh process—constant density.

shown in Figure 7.10 on an h-S diagram, and plots exactly as all other processes on the generalized R-Γ_i diagram of Figure 7.6.

Darcy-Weisbach Equation. The usual form of the Darcy-Weisbach equation is that of (7.6), in terms of the static pressure head. However, as pointed out, this form is not general because it overlooks changes in elevation, in area, and in density. Some references go a step further (in the right direction) and set

$$\frac{\Delta p_t}{\rho} = f\frac{L}{D}\frac{V^2}{2g_c}, \qquad (7.30)$$

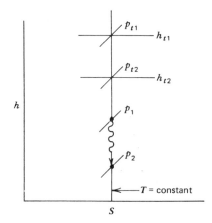

Figure 7.10 Isothermal process—constant density.

which is better, but still is premised on constant density and constant elevation. Much more general is the head loss equation given in terms of a loss coefficient by (7.7).

7.6.2 Generalized Flow Map

The generalized flow function of (7.25) will solve any constant density problem, provided that three of the four terms are known beforehand. The problem often reduces to forming a reliable estimate of p_{t1}/p_{t2}, as in terms of a loss coefficient, K. It is possible to exploit the interior of the R-Γ_i envelope, by showing K and TPR ($=p_{t1}/p_{t2}$) parameters within the Figure 7.6 plot [7], to obtain a generalized flow map as shown in Figure 7.11. The basis of these parametric additions is as follows. For a *constant area* flow

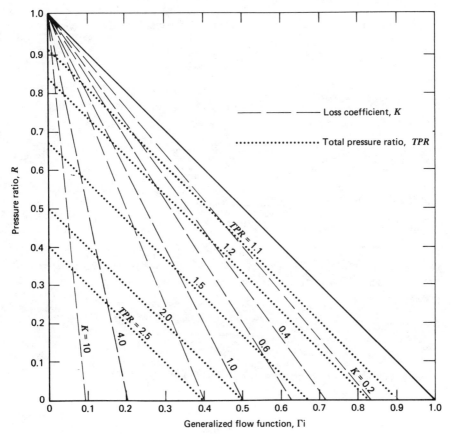

Figure 7.11 Generalized constant density flow map.

286 Flow of Real Liquids in Pipes

passage, such as a pipe, the energy equation of (7.5) reduces to

$$K\frac{V^2}{2g_c} = \frac{p_1 - p_2}{\rho} + \frac{g}{g_c}(Z_1 - Z_2). \qquad (7.31)$$

Dividing through by p_{t1}/ρ, we have

$$K\left(\frac{\rho V^2/2g_c}{p_{t1}}\right) = \frac{p_1 - p_2}{p_{t1}} + \rho \frac{g}{g_c} \frac{(Z_1 - Z_2)}{p_{t1}}$$

or, upon rearrangement,

$$K = \frac{2}{\alpha_i^2}(R_1 - R_{21} + Z_t) = \frac{1}{\Gamma_{i1}}(R_1 - R_{21} + Z_t). \qquad (7.32)$$

In (7.32) we introduced a flow number (α) patterned after (3.68), that is,

$$\alpha_i = \frac{\dot{m}}{Ap_{t1}}\left(\frac{p_{t1}}{\rho g_c}\right)^{1/2}. \qquad (7.33)$$

This incompressible total flow number also can be given in the alternative forms

$$\alpha_i = \frac{V}{(p_{t1} g_c/\rho)^{1/2}} \qquad (7.34)$$

and, in terms of the inlet pressure ratio,

$$\alpha_i = [2(1 - R_1)]^{1/2}. \qquad (7.35)$$

Of course, α_i is a pure number. Also, in (7.32), we introduced the hybrid pressure ratio (R_{21}), which relates the static pressure at 2 to the total pressure at 1, that is,

$$R_{21} = \frac{p_2}{p_{t1}}. \qquad (7.36)$$

Finally, (7.32) also can be solved for R_1 to yield

$$R_1 = \frac{R_{21} + K - Z_t}{1 + K}. \qquad (7.37)$$

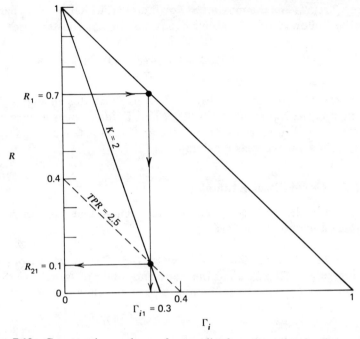

Figure 7.12 Construction and use of generalized constant density flow map.
Given: $R_1 = 0.7, TPR = 2.5, K = 2$.
By (7.38) and (7.40):

$$\Gamma_{i1} = 1 - R_1 = 0.3,$$

$$R_{21} = -\Gamma_{i1} + \frac{1}{TPR} = 0.1.$$

Constant TPR intercepts: at $P = 0$, $\quad \Gamma_i = \frac{1}{TPR} = 0.4;$

at $\Gamma_i = 0$, $\quad R = \frac{1}{TPR} = 0.4.$

Constant K intercepts: at $R = 0$, $\quad \Gamma_i = \frac{1}{1+K} = 0.333;$

at $\Gamma_i = 0$, $\quad R = 1.$

The generalized flow function (Γ_{i1}) of (7.25) can be given in the alternative forms

$$\Gamma_{i1} = 1 - R_1 = \left(\frac{\alpha_i}{\alpha_i^*}\right)^2 \tag{7.38}$$

by making use of (7.35) and the definition

$$\alpha_i^* = \sqrt{2} \tag{7.39}$$

288 Flow of Real Liquids in Pipes

where α_i^* represents the maximum flow number, which occurs when $R_1 = 0$. It further follows from the above equations that, at constant area,

$$R_{21} = -\Gamma_{i1} + \frac{p_{t2}}{p_{t1}} = -(K+1)\Gamma_{i1} + 1. \tag{7.40}$$

Equation 7.40 is the basis of Figure 7.11, as illustrated in Figure 7.12, for note that p_{t2}/p_{t1} represents both the R and Γ_i intercepts for constant TPR parameters, and furthermore, for constant K parameters, the Γ_i intercept is simply $1/(1 + K)$, whereas the R intercept is 1.

7.6.3 Generalized Examples

Several examples will be given in this section that are solved by the generalized methods just developed.

Example 7.6. Consider the flow of water ($\rho = 62.4$ lbm/ft^3) from an inlet total pressure of 50 psia. Find the static and total pressure drops through a constant area ($A = 5$ in.^{2}) horizontal pipe if the flow rate is 100 lbm/sec and the loss coefficient is 1.

Solution: The flow number is by (7.33)

$$\alpha_i = \frac{100}{5 \times 50} \left(\frac{50 \times 144}{62.4 \times 32.174} \right)^{1/2} = 0.7575.$$

But by (7.35):

$$R_1 = 1 - \frac{\alpha_i^2}{2} = 1 - \frac{0.7575^2}{2} = 0.7131,$$

and hence $p_1 = p_{t1} \times 0.7131 = 35.655$ psia.
The generalized flow function is by (7.38)

$$\Gamma_{i1} = 1 - R_1 = 1 - 0.7131 = 0.2869.$$

By (7.40):

$$R_{21} = -(K+1)\Gamma_{i1} + 1 = -2 \times 0.2869 + 1 = 0.4262$$

and hence

$$p_2 = p_{t1} \times R_{21} = 50 \times 0.4262 = 21.31 \text{ psia.}$$

Also by (7.40):

$$\frac{p_{t2}}{p_{t1}} = R_{21} + \Gamma_{i1} = 0.4262 + 0.2869 = 0.7131$$

and hence

$$p_{t2} = p_{t1} \times 0.7131 = 35.655 \text{ psia}.$$

Thus the static pressure drop is given by

$$p_1 - p_2 = 35.655 - 21.310 = 14.345 \text{ psi},$$

and the total pressure drop is given by

$$p_{t1} - p_{t2} = 50 - 35.655 = 14.345 \text{ psi}.$$

Of course, the static and total pressure drops are equal, as they must be in any constant area, constant elevation, constant density flow.

Although not called for, these same results could be obtained from the generalized flow map of Figure 7.11.

Example 7.7. Consider the leakage of oil ($\rho = 50$ lbm/ft^3) from a large tank where the total pressure is 29.4 psia. Find the flow rate and the static and total pressure drops through a small orifice-like passage which discharges to the atmosphere ($p_2 = 14.7$ psia) if the exit area is 1 in.^{2} and the loss coefficient is estimated at 0.2.

Solution: At $R_{21} = p_2/p_{t1} = 14.7/29.4 = 0.5$, we have from (7.40)

$$\Gamma'_i = \frac{1 - R_{21}}{K + 1} = \frac{1 - 0.5}{1.2} = 0.41667$$

and

$$\frac{p_{t2}}{p_{t1}} = \Gamma'_i + R_{21} = 0.41667 + 0.5 = 0.91667.$$

It is important to note that in a variable area problem, such as we are dealing with here, (7.40), which was derived for constant area flow, applies only to the *exit area*. Hence we use the primed quantity (Γ'_i) rather than the Γ_{i1} of a constant area problem. The same can be said for α_i, so that we have by (7.38)

$$\alpha'_i = \alpha_i^* \sqrt{\Gamma'_i} = \sqrt{2 \times 0.41667} = 0.91287.$$

By (7.33) this leads immediately to the flow rate, that is,

$$\dot{m} = \frac{\alpha'_i A_2 p_{t1}}{(p_{t1}/(\rho g_c))^{1/2}} = \frac{0.91287 \times 1 \times 29.4}{[29.4 \times 144/(50 \times 32.174)]^{1/2}} = 16.544 \text{ lbm/sec}.$$

Taking $\Gamma_{i1} \simeq 0$, $R_1 \simeq 0$, and $A_2/A_1 \simeq 0$, because we are going from a tank of relatively large area to a pipe of much smaller area, we have

$$p_1 - p_2 = p_{t1} - p_2 = 29.4 - 14.7 = 14.7 \text{ psi}$$

and

$$p_{t1} - p_{t2} = 29.4 - 29.4 \times 0.91667 = 2.45 \text{ psi}$$

REFERENCES

1. J. K. Vennard and R. L. Street, *Elementary Fluid Mechanics*, 5th ed., John Wiley, 1975.
2. R. P. Benedict, "On the determination and combination of loss coefficients for compressible fluid flows," *Trans. ASME J. Eng. Power*, January 1966, p. 67.
3. R. P. Benedict, N. A. Carlucci, and S. D. Swetz, "Flow losses in abrupt enlargements and contractions," *Trans. ASME, J. Eng. Power*, January 1966, p. 73.
4. R. P. Benedict, J. S. Wyler, J. A. Dudek, and A. R. Gleed, "Generalized flow across an abrupt enlargement," *Trans. ASME J. Eng. Power*, July 1976, p. 327.
5. J. G. Knudsen and D. L. Katz, *Fluid Dynamics and Heat Transfer*, McGraw-Hill, 1958, p. 182.
6. W. G. Steltz and R. P. Benedict, "Some generalizations in one-dimensional constant density fluid dynamics," *Trans. ASME, J. Eng. Power*, January 1962, p. 44.
7. R. P. Benedict and N. A. Carlucci, "Flow with losses," *Trans. ASME, J. Eng. Power*, January 1965, p. 37.

NOMENCLATURE

Roman

- A cross-sectional area
- C specific heat capacity, constant
- D pipe diameter
- e absolute pipe roughness
- f friction factor
- F frictional head
- g local acceleration of gravity
- g_c gravitational constant
- h_{loss} head loss

K loss coefficient
K.E. kinetic energy
L length
$\dot{m}$ mass flow rate
p static pressure
p_t total pressure
Q heat transferred, volumetric flow rate
R static/total pressure ratio
R_D pipe Reynolds number
S entropy
S.G. specific gravity
t time
T temperature
u internal energy
v specific volume
V volumetric average velocity
W work
Z vertical displacement
Z_t elevation factor

Greek

α kinetic energy coefficient, flow number
β diameter ratio
Γ generalized flow function
Δ finite difference
μ dynamic viscosity
ν kinematic viscosity
ρ fluid density

Subscripts

1, 2 axial stations
i incompressible
L laminar
R reference
T turbulent

8

Flow of Real Gases in Pipes

> From the top of the arch branches go to the head and brain. Below, branches go to the lungs, sac of heart, diaphragm, liver, bowel, spleen, pancreas, right and left a stout branch goes to each kidney: branch, branch, branch, at last those distant ones that deliver to a man's toes... —*Gustav Eckstein*

8.1 GENERAL REMARKS

In this chapter we admit that real gases flow in pipes with losses introduced by the viscosity of the fluid, and by other loss mechanisms already mentioned. Again, the conservation equations of continuity and energy are used, with the head loss term now added to the general energy equation of ideal flow. An application of real gas flow involving the conservation of momentum also is given to illustrate its use. The friction factor of Chapter 6 is carried over to the real gas flows of this chapter, as it was in the real liquid flows of Chapter 7. Finally, a generalized compressible pipe flow solution is given in terms of an isentropic flow function and various generalized pipe flow maps.

8.2 THE ONE-DIMENSIONAL ENERGY EQUATION WITH LOSSES

Combining the steady flow general energy equation of (3.19) and the first law of thermodynamics of (3.13), we obtain for one-dimensional gas flow

$$\delta W = \delta F + \frac{dp}{\rho} + \frac{V\,dV}{g_c}. \tag{8.1}$$

In connection with (8.1), note that we have already dispensed with the kinetic energy coefficient, as discussed in Section 7.2, and with the elevation term, as discussed in Section 3.2.7. Equation 8.1 is called the one-dimensional energy equation with losses, and applies to real gas flow in pipes, with or without heat transfer.

8.3 THE HEAD LOSS EQUATION

As discussed in Section 7.3, the thermodynamic loss term (δF) is usually called the head loss (h_{loss}) in pipe flow work, and is defined, from (8.1), for workless gas flow as

$$\int_1^2 \delta F = (h_{\text{loss}})_{1,2} = \int_2^1 \frac{dp}{\rho} + \int_2^1 \frac{V\,dV}{g_c}. \tag{8.2}$$

In our work in this chapter, we must be careful to *avoid* applying the usual Darcy-Weisbach equation of (4.8), (6.1), and (7.6) to a gas flow, even in the constant-diameter case, because of density variations and accompanying fluid accelerations in all but the shortest pipe runs. However, in its differential form:

$$\delta F = f \frac{dx}{D} \frac{V^2}{2g_c} \tag{8.3}$$

we can make use of this modified Darcy-Weisbach equation with some fruitful results.

8.4 COMPRESSIBLE LOSS COEFFICIENTS

A compressible loss coefficient can be defined, in general, in terms of (8.3) as

$$dK = f \frac{dx}{D}. \tag{8.4}$$

Much is to be gained by maintaining the *incompressible* definition of f in (8.4) because of the bulk of information already published on this friction factor. We further assume that f does not vary with length. In fact, if f is found to be a function of length, an average value of the friction factor may be used. However, most authorities agree that this is not a significant variation.

294 Flow of Real Gases in Pipes

To further particularize this loss coefficient, it is necessary to specify the thermodynamic flow process involved.

8.4.1 Adiabatic Loss Coefficient

Combining (8.2), (8.3), and (8.4), we obtain

$$dK = \frac{-2g_c \, dp}{\rho V^2} - 2\frac{dV}{V}. \tag{8.5}$$

In (8.5), density can be replaced, according to continuity, by

$$\rho = \frac{\dot{m}}{AV}. \tag{8.6}$$

It is further convenient to introduce here the total flow number of (3.68), in the form [1]

$$\alpha = \frac{\dot{m}}{Ap_{t1}}\left(\frac{p_t}{\rho_t g_c}\right)^{1/2}, \tag{8.7}$$

where α is a constant for a given area (A) and given inlet total pressure (p_{t1}), and also can be expressed in terms of R_1, via (3.68), as

$$\alpha = \left\{\left(\frac{2\gamma}{\gamma-1}\right)R_1^{2/\gamma}\left[1 - R_1^{(\gamma-1)/\gamma}\right]\right\}^{1/2}. \tag{8.8}$$

The remaining variable to be replaced in (8.5) is the velocity (V). Beginning with the general compressible relation of (3.67), namely,

$$V^2 = \left(\frac{2g_c\gamma}{\gamma-1}\right)\left(\frac{p_t}{\rho_t} - \frac{p}{\rho}\right), \tag{8.9}$$

and replacing ρ by (8.6), we obtain the quadratic

$$V^2 + \left(\frac{2g_c\gamma}{\gamma-1}\right)\left(\frac{pA}{\dot{m}}\right)V - \left(\frac{2g_c\gamma}{\gamma-1}\right)\frac{p_t}{\rho_t} = 0,$$

which has the real solution [2]

$$V = \frac{2(p_t/\rho_t g_c)^{1/2}}{[(R/\alpha)^2 + 2(\gamma-1)/\gamma]^{1/2} + R/\alpha}. \tag{8.10}$$

In (8.10), R is the point static-total pressure ratio.

Compressible Loss Coefficients

When (8.6), (8.7), and (8.10) are introduced in (8.5), there results

$$dK = -\frac{1}{\alpha}\left\{\left[\left(\frac{R}{\alpha}\right)^2 + \frac{2(\gamma-1)}{\gamma}\right]^{1/2} + \frac{R}{\alpha}\right\} dR - 2\frac{dV}{V}, \qquad (8.11)$$

which can be integrated between any two stations in the pipe by making use of the general form

$$\int (ax^2+b)^{1/2} dx = \frac{x}{2}(ax^2+b)^{1/2} + \frac{b}{2\sqrt{a}} \ln\left[(ax^2+b)^{1/2} + x\sqrt{a}\,\right]. \qquad (8.12)$$

It is important to note, in connection with the integration, that the quantity p_t/ρ_t is given by the perfect gas relation:

$$\frac{p_t}{\rho_t} = \bar{R} T_t, \qquad (8.13)$$

which is a constant for the *adiabatic* case under consideration here. There results

$$(K_{\text{adi}})_{1,2} = \frac{1}{2}\left(\frac{R}{\alpha}\right)\left[\left(\frac{R}{\alpha}\right)^2 + \frac{2(\gamma-1)}{\gamma}\right]^{1/2}\bigg|_2^1 + \frac{1}{2}\left(\frac{R}{\alpha}\right)^2\bigg|_2^1 \qquad (8.14)$$

$$-\left(\frac{\gamma+1}{\gamma}\right)\ln\left\{\left[\left(\frac{R}{\alpha}\right)^2 + \frac{2(\gamma-1)}{\gamma}\right]^{1/2} + \frac{R}{\alpha}\right\}\bigg|_2^1$$

where, for the upper limit, R takes on the value $R_1 = p_1/p_{t1}$, a point pressure ratio, while for the lower limit we have $R_{21} = p_2/p_{t1}$, a *hybrid* pressure ratio.

The term K_{adi} of (8.14) maximizes when the value of R_{21} at the critical state (i.e., where the flow maximizes) is used for the lower limit. This critical R_{21} is given in terms of the initial pressure ratio by

$$(R_{21})^*_{\text{adi}} = \frac{2R_1^{1/\gamma}[1 - R_1^{(\gamma-1)/\gamma}]^{1/2}}{[(\gamma+1)(\gamma-1)]^{1/2}}. \qquad (8.15)$$

The significance of K^*_{adi} is that it indicates the maximum pipe length possible for the given inlet pressure ratio, R_1. Incidentally, it follows from (3.125) and (8.8) that the adiabatic critical flow number is given by

$$\alpha^*_{\text{adi}} = \left[\gamma\left(\frac{2}{\gamma+1}\right)^{(\gamma+1)/(\gamma-1)}\right]^{1/2}. \qquad (8.16)$$

When (8.8) and (8.14) are solved simultaneously, we get $\alpha = f(R_{21})$, the parameter being constant K's. Such solutions represent exactly the adiabatic flow of a compressible fluid with losses in constant area passages.

296 Flow of Real Gases in Pipes

In terms of the Mach number, this same adiabatic loss coefficient has been given [3-5] by

$$(K_{adi})_{1,2} = \frac{1}{\gamma}\left(\frac{1}{M_1^2} - \frac{1}{M_2^2}\right) + \left(\frac{\gamma+1}{2\gamma}\right)\ln\left\{\frac{M_1^2}{M_2^2}\left[\frac{2+(\gamma-1)M_2^2}{2+(\gamma-1)M_1^2}\right]\right\}, \quad (8.17)$$

which attains the maximum value, when $M_2 = 1$, of

$$(K_{adi})_{1,*} = \frac{1}{\gamma}\left(\frac{1}{M_1^2} - 1\right) + \left(\frac{\gamma+1}{2\gamma}\right)\ln\left[\frac{(\gamma+1)M_1^2}{2+(\gamma-1)M_1^2}\right]. \quad (8.18)$$

Examples, tables, and graphs involving K_{adi} will be given in the next few sections of this chapter.

Example 8.1. Find the adiabatic compressible loss coefficient if the pressure ratios, $R_1 = 0.9$ and $R_{21} = 0.76504$, are determined experimentally in air at $\gamma = 1.4$.

Solution: By (8.8):

$$\alpha = \left[\left(\frac{2 \times 1.4}{0.4}\right)(0.9)^{2/1.4}(1 - 0.9^{0.4/1.4})\right]^{1/2} = 0.422581.$$

By (8.14):

$$(K_{adi})_{1,2} = \frac{1}{2}\left(\frac{R_1}{\alpha}\right)\left[\left(\frac{R_1}{\alpha}\right)^2 + \frac{2(\gamma-1)}{\gamma}\right]^{1/2} + \frac{1}{2}\left(\frac{R_1}{\alpha}\right)^2$$

$$- \left(\frac{\gamma+1}{\gamma}\right)\ln\left\{\left[\left(\frac{R_1}{\alpha}\right)^2 + \frac{2(\gamma-1)}{\gamma}\right]^{1/2} + \frac{R_1}{\alpha}\right\}$$

$$- \frac{1}{2}\left(\frac{R_{21}}{\alpha}\right)\left[\left(\frac{R_{21}}{\alpha}\right)^2 + \frac{2(\gamma-1)}{\gamma}\right]^{1/2} - \frac{1}{2}\left(\frac{R_{21}}{\alpha}\right)^2$$

$$+ \left(\frac{\gamma+1}{\gamma}\right)\ln\left\{\left[\left(\frac{R_{21}}{\alpha}\right)^2 + \frac{2(\gamma-1)}{\alpha}\right]^{1/2} + \frac{R_{21}}{\alpha}\right\}$$

$$= \left(\frac{0.9}{2 \times 0.422581}\right)\left[\left(\frac{0.9}{0.422581}\right)^2 + \frac{2 \times 0.4}{1.4}\right]^{1/2}$$

$$+ 2.267958 - 2.535881$$

$$- \left(\frac{0.76504}{2 \times 0.422581}\right)\left[\left(\frac{0.76504}{0.422581}\right)^2 + 0.5714286\right]^{1/2}$$

$$- 1.6387713 + 2.276021$$

$$= 1.000014.$$

Example 8.2. For the same conditions as given in Example 8.1, find the adiabatic loss coefficient by the Mach number formulation of (8.17).

Solution: Equation 3.64 can be rewritten as

$$\mathbf{M} = \left\{ \left[R^{(1-\gamma)/\gamma} - 1 \right] \frac{2}{\gamma - 1} \right\}^{1/2}. \tag{8.19}$$

Hence

$$\mathbf{M}_1 = \left[(0.9^{-0.4/1.4} - 1) \frac{2}{0.4} \right]^{1/2} = 0.390901.$$

To get $\mathbf{M}_2$ it is necessary to find $R_2 = p_2/p_{t2}$. By an iterative procedure that we will discuss in Section 8.5.2 (and the results of which we will present in tabular form), the total pressure ratio for this problem is found to be $p_{t1}/p_{t2} = 1.13244$. It follows that

$$R_2 = R_{21} \left(\frac{p_{t1}}{p_{t2}} \right) = 0.76504 \times 1.13244 = 0.866362.$$

From (8.19):

$$\mathbf{M}_2 = \left[(0.866362^{-0.4/1.4} - 1) \frac{2}{0.4} \right]^{1/2} = 0.457373.$$

By (8.17):

$$(K_{\text{adi}})_{1,2} = \frac{1}{1.4} \left(\frac{1}{0.390901^2} - \frac{1}{0.457373^2} \right)$$

$$+ \frac{2.4}{2.8} \ln \left[\frac{0.390901^2}{0.457373^2} \left(\frac{2 + 0.4 \times 0.457373^2}{2 + 0.4 \times 0.390901^2} \right) \right]$$

$$= 1.26000503 + 0.85714286 \ln \left(\frac{0.318393177}{0.431166117} \right)$$

$$= 1.000114,$$

which closely checks the result of Example 8.1.

Example 8.3. For the same inlet conditions as in Examples 8.1 and 8.2, find the maximum K_{adi} possible by the pressure ratio formulation.

298 Flow of Real Gases in Pipes

Solution: By (8.15):

$$(R_{21})^*_{adi} = \frac{2 \times 0.9^{1/1.4} \times (1 - 0.9^{0.4/1.4})^{1/2}}{(2.4 \times 0.4)^{1/2}} = 0.3260282.$$

By (8.14), using the results of Example 8.1 for the upper limit of integration, we have

$$(K_{adi})_{1,*} = 2.406579 + 2.267958 - 2.535881$$
$$- 0.4166664 - 0.29761883 + 1.0561227$$
$$= 2.480494.$$

Example 8.4. Find the maximum K_{adi} of Example 8.3 by Mach number relations.

Solution: By (8.18):

$$(K_{adi})_{1,*} = \frac{1}{1.4}(6.5443489 - 1) + 0.85714286 \ln\left(\frac{2.4 \times 0.15280359}{2 + 0.4 \times 0.15280359}\right)$$
$$= 2.480492,$$

which closely checks the result of Example 8.3.

8.4.2 Isothermal Loss Coefficient

Beginning with (8.5), we multiply the numerator and the denominator of the second term by the perfect gas relation:

$$\rho = \frac{p}{\bar{R}T} \tag{8.20}$$

to obtain

$$dK = \frac{-2g_c p \, dp}{\rho^2 V^2 \bar{R} T} - 2\frac{dV}{V}. \tag{8.21}$$

This can be integrated at once, noting that $\rho^2 V^2 = G^2 = (\dot{m}/A)^2 =$ constant. There results

$$K_{1,2} = \frac{g_c}{G^2 \bar{R} T}(p_1^2 - p_2^2) + 2\ln\left(\frac{V_1}{V_2}\right). \tag{8.22}$$

But, in the *isothermal* case,

$$\frac{V_1}{V_2} = \frac{\rho_2}{\rho_1} = \frac{p_2}{p_1} = \frac{R_{21}}{R_1}. \tag{8.23}$$

We also introduce the total flow number patterned after (8.7), but based on inlet conditions, as

$$\alpha_1^2 = \left(\frac{\dot{m}}{A}\right)^2 \left(\frac{p_{t1}}{\rho_{t1} g_c}\right) \frac{1}{p_{t1}^2} = \frac{G^2 \bar{R} T_{t1}}{p_{t1}^2 g_c}. \tag{8.24}$$

When (8.23) and (8.24) are combined with (8.22), we have

$$(K_{\text{iso}})_{1,2} = \frac{R_1^{(\gamma+1)/\gamma}}{\alpha_1^2} \left[1 - \left(\frac{R_{21}}{R_1}\right)^2\right] + \ln\left(\frac{R_{21}}{R_1}\right)^2. \tag{8.25}$$

The term K_{iso} of (8.25) maximizes when the value of R_{21} at the critical state is used. This critical R_{21} is given in terms of the initial pressure ratio by

$$(R_{21})_{\text{iso}}^* = \left\{\left(\frac{2\gamma}{\gamma-1}\right) R_1^{(\gamma+1)/\gamma} \left[1 - R_1^{(\gamma-1)/\gamma}\right]\right\}^{1/2}. \tag{8.26}$$

Incidentally, it follows from (8.8) and the isothermal critical pressure ratio:

$$R_{\text{iso}}^* = \left(\frac{2\gamma}{3\gamma-1}\right)^{\gamma/(\gamma-1)} \tag{8.27}$$

that the isothermal critical flow number is given by

$$\alpha_{\text{iso}}^* = \left(\frac{2\gamma}{3\gamma-1}\right)^{(\gamma+1)/[2(\gamma-1)]}. \tag{8.28}$$

The maximum value of K_{iso} is given by

$$(K_{\text{iso}})_{1,*} = \frac{R_1^{(\gamma+1)/\gamma}}{\alpha_1^2} \left\{1 - \left(\frac{2\gamma}{\gamma-1}\right)\left[R_1^{(1-\gamma)/\gamma} - 1\right]\right\}$$
$$+ \ln\left\{\left(\frac{2\gamma}{\gamma-1}\right)\left[R_1^{(1-\gamma)/\gamma} - 1\right]\right\}. \tag{8.29}$$

When (8.8) and (8.25) are solved simultaneously, we get $\alpha = f(R_{21})$, the parameter being constant K's. Such solutions represent exactly the isothermal flow of a compressible fluid with losses in constant area passages.

300 Flow of Real Gases in Pipes

In terms of the Mach number, this same isothermal loss coefficient has been given [3–5] by

$$(K_{iso})_{1,2} = \frac{1-(M_1/M_2)^2}{\gamma M_1^2} + \ln\left(\frac{M_1}{M_2}\right)^2, \qquad (8.30)$$

which attains the maximum value, when $M_2 = 1/\sqrt{\gamma}$, of

$$(K_{iso})_{1,*} = \frac{1-\gamma M_1^2}{\gamma M_1^2} + \ln(\gamma M_1^2). \qquad (8.31)$$

Examples, tables, and graphs of K_{iso} will be given in the next few sections of this chapter.

Example 8.5. Find $(K_{iso})_{1,2}$ by pressure ratio and Mach number relations for the pressure ratios of Example 8.1.

Solution: By (8.25):

$$(K_{iso})_{1,2} = \frac{0.9^{2.4/1.4}}{\alpha_1^2}\left[1-\left(\frac{0.76504}{0.9}\right)^2\right] + \ln\left(\frac{0.76504}{0.9}\right)^2$$

$$= \frac{0.834754}{0.178575}(1-0.7225756) + \ln(0.7225756)$$

$$= 0.971898.$$

By (8.30) we have M_1, but M_2 is not known yet since R_2 is not necessarily the same in the adiabatic and isothermal cases. By an iteration that will be discussed in Section 8.5.2, we find that $p_{t1}/p_{t2} = 1.1307067$ and hence $(R_2)_{iso} = 0.865036$. Then, by (8.19), $M_2 = 0.459859$, and

$$(K_{iso})_{1,2} = \frac{1-0.722577}{1.4 \times 0.390901^2} + \ln(0.722577)$$

$$= 0.971891,$$

which checks the pressure result closely.

Example 8.6. For the same inlet conditions of Example 8.5, find the maximum isothermal loss coefficient by pressure ratio and Mach number relations.

Solution: By (8.29):

$$(K_{iso})_{1,*} = \frac{0.9^{2.4/1.4}}{\alpha_1^2}\left[1-\left(\frac{2.8}{0.4}\right)(0.9^{-0.4/1.4}-1)\right]+\ln\left[\frac{2.8}{0.4}(0.9^{-0.4/1.4}-1)\right]$$

$$= \frac{0.83475415}{0.1785747}[1-7(0.03056068)]+\ln(7\times 0.03056068)$$

$$= 2.132410.$$

By (8.31):

$$(K_{iso})_{1,*} = \frac{1-1.4\times 0.390901^2}{1.4\times 0.390901^2}+\ln(1.4\times 0.390901^2)$$

$$= 2.132405.$$

These results check very closely.

8.4.3 Comparison of Compressible and Incompressible Loss Coefficients

The *incompressible* loss coefficient, as defined by the Darcy-Weisbach equation of (4.8), (6.1), and (7.6), can be applied to compressible flows to obtain an approximation. However, we must first decide whether the static or total pressure drop is of interest. Furthermore, we must choose which density and velocity are to be used. Although inlet conditions, exit conditions, and mean conditions all have been used, we will settle here, for the sake of illustration, on the exit conditions, believing that these are used most frequently.

Darcy Static Approximation. Starting with

$$\frac{\Delta p}{\rho_2} = K_s \frac{V_2^2}{2g_c}, \qquad (8.32)$$

where the subscript s signifies a static pressure drop, we rearrange this to

$$K_s = \frac{(p_1-p_2)2g_c}{\rho_2 V_2^2}. \qquad (8.33)$$

The total flow number of (8.7) can be written as

$$\alpha_2^2 = \frac{\rho_2^2 V_2^2 \bar{R} T_t}{\rho_{t1}^2 g_c}. \qquad (8.34)$$

302 Flow of Real Gases in Pipes

The density of a gas is simply

$$\rho_2 = \frac{p_2}{\bar{R}T_2}, \tag{8.35}$$

and the temperature ratio at 2 is

$$\frac{T_2}{T_t} = R_2^{(\gamma-1)/\gamma}. \tag{8.36}$$

When the last three equations are combined with (8.33), we obtain [6]

$$K_s = \frac{2}{\alpha_2^2} \left[\frac{(R_1 - R_{21})R_{21}}{R_2^{(\gamma-1)/\gamma}} \right]. \tag{8.37}$$

Darcy Total Approximation. Starting with

$$\frac{\Delta p_t}{\rho_2} = K_t \frac{V_2^2}{2g_c}, \tag{8.38}$$

we obtain by a similar development [6]

$$K_t = \frac{2}{\alpha_2^2} \left[\frac{(1 - 1/TPR)R_{21}}{R_2^{(\gamma-1)/\gamma}} \right], \tag{8.39}$$

where *TPR* signifies the total pressure ratio, p_{t1}/p_{t2}.

For given inlet conditions and for measured pressure ratios, R_1 and R_{21}, we can compare the compressible loss coefficients with the static and total Darcy approximations given above.

Example 8.7. Given the conditions of Examples 8.1 and 8.2, find the degree of approximation made by using the incompressible relations of (8.37) and (8.39).

Solution: By (8.37):

$$K_s = \left(\frac{2}{0.422581^2} \right) \left[\frac{(0.9 - 0.76504) \times 0.76504}{0.866362^{0.4/1.4}} \right] = 1.20476.$$

By (8.39):

$$K_t = \left(\frac{2}{0.422581^2} \right) \left[\frac{(1 - 1/1.13244) \times 0.76504}{0.866362^{0.4/1.4}} \right] = 1.04400.$$

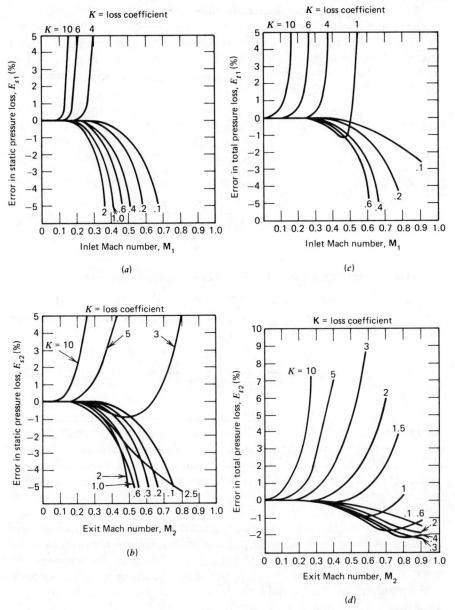

Figure 8.1 Darcy-Weisbach incompressible approximations compared with the adiabatic compressible loss coefficient. (*a*) Error in static-pressure approximation based on inlet conditions. (*b*) Error in static-pressure approximations based on exit conditions. (*c*) Error in total-pressure approximations based on inlet conditions. (*d*) Error in total-pressure approximations based on exit conditions.

304 Flow of Real Gases in Pipes

We conclude that the incompressible Darcy total approximation is higher than the compressible adiabatic loss coefficient of 1 by 4.4%, whereas the incompressible Darcy static approximation is higher by about 20.5%. These results will vary with the exit Mach number and the magnitude of the loss coefficient involved. Figure 8.1 gives a useful summary [2] of such approximations in terms of error functions (E), defined as

$$E_s = \frac{(p_2/p_1)_{\text{adi}} - (p_2/p_1)_{\text{Darcy}}}{(p_2/p_1)_{\text{adi}}} \qquad (8.40)$$

and

$$E_t = \frac{(p_{t2}/p_{t1})_{\text{adi}} - (p_{t2}/p_{t1})_{\text{Darcy}}}{(p_{t2}/p_{t1})_{\text{adi}}}. \qquad (8.41)$$

8.4.4 Combining Compressible Loss Coefficients

As already discussed in Section 7.4.2, we know that in the constant density case loss coefficients are additive when adjusted to a common reference area (A_R), as indicated by (7.9), that is,

$$(K_{0,N})_{\text{inc}} = \left[(K_{0,1})_{\text{inc}} \left(\frac{A_R}{A_1}\right)^2 + (K_{1,2})_{\text{inc}} \left(\frac{A_R}{A_2}\right)^2 + \cdots + K_{N-1,N} \right]. \qquad (8.42)$$

For the compressible case under consideration here, examination of the adiabatic loss coefficient of (8.14) or the isothermal loss coefficient of (8.25) reveals that area is inexorably involved in the definition of K_{adi} and K_{iso} (through the flow number, α), and hence resists factoring, which was possible in deriving (8.42). Thus we can expect no simple combining relation for compressible flow loss coefficients such as was given by (7.9) and (8.42) for the constant density case.

However, the problem becomes tractable if all compressible loss coefficients are *initially* adjusted to a common reference area. This entails an evaluation of K according to (8.14) or (8.25), based on the pertinent exit area, and then a reevaluation of K based on the common reference area while maintaining the original total pressure drop and, of course, the same flow rate.

To illustrate this idea, consider the adiabatic loss coefficient of (8.14). In terms of a common reference area we have

$$(K_{x,y})_{\text{ref}} = f(p_x) - f(p_y) - \left(\frac{\gamma+1}{\gamma}\right) \ln \frac{F(p_x)}{F(p_y)}. \qquad (8.43)$$

Compressible Loss Coefficients

The *overall* loss coefficient for an arbitrary series installation is

$$(K_{0,N})_{\text{ref}} = f(p_0) - f(p_N) - \left(\frac{\gamma+1}{\gamma}\right) \ln \frac{F(p_0)}{F(p_N)}. \tag{8.44}$$

The loss coefficients of the various components of the installation are

$$(K_{0,1})_{\text{ref}} = f(p_0) - f(p_1) - \left(\frac{\gamma+1}{\gamma}\right) \ln \frac{F(p_0)}{F(p_1)},$$

$$(K_{1,2})_{\text{ref}} = f(p_1) - f(p_2) - \left(\frac{\gamma+1}{\gamma}\right) \ln \frac{F(p_1)}{F(p_2)}, \tag{8.45}$$

$$(K_{N-1,N})_{\text{ref}} = f(p_{N-1}) - f(p_N) - \left(\frac{\gamma+1}{\gamma}\right) \ln \frac{F(p_{N-1})}{F(p_N)}.$$

A comparison of (8.44) and (8.45) indicates that

$$(K_{0,N})_{\text{ref}} = (K_{0,1})_{\text{ref}} + (K_{1,2})_{\text{ref}} + \cdots + (K_{N-1,N})_{\text{ref}}. \tag{8.46}$$

Thus we conclude in the compressible case also that loss coefficients are additive when adjusted to a common reference area.

This concept of reevaluating K based on a common reference area will now be illustrated by a numerical example.

Example 8.8. From Examples 8.1 and 8.2 we found that $R_1 = 0.9$, $R_{21} = 0.76504$, $(\alpha_1)_{\text{point}} = 0.422581$, $K_{\text{adi}} = 1.000$, $R_2 = 0.866362$, and $TPR_{1,2} = 1.13244$. Find the overall loss coefficient if this loss element, at $A_1 = A_2 = 2$ in.^{2}, is to be combined with another loss element whose inlet conditions just match the exit conditions of the first loss element, and whose exit conditions are at $R_{32} = 0.61579$ and at $A_3 = 1.5$ in.^{2}.

Solution: According to Figure 8.2, the hybrid flow number based on the exit area is

$$(\alpha_3)_{\text{hybrid}} = \frac{\dot{m}}{A_3 p_{t2}} \sqrt{\frac{\bar{R} T_t}{g_c}} = (\alpha_1)_{\text{point}} \times TPR_{1,2} \times \frac{A_2}{A_3}$$

$$= 0.422581 \times 1.13244 \times \frac{2}{1.5} = 0.638064.$$

This yields, for an *effective* inlet pressure ratio to loss element 2, based on the exit area, $R_2' = 0.70042$. The values $(\alpha_3)_{\text{hybrid}}$, R_2', and R_{32}, for loss element 2, lead to $(K_{\text{adi }2,3})_3 = 0.1$, by calculations similar to those used in

306 Flow of Real Gases in Pipes

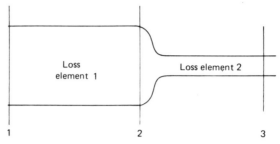

Figure 8.2 Combining two loss elements in series.

Examples 8.1 and 8.2. There also result $TPR_{2,3} = 1.04264$, $R_3 = 0.64205$, and $(\alpha_3)_{point} = 0.664825$.

The question is: How do we combine these two loss coefficients to get one overall coefficient? This is no trivial problem without the aid of a computer or tables, which will be discussed later. For now, we must refer both loss coefficients to a common reference area, and the most convenient one is the exit area of loss element 2. This means that we must reexpress $(K_{1,2})_2$ as $(K_{1,2})_3$:

$$(\alpha_1)_{hybrid} = \frac{\dot{m}}{A_3 P_{t1}} \sqrt{\frac{\overline{R} T_t}{g_c}} = (\alpha_1)_{point} \times \frac{A_2}{A_3}$$

$$= 0.422581 \times \frac{2}{1.5} = 0.56344.$$

This yields for an *effective* inlet pressure ratio to loss element 1, based on the exit area, $R'_1 = 0.79726$. But R'_2 has already been determined to be $R'_2 = 0.70042$, which leads to $R'_{21} = R'_2 \times TPR_{1,2} = 0.79318$. By usual means, R'_1, R'_{21}, and $(\alpha_1)_{hybrid}$ lead to $(K_{1,2})_3 = 0.0128$.

Hence the overall loss coefficient for elements 1 and 2, based on A_3, is simply

$$(K_{1,3})_3 = (K_{1,2})_3 + (K_{2,3})_3 = 0.1128.$$

As a check, we have, from initial calculations,

$$TPR_{1,3} = TPR_{1,2} \times TPR_{2,3}$$
$$= 1.13244 \times 1.04264 = 1.18073,$$

whereas

$$TPR_{1,3} = \frac{\alpha_3}{\alpha_{1, hybrid}} = \frac{0.664825}{0.56344} = 1.17994,$$

which agrees closely.

Table 8.1 Series Combination Characteristics of Compressible Flow Loss Coefficients[a]

$K_{0,2}$			$K_{2,3}$			$K_{3,4}$			$(K_{0,4})_{by\ sum}$			$(K_{0,4})_{overall}$			Exit Mach Number
EXACT	STATIC	TOTAL	EXACT	STATIC	TOTAL	EXACT	STATIC	TOTAL	EXACT	STATIC	TOTAL	EXACT	STATIC	TOTAL	
1.60	2.57	1.61	18.28	17.58	17.48	25.88	23.86	23.70	45.76	43.91	42.79	45.78	40.47	39.50	0.114
1.26	2.22	1.26	15.84	14.94	14.78	22.00	18.82	18.52	39.10	35.98	34.66	39.16	30.77	29.79	0.182
1.24	2.21	1.25	15.81	14.57	14.39	21.79	16.10	14.33	38.84	32.88	29.97	38.76	25.04	24.04	0.283
1.27	2.23	1.27	16.10	14.70	14.49	21.95	14.40	13.80	39.32	31.33	29.56	39.39	21.66	20.64	0.367
1.28	2.25	1.29	16.38	14.88	14.67	22.30	13.17	12.38	39.96	30.30	28.34	40.31	19.28	18.24	0.449
1.22	2.16	1.23	15.38	14.41	14.26	8.04	7.80	7.68	24.64	24.37	23.17	24.66	22.28	21.33	0.157
1.19	2.16	1.20	15.80	14.16	13.92	7.99	7.49	7.28	24.98	23.81	22.40	25.02	20.12	19.13	0.226
1.22	2.20	1.23	16.26	13.88	13.56	8.16	7.10	6.73	25.64	23.18	21.52	25.58	17.03	16.02	0.338
1.21	2.19	1.22	16.13	13.40	13.10	8.13	6.57	6.07	25.47	22.16	20.39	25.46	14.60	13.57	0.443
1.18	2.15	1.18	16.12	13.28	12.90	7.95	6.02	5.41	25.25	21.45	19.49	25.17	12.76	11.69	0.550
1.22	2.18	1.22	15.96	15.01	14.86	5.60	5.54	5.48	22.78	23.73	21.56	22.78	21.32	20.34	0.142
1.14	2.14	1.15	15.28	13.48	13.21	5.29	5.15	4.97	21.71	20.77	19.33	21.76	17.76	16.75	0.233
1.23	2.22	1.24	16.42	13.56	14.30	5.71	5.28	4.96	23.36	21.06	20.50	23.30	15.68	14.66	0.354
1.22	2.21	1.23	16.55	13.22	12.81	5.63	4.97	4.52	23.40	20.40	18.56	23.35	13.56	12.51	0.462
1.18	2.17	1.19	16.24	12.75	12.32	5.42	4.58	4.02	22.84	19.50	17.53	22.78	11.69	10.61	0.580

[a] Exact K's are based on (8.14). Static K's are based on (8.37). Total K's are based on (8.39). Note the progressive deterioration of the Darcy approximations as compressibility (indicated by the exit Mach number) becomes more significant.

308 Flow of Real Gases in Pipes

A comparison of several incompressible Darcy approximations of compressible loss coefficients with the more exact compressible adiabatic loss coefficients is given in Table 8.1 for various combinations of loss elements in series. Of course, all loss coefficients are referred to a common reference area. Note that the deviations between the compressible results and the various incompressible approximations vary directly with the exit Mach number.

8.4.5 Deriving a Loss Coefficient

As indicated in Section 7.4.3, very few loss coefficients can be derived analytically. This means that most loss coefficients must be determined empirically, as by experiment. However, just as the loss coefficient of an abrupt enlargement was derived analytically for the constant density case in Section 7.4.3, so here we will derive the abrupt enlargement loss coefficient analytically for compressible fluids.

Conservation of Mass. Here we presume *one-dimensional velocity* distributions at stations 1 and 2 of Figure 7.1. This is quite reasonable since the streamlines are straight and parallel at both these stations. We have

$$\dot{m} = \rho_1 A_1 V_1 = \rho_2 A_2 V_2. \tag{8.47}$$

Conservation of Momentum. Here we presume there is *free shear flow* from station 1 to station 2 of Figure 7.1. This means quite simply that there is an absence of wall shear stresses between stations 1 and 2. There results

$$p_1 A_1 + p_1'(A_2 - A_1) - p_2 A_2 = \frac{\dot{m}}{g_c}(V_2 - V_1), \tag{8.48}$$

where p_1' here represents the static pressure outside the jet issuing from the step.

Conservation of Energy. Here we presume *adiabatic* flow, and obtain

$$\frac{T_1}{T_2} = \frac{2 + (\gamma - 1)M_2^2}{2 + (\gamma - 1)M_1^2}. \tag{8.49}$$

When (8.47), (8.48), and (8.49) are combined in terms of the perfect gas equation of state, that is, $pv = \overline{R}T$, there results

$$\frac{M_2[2 + (\gamma - 1)M_2^2]^{1/2}}{1 + \gamma M_2^2} = \frac{M_1[2 + (\gamma - 1)M_1^2]^{1/2}}{1 + \gamma M_1^2 + \left(\frac{p_1'}{p_1}\right)[(1 - \varphi)/\varphi]}, \tag{8.50}$$

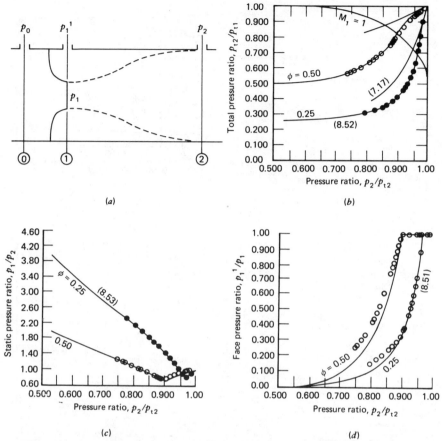

Figure 8.3 Experimental results obtained in air for area ratios of 0.25 and 0.5 (after Benedict et al. [7]). (a) Test setup. (b) Total pressure loss. (c) Static pressure recovery. (d) Face pressure variation.

where $\varphi = A_1/A_2$, the area ratio [7, 8].

Subsonic Case. The pressure ratio (p_1'/p_1) is considered unity by most experimenters for the subsonic case [7, 9–12].

Supercritical Case. When the Mach number at station 1 is unity, (8.50) yields p_1'/p_1 for every value of M_2, that is, (8.50) can be rearranged to

$$\left(\frac{p_1'}{p_1}\right)_{\text{supersonic}} = \left(\frac{\varphi}{1-\varphi}\right)(1+\gamma)^{1/2}\left\{\frac{1+\gamma M_2^2}{M_2[2+(\gamma-1)M_2^2]^{1/2}} - (1+\gamma)^{1/2}\right\}.$$

(8.51)

310 Flow of Real Gases in Pipes

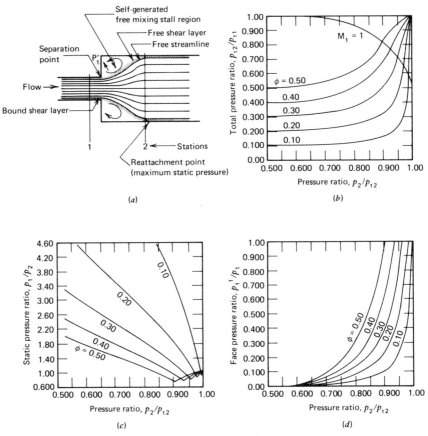

Figure 8.4 Design curves for abrupt enlargements of various area ratios in air (after Benedict et al. [7]). (*a*) Abrupt enlargement. (*b*) Total pressure loss. (*c*) Static pressure recovery. (*d*) Face pressure variation.

With p_1'/p_1 defined for all compressible situations, it follows that the total pressure ratio across the abrupt enlargement is given by

$$\frac{p_{t2}}{p_{t1}} = \varphi\left(\frac{M_1}{M_2}\right)\left[\frac{2+(\gamma-1)M_2^2}{2+(\gamma-1)M_1^2}\right]^{(\gamma+1)/[2(\gamma-1)]}. \tag{8.52}$$

This is to be compared with the constant density result of (7.17). Similarly, the static pressure ratio is given by

$$\frac{p_1}{p_2} = \frac{1}{\varphi}\left(\frac{M_2}{M_1}\right)\left[\frac{2+(\gamma-1)M_2^2}{2+(\gamma-1)M_1^2}\right]^{1/2}. \tag{8.53}$$

This is to be compared with the constant density result of (7.19).

Experimental results, obtained in air for abrupt area changes of 0.25 and 0.5, are presented in Figure 8.3. Design curves for abrupt enlargements of various area ratios in air are given in Figure 8.4.

8.5 SOLUTION OF REAL FLOW OF GAS IN PIPES

8.5.1 Solution Steps

Some general steps can be outlined which lead to solutions to compressible pipe flow problems involving losses.

1. We determine the pipe Reynolds number, according to (5.1) and (7.20). Note that for steady flow in a constant diameter pipe there is essentially only one Reynolds number for the pipe, neglecting only the minor variations in viscosity with temperature.
2. By usual means, according to the methods of Chapter 6, we determine the friction factor of the pipe.
3. We determine the head loss, via (8.3) and (8.4), as

$$(h_{\text{loss}})_{1,2} = f \frac{L}{D} \frac{V^2}{2g_c} = K_x \frac{V_x^2}{2g_c}. \tag{8.54}$$

4. From the energy equation of (8.2), we can now get either Δp or ΔK.E., depending on the unknown in the pipe flow problem, where

$$\Delta \text{K.E.} = \int_1^2 \frac{V\,dV}{g_c} = \left(\frac{\gamma}{\gamma-1}\right)\left(\frac{p_1}{\rho_1} - \frac{p_2}{\rho_2}\right) \tag{8.55}$$

according to (3.62) for a perfect gas. Note that the velocities called for in (8.54) and (8.55) are the volumetric average velocities of the pipe as defined by the continuity equation. As before noted, if the pipe is not circular in cross section, (7.22) can be used to define an equivalent diameter.

8.5.2 Numerical Examples

First, a word on obtaining TPR, given R_1 and R_{21}, is in order.

Adiabatic Flow. Here we guess a value of R_2. But TPR is given by

$$TPR_A = \frac{R_2}{R_{21}} \tag{8.56}$$

and by

$$(TPR_B)_{\text{Fanno}} = \left\{ \left(\frac{R_2}{R_1} \right)^{2/\gamma} \left[\frac{1 - R_2^{(\gamma-1)/\gamma}}{1 - R_1^{(\gamma-1)/\gamma}} \right] \right\}^{1/2}. \tag{8.57}$$

When the alternative forms, (8.56) and (8.57), for TPR converge within the accuracy required, we have obtained both R_2 and TPR.

Isothermal Flow. Once more, we guess a value of R_2. But TPR_A is given by (8.56), while

$$(TPR_B)_{\text{iso}} = \left\{ \left(\frac{R_2}{R_1} \right)^{(\gamma+1)/\gamma} \left[\frac{1 - R_2^{(\gamma-1)/\gamma}}{1 - R_1^{(\gamma-1)/\gamma}} \right] \right\}^{1/2}. \tag{8.58}$$

When the alternative forms, (8.56) and (8.58), for TPR converge within the accuracy required, we have obtained both R_2 and TPR.

Example 8.9. For the conditions given in Example 8.5, namely, $R_1 = 0.9$ and $R_{21} = 0.76504$, find $(TPR)_{\text{iso}}$ and $(R_2)_{\text{iso}}$.

Solution: We arbitrarily guess $R_{2,0} = 0.874071$ (where $R_{2,0}$ indicates initial guess of R_2), and find

$$TPR_A = \frac{R_2}{R_{21}} = \frac{0.874071}{0.76504} = 1.142517$$

and, by (8.58), $TPR_B = 1.10000$, yielding an error of $e_0 = 0.042517$. A second guess of $R_{2,1}$ is based arbitrarily on the adiabatic result of Example 8.2 as 0.866362. This yields $TPR_A = 1.13244$ and $TPR_B = 1.126317$, for an error of $e_1 = 0.006123$. This is seen to be closer to zero, but perhaps we require five-place decimal accuracy. The third guess of R_2 need not be arbitrary. Instead, the Newton-Raphson scheme, already given in Example 3.1, can be used. Thus

$$R_2 = R_{2,1} - e_1 \left(\frac{R_{2,0} - R_{2,1}}{e_0 - e_1} \right) \tag{8.59}$$

$$= 0.866362 - 0.006123 \left(\frac{0.874071 - 0.866362}{0.042517 - 0.006123} \right).$$

There results $R_2 = 0.865065$, which leads to a new error 1 of $e_1 = 0.0001346$. A final calculation, using (8.59), yields $R_2 = 0.865036$ and $(TPR)_{\text{iso}} = 1.1307067$, at a negligible error. These are the values used in Example 8.5.

Solution of Real Flow of Gas in Pipes 313

Example 8.10. Air flows adiabatically through a 10 in. duct of 200 ft length with an average friction factor of $f=0.01589$. The measured exit conditions are $p_2=94$ psia, $p_{t2}=100$ psia, $T_t=540°R$. Find the inlet pressure ratio.

Solution: At $R_2=p_2/p_{t2}=0.94$, $K_{2,*}=5.3646$ via (8.14) and (8.15). But

$$K_{1,2}=f\frac{L}{D}=\frac{0.01589\times 200}{10/12}=3.8136$$

via (8.54). Thus, according to Figure 8.5,

$$K_{1,*}=K_{1,2}+K_{2,*}=9.1781.$$

Once more applying (8.14), with R_1 the unknown and with R_{21} given by (8.15), we obtain, for $K_{1,*}=9.1781$, the result $R_1=0.96$.

Example 8.11. Air flows adiabatically through a duct of area $=25$ in.^{2} at a rate of 11.559 lbm/sec. Inlet total conditions are $p_{t1}=50$ psia and $T_t=540°R$. For a loss coefficient of 2, find the static and total pressure drops through the duct.

Solution: By (8.7):

$$\alpha=\frac{11.559}{25\times 50}\left(\frac{53.35\times 540}{32.174}\right)^{1/2}=0.27671.$$

By (8.8), solving for R_1, we obtain

$$R_1=0.96 \quad \text{and hence} \quad p_1=48 \text{ psia.}$$

According to the iteration scheme outlined at the beginning of this section and illustrated in Example 8.9, we guess R_2 and get $(TPR)_{\text{adi}}$. These results

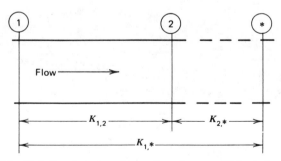

Figure 8.5 Relation between the various loss coefficients in a pipe.

314 Flow of Real Gases in Pipes

are as follows:

$$TPR_{1,2} = 1.09461 \quad \text{and hence} \quad p_{t2} = 45.678 \text{ psia},$$

$$R_{21} = 0.86936 \quad \text{and hence} \quad p_2 = 43.468 \text{ psia}.$$

It follows that

$$p_1 - p_2 = 48 - 43.468 = 4.532 \text{ psi},$$

$$p_{t1} - p_{t2} = 50 - 45.678 = 4.322 \text{ psi}.$$

We note that the static pressure drop always exceeds the total pressure drop in constant area adiabatic compressible flow because of fluid accelerations.

8.6 GENERALIZED COMPRESSIBLE PIPE FLOW

8.6.1 Generalized Flow Function

A generalized compressible flow equation can be given [13], in terms of continuity, energy, and the perfect gas equation of state, by combining the relations

$$\dot{m} = \rho A V = \text{constant}, \tag{8.6}$$

$$V^2 = \left(\frac{2g_c \gamma}{\gamma - 1}\right)\left(\frac{p_t}{\rho_t} - \frac{p}{\rho}\right), \tag{8.9}$$

and

$$\frac{p}{\rho} = \bar{R}T. \tag{8.13}$$

There results

$$\left(\frac{T_{t2}}{T_{t1}}\right)^{1/2}\left(\frac{p_{t1}}{p_{t2}}\right)\left(\frac{A_1}{A_2}\right)\left\{\left(\frac{p_1}{p_{t1}}\right)^{1/\gamma}\left[1 - \left(\frac{p_1}{p_{t1}}\right)^{(\gamma-1)/\gamma}\right]^{1/2}\right\}$$

$$= \left(\frac{p_2}{p_{t2}}\right)^{1/\gamma}\left[1 - \left(\frac{p_2}{p_{t2}}\right)^{(\gamma-1)/\gamma}\right]^{1/2} \tag{8.60}$$

or, more simply,

$$\frac{(TPR)(AR)}{(TTR)^{1/2}} P_1 = P_2, \qquad (8.61)$$

where P is a pressure function whose definition is obvious from (8.60). Equations (8.60) and (8.61) are entirely general dimensionless expressions for the flow of an ideal compressible fluid.

When both sides of (8.60) and/or (8.61) are referred to a function similar to P, but based on the isentropic critical pressure ratio of (3.125), namely,

$$R^* = \frac{p^*}{p_t} = \left(\frac{2}{\gamma+1}\right)^{\gamma/(\gamma-1)}, \qquad (8.62)$$

the general compressible flow equation can be written as

$$\frac{TPR \times AR}{TTR^{1/2}} \Gamma_1 = \Gamma_2, \qquad (8.63)$$

where Γ represents the generalized compressible flow function, which is defined uniquely in terms of the point pressure ratio, $R = p/p_t$, as

$$\Gamma = \left\{ \frac{R^{2/\gamma}[1 - R^{(\gamma-1)/\gamma}]}{[2/(\gamma+1)]^{2/(\gamma-1)}[(\gamma-1)/(\gamma+1)]} \right\}^{1/2}. \qquad (8.64)$$

Since the compressible flow function (Γ) varies between 0 and 1, as does R, for all flow processes, these parameters are said to be normalized. The envelope of all possible values of R and Γ is shown in Figure 8.6. Because of its generality, any compressible process can be shown on this envelope as from state 1 to 2 in Figure 8.7.

A skeleton table of the Γ function is given in Table 8.2 for $\gamma = 1.4$.

Example 8.12. If air flows with loss and heat transfer through a convergent passage whose area ratio, $AR = A_1/A_2$, is 2, such that the total pressure ratio, $TPR = p_{t1}/p_{t2}$, is 1.09138, the total temperature ratio, $1/TTR = T_{t2}/T_{t1}$, is 1.2, and the inlet pressure ratio, $R_1 = p_1/p_{t1}$, is 0.98, find the exit Mach number.

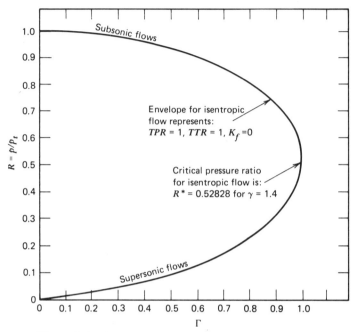

Figure 8.6 Generalized gas dynamics map ($\gamma = 1.4$).

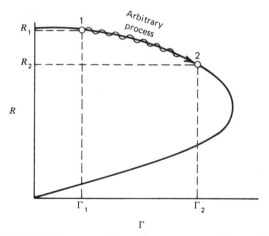

Figure 8.7 Arbitrary gas dynamics process on the $R - \Gamma$ plot.

Table 8.2 Generalized Compressible Flow Table ($\gamma = 1.4$)

Pressure Ratio, p/p_t	Temperature Ratio, T/T_t	Mach Number, M	Isentropic and Fanno, Γ	Rayleigh, T_t/T_t^*	Isothermal, T_t/T_t^*	Normal Shock, p_{t1}/p_{t2}
1.00000	1.00000	0.	0.	0.	0.87500	0
0.98000	0.99424	0.17013	0.28894	0.12907	0.88007	0
0.96000	0.98840	0.24220	0.40412	0.24327	0.88527	0
0.94000	0.98248	0.29863	0.48938	0.34434	0.89061	0
0.92000	0.97646	0.34720	0.55858	0.43380	0.89610	0
0.90000	0.97035	0.39090	0.61715	0.51294	0.90174	0
0.88000	0.96414	0.43127	0.66789	0.58290	0.90755	0
0.86000	0.95782	0.46922	0.71249	0.64468	0.91353	0
0.84000	0.95141	0.50536	0.75203	0.69914	0.91969	0
0.92000	0.94488	0.54009	0.78729	0.74706	0.92605	0
0.80000	0.93823	0.57372	0.81880	0.78911	0.93260	0
0.78000	0.93147	0.60650	0.84701	0.82589	0.93937	0
0.76000	0.92458	0.63862	0.87222	0.85791	0.94637	0
0.74000	0.91757	0.67022	0.89469	0.88565	0.95361	0
0.72000	0.91041	0.70144	0.91464	0.90953	0.96110	0
0.70000	0.90311	0.73239	0.93222	0.92990	0.96887	0
0.68000	0.89566	0.76318	0.94756	0.94710	0.97693	0
0.66000	0.88806	0.79389	0.96079	0.96142	0.98530	0
0.64000	0.88028	0.82461	0.97199	0.97312	0.99400	0
0.62665	0.87500	0.84515	0.97837	0.97959	1.00000	0
0.62000	0.87234	0.85542	0.98123	0.98244	1.00305	0
0.60000	0.86420	0.88639	0.98858	0.98959	1.01250	0
0.58000	0.85587	0.91761	0.99409	0.99474	1.02235	0
0.56000	0.84733	0.94914	0.99778	0.99808	1.03265	0
0.54000	0.83857	0.98107	0.99970	0.99974	1.04344	0
0.52828	0.83333	1.00000	1.00000	1.00000	1.05000	1.00000
0.52000	0.82958	1.01348	0.99985	0.99988	1.05475	1.00000
0.50000	0.82034	1.04645	0.99825	0.99859	1.06664	1.00012
0.48000	0.81082	1.08008	0.99490	0.99600	1.07915	1.00057
0.46000	0.80102	1.11446	0.98979	0.99219	1.09235	1.00157
0.44000	0.79091	1.14969	0.98291	0.98726	1.10631	1.00330
0.42000	0.78047	1.18591	0.97424	0.98126	1.12112	1.00596

318 Flow of Real Gases in Pipes

Table 8.2 Continued

Pressure Ratio, p/p_t	Temperature Ratio, T/T_t	Mach Number, M	Isentropic and Fanno, Γ	Rayleigh, T_t/T_t^*	Isothermal, T_t/T_t^*	Normal Shock, p_{t1}/p_{t2}
0.40000	0.76967	1.22324	0.96375	0.97428	1.13685	1.00973
0.38000	0.75847	1.26183	0.95139	0.96636	1.15364	1.01480
0.36000	0.74684	1.30186	0.93712	0.95756	1.17160	1.02140
0.34000	0.73475	1.34353	0.92088	0.04790	1.19089	1.02977
0.32000	0.72213	1.38707	0.90258	0.93742	1.21170	1.04020
0.30000	0.70893	1.43277	0.88214	0.92614	1.23425	1.05303
0.28000	0.69510	1.48096	0.85945	0.91407	1.25882	1.06871
0.26000	0.68053	1.53205	0.83438	0.90121	1.28576	1.08777
0.24000	0.66515	1.58655	0.80677	0.88754	1.31550	1.11090
0.22000	0.64882	1.64510	0.77642	0.87304	1.34861	1.13906
0.20000	0.63139	1.70853	0.74311	0.85766	1.38584	1.17348
0.18000	0.61266	1.77795	0.70652	0.84134	1.42819	1.21593
0.16000	0.59239	1.85484	0.66630	0.82397	1.47707	1.26898
0.14000	0.57021	1.94130	0.62194	0.80539	1.53451	1.33647
0.12000	0.54564	2.04046	0.57280	0.78540	1.60361	1.42457
0.10000	0.51795	2.15719	0.51795	0.76364	1.68936	1.54383
0.08000	0.48596	2.29978	0.45606	0.73958	1.80057	1.71411
0.06000	0.44761	2.48403	0.38495	0.71224	1.95482	1.97848
0.04000	0.39865	2.74634	0.30066	0.67966	2.19492	2.45413
0.02000	0.32703	3.20769	0.19386	0.63639	2.67562	3.64440
0.	0.	∞	0.	0.48980	∞	∞

Solution: From the generalized flow table (Table 8.2), at $R_1 = 0.98$, we find $\Gamma_1 = 0.28894$. Thus by (8.63), we have

$$(1.2)^{1/2} \times 1.09138 \times 2 \times 0.28894 = \Gamma_2 = 0.69088.$$

At this Γ_2 we find that $M_2 = 0.45051$ and $R_2 = 0.87$.

8.6.2 Compressible Processes

Note that Table 8.2 has columns labeled "Isentropic and Fanno," "Rayleigh," "Isothermal," and "Normal Shock." These have all been discussed previously, but in this section we clarify the use of these columns.

Generalized Compressible Pipe Flow 319

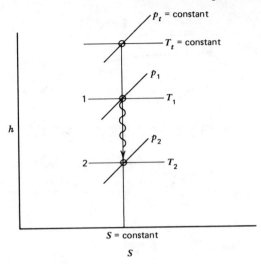

Figure 8.8 Compressible isentropic process.

Isentropic Process. Since $TTR = TPR = 1$ for adiabatic flow without loss, it follows from (8.63) that

$$\left(\frac{A_1}{A_2}\right)\Gamma_1 = \Gamma_2. \tag{8.65}$$

Thus one enters the generalized compressible flow table (Table 8.2) at R_1, M_1, or Γ_1, multiplies Γ_1 by the area ratio, and obtains Γ_2, M_2, and R_2. On an h-S diagram the isentropic process for a perfect gas plots as shown in Figure 8.8. On the generalized R-Γ diagram the isentropic process, of course, plots as shown in Figure 8.7.

Fanno Type Process.* Since $TTR = 1$ for any adiabatic flow, it follows from (8.63) that for adiabatic flow with loss

$$\left(\frac{p_{t1}}{p_{t2}}\right)\left(\frac{A_1}{A_2}\right)\Gamma_1 = \Gamma_2. \tag{8.66}$$

Thus one enters Table 8.2 at R_1, M_1, or Γ_1, multiplies Γ_1 by the area ratio and the total pressure ratio, and obtains Γ_2, M_2, and R_2. A Fanno process is shown on the h-S diagram of Figure 8.9.

*So called after Fanno, a student who first described this type of flow in his thesis at the Eidgen. Techn. Hoshschule in 1904.

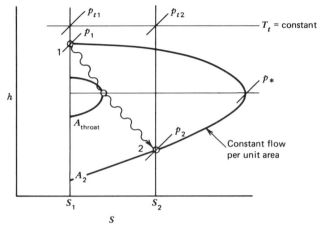

Figure 8.9 Compressible Fanno-type process.

Rayleigh Process. In this *constant area* heat transfer process neither the total temperature nor the total pressure remains constant, so it follows from (8.63) for diabatic flow without loss that

$$\left(\frac{T_{t2}}{T_{t1}}\right)^{1/2}\left(\frac{p_{t1}}{p_{t2}}\right)\Gamma_1 = \Gamma_2. \tag{8.67}$$

Now one enters Table 8.2 at R_1, M_1, or Γ_1 and obtains $(T_{t1}/T_{t*})_{\text{Rayleigh}}$. This is multiplied by the reciprocal of the total temperature ratio (T_{t2}/T_{t1}) to obtain $(T_{t2}/T_{t*})_{\text{Rayleigh}}$ according to the elementary relation

$$\left(\frac{T_{t2}}{T_{t1}}\right)\left(\frac{T_{t1}}{T_{t*}}\right) = \left(\frac{T_{t2}}{T_{t*}}\right). \tag{8.68}$$

On the same row in the table that (T_{t2}/T_{t*}) appears, one finds Γ_2, M_2, and R_2. A Rayleigh process is shown as the h-S diagram of Figure 8.10.

Isothermal Process. In this *constant area*, constant static temperature heat transfer process with loss, neither *TTR* nor *TPR* remains constant, so it follows from (8.63) that (8.67) applies. Similarly to the method for Rayleigh flow, one enters Table 8.2 at R_1, M_1, or Γ_1 and obtains $(T_{t1}/T_{t*})_{\text{iso}}$. This is multiplied by the reciprocal of total temperature (T_{t2}/T_{t1}) to obtain $(T_{t2}/T_{t*})_{\text{iso}}$ according to (8.68). On this same row, one finds Γ_2, M_2, and R_2. An isothermal process is shown on the h-S diagram of Figure 8.11, and plots exactly as all other processes on the generalized R-Γ diagram of Figure 8.7.

Generalized Compressible Pipe Flow 321

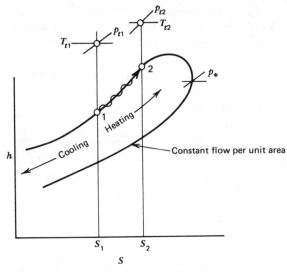

Figure 8.10 Compressible Rayleigh process.

Shock Process. The last column in Table 8.2 concerns the normal shock process, wherein the fluid, initially at a supersonic state, with $M_1 > 1$, experiences an abrupt change to a subsonic state. The equation is that of a constant area Fanno process, that is,

$$\left(\frac{p_{t1}}{p_{t2}}\right)\Gamma_1 = \Gamma_2. \tag{8.69}$$

One enters Table 8.2 with R_1, M_1, or Γ_1 to obtain p_{t1}/p_{t2}, the drop in total

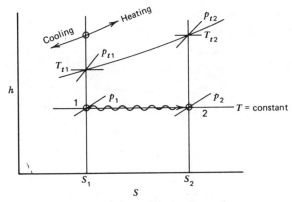

Figure 8.11 Compressible isothermal process.

322 Flow of Real Gases in Pipes

pressure across the normal shock. If conditions at 2 are required, one simply multiplies Γ_1 by this *TPR* to obtain Γ_2, M_2, and R_2. The normal shock process is shown on the h-S and R-Γ diagrams of Figure 8.12.

Example 8.13. If air flows at constant area through a combustor such that $TTR = 2.03038$ from an inlet pressure ratio of 0.94, find the exit Mach number and the total pressure ratio across the combustor.

Solution: At $R_1 = 0.94$, we find, from Table 8.2, that

$$\Gamma_1 = 0.48938, \qquad \left(\frac{T_{t1}}{T_{t*}}\right)_{\text{Ray}} = 0.34434.$$

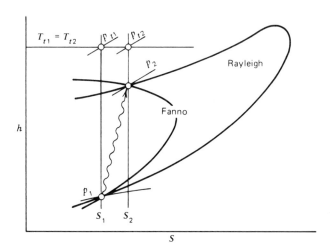

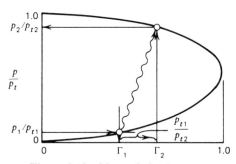

Figure 8.12 Normal shock process.

Thus, according to (8.68),

$$\frac{T_{t2}}{T_{t*}} = \left(\frac{T_{t2}}{T_{t1}}\right)\left(\frac{T_{t1}}{T_{t*}}\right) = 2.03038 \times 0.34434 = 0.69914.$$

On this same row in the table, we read:

$$R_2 = 0.84, \quad M_2 = 0.50536, \quad \Gamma_2 = 0.75203.$$

Hence the exit Mach number is 0.50536. The total pressure ratio is determined from (8.67) as

$$\frac{p_{t1}}{p_{t2}} = \frac{\Gamma_2}{\Gamma_1 \times (1/TTR)^{1/2}}$$

or

$$\frac{p_{t1}}{p_{t2}} = \frac{0.75203}{0.48938 \times (2.03038)^{1/2}} = 1.07845.$$

8.6.3 Generalized Flow Maps

Having presented the general compressible flow equation of (8.63), one would think that the goal of solving the general compressible pipe flow problem has been accomplished. However, while AR can be specified, TPR and TTR are usually the very unknowns we seek in a pipe flow problem. We ask: What is the pressure loss across this piping system, or what is the temperature rise across this piping component? Such questions require answers, and indicate that we are in need of more information before we really can apply (8.63).

One method for establishing TPR is to calculate or estimate the loss coefficient (K_f) for the process. This, of course, we have already discussed in great detail in this chapter. And a method for establishing TTR is to calculate or estimate the heat transfer coefficient (K_q), which will be discussed in this section.

The innovation introduced here is to exploit the interior of the envelope of Figure 8.6, through the use of empirical loss and heat transfer coefficients. When these empirical coefficients are plotted within the generalized envelope of Figure 8.6, we will have a *generalized flow map* for each of the common flow processes that have been discussed.

324 Flow of Real Gases in Pipes

Fanno Flow Map [14, 15] Briefly, the loss coefficient of (8.14) can be expressed in terms of R-Γ through the simple substitution of

$$\alpha = \Gamma \alpha^*, \tag{8.70}$$

where Γ is given by (8.64) and α^* by (8.16). K_f has the coordinates $R_2 = 1$ at $\Gamma_1 = 0$, and maximizes when R_{21}^* of (8.15) is used for limit 2 in (8.14). Also within the generalized envelope can be plotted TPR parameters, where TPR_Fanno has been given by (8.57). When R_1 is unity, we get the maximum R_{21} value for a given TPR as

$$(R_{21})_\text{max} = \frac{1}{TPR}. \tag{8.71}$$

This represents the $\Gamma_1 = 0$ intercept of any TPR parameter. The *minimum* R_{21} value for a given TPR is found to be

$$(R_{21})_\text{min} = \frac{1}{TPR}\left(\frac{2}{\gamma+1}\right)^{\gamma/(\gamma-1)} \tag{8.72}$$

based on $(\Gamma_1)_\text{max} = 1/TPR$, since then Γ_2 attains its maximum value of $\Gamma_2^* = 1$.

With K_f and TPR completely defined, the Fanno flow map of Figure 8.13 results. Briefly, one enters the outer envelope with R_1 to obtain Γ_1, which intercepts the chosen K_f parameter. At this intercept also can be located R_{21} and TPR, which in turn lead to Γ_2 and the complete solution. Note that, of the three parameters of interest, K_f and TPR are now found inside the envelope, while TTR is still found on the envelope, it being unity in this process. Such a Fanno solution is well adapted to flow in short ducts and piping networks, where the flow is essentially adiabatic.

Rayleigh Flow Map [15] The required TTR for Rayleigh flow can be obtained by means of the empirical heat transfer coefficient:

$$K_q = \frac{h_c L}{900 G c_p D}. \tag{8.73}$$

The basis of this equation is as follows. The convective heat transfer coefficient, h_c, is determined by usual means (see Figure 8.14, e.g.). The steady flow energy equation of (3.19) combines with the definition of total enthalpy for a perfect gas as

$$\delta Q = dh_t = c_p dT_t. \tag{8.74}$$

Generalized Compressible Pipe Flow

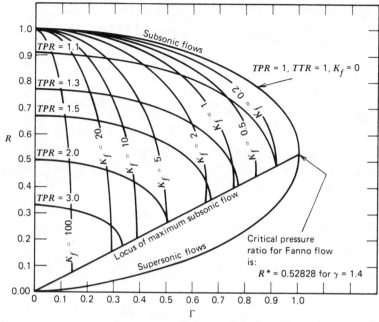

Figure 8.13 Generalized Fanno flow map.

Newton's law of convective heat transfer, modified for moving fluid effects, can be given for a pipe as

$$\dot{m}\,\delta Q = h_c(T_2 - T_t)\pi D\,dx, \tag{8.75}$$

where the adiabatic temperature could be used in place of T_t if the recovery factor (see Chapter 12) differed significantly from 1. Combining (8.74) and (8.75), we obtain

$$\frac{dT_t}{T_w - T_t} = \frac{h_c\,dx}{900\,Gc_p D} = dK_q. \tag{8.76}$$

This equation can be integrated directly if the flow element is of a length small enough so that T_w and h can be considered constants. Otherwise, a stepwise approach must be used. The result, simplified, gives the required TTR as

$$\frac{1}{TTR} = \frac{T_2}{T_{t1}}\left(1 - \frac{1}{e^{K_q}}\right) + \frac{1}{e^{K_q}}. \tag{8.77}$$

Thus TTR can be plotted as a simple plane in terms of T_w/T_{t1} and e^{K_q}, as shown in Figure 8.15.

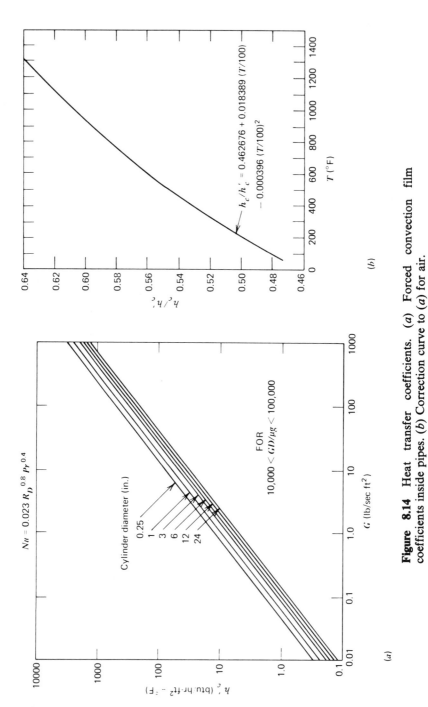

Figure 8.14 Heat transfer coefficients. (*a*) Forced convection film coefficients inside pipes. (*b*) Correction curve to (*a*) for air.

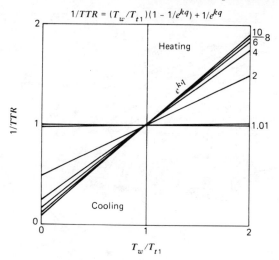

Figure 8.15 Total temperature ratio as a function of wall and fluid temperature and the heat transfer coefficient.

Also within the generalized envelope can be plotted *TPR* parameters, where TPR_{Rayleigh} can be given by

$$TPR_{\text{Ray}} = \frac{R_2[2\gamma/(\gamma-1)](R_2^{1/\gamma} - R_2)}{R_1 + [2\gamma/(\gamma-1)](R_1^{1/\gamma} - R_1)}. \tag{8.78}$$

With K_q, *TTR*, and *TPR* completely defined, the Rayleigh flow map of Figure 8.16 results. Briefly, one enters the outer envelope with R_1 to obtain Γ_1, which intercepts the chosen *TTR* parameter. At this intercept also can be located R_{21} and *TPR*, which in turn lead to Γ_2 and the complete solution. Note that, of the three parameters of interest, *TTR* and *TPR* are now found inside the envelope, while K_f is found on the envelope, it being zero in this process. Although a Rayleigh flow is difficult to achieve in practice because of the close relationships between frictional effects and heating effects, such solutions are well adapted to flow processes involving large external heat transfer, where the wall temperature differs greatly from the fluid temperature.

Isothermal Flow Map [15] The required *TPR* for isothermal flow already has been given by (8.58). The isothermal loss coefficient also has been given by (8.25). Both of these can be expressed in terms of R-Γ through the use of (8.70). For the heat transfer aspects of the isothermal problem, we again apply Figure 8.14 and (8.73) and (8.77), noting that K_f and K_q cannot both be specified unless flow rate is the unknown parameter.

328 Flow of Real Gases in Pipes

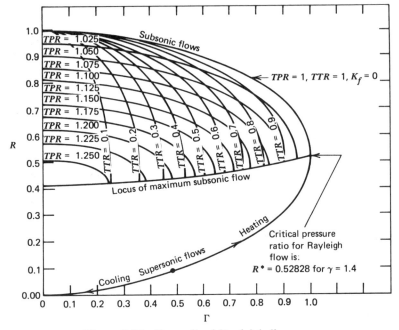

Figure 8.16 Generalized Rayleigh flow map.

With K_q and TTR, and K_f and TPR, complete defined, the isothermal flow map of Figure 8.17 results.

Of course, in the Fanno and isothermal processes the location of the K_f parameters is different, but the meaning of K_f is the same. Briefly, one enters the outer envelope of Figure 8.17 with R_1 to obtain Γ_1, which intercepts the chosen K_f parameter. At this intercept also can be located R_{21}, TPR, and TTR, which in turn lead to Γ_2 and the complete solution. Note that all three parameters of interest, that is, K_f, TPR, and TTR, are now found inside the envelope, and the circle that began with isentropic flow (with all three parameters on the envelope) is now complete. Such an isothermal solution is well adapted to flow in extremely long ducts, where there is sufficient area for heat transfer to keep the fluid temperature uniform.

Such generalized flow maps as those given by Figures 8.13, 8.16, and 8.17 are of value because, in addition to their utility in providing graphical solutions, they exhibit a certain beauty of symmetry and continuity.

Generalized skeleton tables for the Fanno, Rayleigh, and isothermal processes are presented in Tables 8.3, 8.4, and 8.5, respectively.

Generalized examples, based on the use of these flow maps and these tables, follow in the next section.

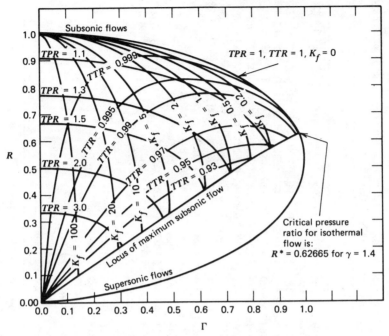

Figure 8.17 Generalized isothermal flow map.

8.6.4 Generalized Examples

Example 8.14: Isentropic Flow. In the subsonic expansion of air, find the mass flow rate per unit area at exit, and the inlet static pressure, if $\gamma = 1.4$, $\bar{R} = 53.35$ ft-lbf/lbm °R, $\alpha^* = 0.68473$, $D_1 = 3$ in., $D_2 = 1$ in., $T_t = 540°$R, $p_t = 20$ psia, and $p_2 = 15$ psia (see Figure 8.18).

Solution:

$$R_2 = \frac{p_2}{p_t} = \frac{15}{20} = 0.75,$$

$\Gamma_2 = 0.88378$ from Table 8.2,

$$\Gamma_1 = \frac{A_2}{A_1}\Gamma_2 = \left(\frac{D_2}{D_1}\right)^2 \Gamma_2 = \tfrac{1}{9} \times 0.88378 = 0.0982,$$

$$R_1 = \frac{p_1}{p_t} = 0.9977, \qquad p_1 = 19.954 \text{ psia},$$

$$\alpha_2 = \alpha^*\Gamma_2 = 0.68473 \times 0.88378 = 0.60515,$$

$$G_2 = \frac{\dot{m}}{A_2} = \frac{p_{t1}\alpha_2}{\sqrt{\bar{R}T_t/g}} = \frac{20 \times 144 \times 0.60515}{\sqrt{53.35 \times 540/32.174}} = 58.243 \text{ lbm/ft}^2 \text{ sec.}$$

Table 8.3 Generalized Fanno Flow Table ($\gamma = 1.4$)[a]

Press Ratio p/p_t	Mach Number, M	Temperature Ratio T/T_t	Flow Factor Γ_{isen}	K^*	Exit Static Pressure to Inlet Total Pressure (p_2/p_{t1}), and Inlet Total Pressure to Exit Total Pressure (p_{t1}/p_{t2}), versus Loss Coefficient (K or fL/D)								
					0.1	0.2	0.4	0.6	1	2	4	6	10	20
0.999	0.038	1.000	0.06528	493.5268	0.99890 1.00010	0.99880 1.00020	0.99860 1.00040	0.99840 1.00060	0.99800	0.99700	0.99499	0.99297	0.98893	0.97876
0.99	0.120	0.997	0.20543	45.48177	0.98898 1.00101	0.98796 1.00202	0.98592 1.00405	0.98388 1.00609	0.97978 1.01022	0.96944 1.02077	0.94843 1.04290	0.92691 1.06656	0.88225 1.11923	0.75857 1.29596
0.98	0.170	0.994	0.28894	21.07852	0.97793 1.00203	0.97585 1.00408	0.97169 1.00821	0.96750 1.01239	0.95908 1.02091	0.93765 1.04324	0.89313 1.09280	0.84606 1.15043	0.74204 1.30127	0.32715 2.57883
0.97	0.209	0.991	0.35193	13.09545	0.96684 1.00307	0.96366 1.00618	0.95728 1.01247	0.95085 1.01890	0.93784 1.03213	0.90439 1.06772	0.83293 1.15221	0.75362 1.26216	0.55333 1.64972	
0.96	0.242	0.988	0.40412	9.17812	0.95570 1.00413	0.95138 1.00832	0.94268 1.01686	0.93388 1.02564	0.91599 1.04393	0.86936 1.09461	0.76589 1.22532	0.64123 1.42584		
0.95	0.272	0.985	0.44929	6.87217	0.94452 1.00521	0.93901 1.01051	0.92787 1.02138	0.91656 1.03263	0.89344 1.05638	0.83216 1.12448	0.68841 1.31988	0.48060 1.72254		
0.94	0.299	0.982	0.48938	5.36456	0.93330 1.00631	0.92654 1.01274	0.91283 1.02603	0.89886 1.03990	0.87008 1.06958	0.79218 1.15807	0.59227 1.45313			
0.93	0.324	0.979	0.52555	4.30904	0.92202 1.00742	0.91395 1.01503	0.89753 1.03083	0.88072 1.04746	0.84577 1.08363	0.74853 1.19652	0.44508 1.68330			
0.92	0.347	0.976	0.55858	3.53348	0.91069 1.00856	0.90124 1.01737	0.88196 1.03578	0.86208 1.05537	0.82034 1.09866	0.66973 1.24162				
0.91	0.370	0.973	0.58898	2.94285	0.89929 1.00972	0.88840 1.01976	0.86606 1.04091	0.84288 1.06365	0.79354 1.11485	0.64307 1.29643				
0.90	0.391	0.970	0.61715	2.48050	0.88783 1.01090	0.87541 1.02221	0.84980 1.04622	0.82301 1.07234	0.76504 1.13244	0.57223 1.36726				
0.89	0.411	0.967	0.64338	2.11056	0.87629 1.01210	0.86226 1.02473	0.83313 1.05176	0.80234 1.08153	0.73438 1.15175	0.46172 1.47278				
					0.86468 1.01332	0.84893 1.02728	0.81598 1.05731	0.78076 1.09081	0.70081					

330

0.88	0.431	0.964	0.66790	1.80929	1.01333 0.85298	1.02731 0.83539	1.05752 0.79829	1.09126 0.75804	1.17323 0.66313
0.87	0.451	0.961	0.69088	1.56034	1.01458 0.84118	1.02997 0.82163	1.06354 0.77996	1.10162 0.73390	1.19757 0.61897
0.86	0.469	0.958	0.71249	1.35213	1.01586 0.82927	1.03270 0.80761	1.06984 0.76085	1.11275 0.70792	1.22604 0.56273
0.85	0.487	0.955	0.73284	1.17618	1.01717 0.81725	1.03552 0.79329	1.07648 0.74080	1.12480 0.67948	1.26096 0.46630
0.84	0.505	0.951	0.75203	1.02621	1.01851 0.80509	1.03843 0.77863	1.08350 0.71957	1.13801 0.64750	1.30964
0.83	0.523	0.948	0.77016	0.89741	1.01988 0.79278	1.04144 0.76354	1.09096 0.69683	1.15273 0.60983	
0.82	0.540	0.945	0.78729	0.78610	1.02128 0.78026	1.04458 0.74800	1.09897 0.67205	1.16961 0.56111	
0.81	0.557	0.942	0.80348	0.68937	1.02274 0.76757	1.04783 0.73188	1.10765 0.64437	1.18981 0.46509	
0.80	0.574	0.938	0.81880	0.60491	1.02423	1.05123	1.11718	1.21742	
0.79	0.590	0.935	0.83330	0.53086	0.75464 1.02576 0.74145	0.71507 1.05479 0.69738	0.61199 1.12796 0.57087		
0.78	0.607	0.931	0.84701	0.46572	1.02734 0.72793	10.5853 0.67855	1.14051 0.49854		
0.77	0.623	0.928	0.85997	0.40825	1.02897 0.71403	1.06251 0.65819	1.15680		
0.76	0.639	0.925	0.87222	0.35743	1.03066 0.69965	1.06676 0.63562			
0.75	0.654	0.921	0.88378	0.31239	1.03242 0.68468	1.07137 0.60951			
0.74	0.670	0.918	0.89469	0.27243	1.03425 0.66896	1.07648 0.57674			
0.73	0.686	0.914	0.90497	0.23692	1.03617 0.65222	1.08230 0.52125			

Table 8.3 Generalized Fanno Flow Table ($\gamma = 1.4$)[a] Continued

Press Ratio p/p_t	Mach Number, M	Temperature Ratio T/T_t	Flow Factor Γ_{isen}	K^*	Exit Static Pressure to Inlet Total Pressure (p_2/p_{t1}), and Inlet Total Pressure to Exit Total Pressure (p_{t1}/p_{t2}), versus Loss Coefficient (K or fL/D)							
					0.1	0.2	0.4	0.6	1	2	4	6	10	20
0.72	0.701	0.910	0.91464	0.20536	1.03820	1.08954								
					0.63409									
0.71	0.717	0.907	0.92371	0.17731	1.04035									
					0.61361									
0.70	0.732	0.903	0.93222	0.15237	1.04275									
					0.58939									
0.69	0.748	0.899	0.94016	0.13023	1.04536									
					0.55581									
0.68	0.763	0.896	0.94756	0.11059	1.04842									

[a]Under loss coefficient is tabulated p_2/p_{t1} as upper figure and p_{t1}/p_{t2} as lower figure. Example: At $p_1/p_{t1} = 0.80$, $M_1 = 0.574$, $T_1/T_t = 0.938$, $\Gamma_1 = 0.81880$, $K^* = 0.60491$, and at $K = 0.4$ $p_2/p_{t1} = 0.64437$ while $p_{t1}/p_{t2} = 1.11718$.

Table 8.4 Generalized Rayleigh Flow Table

					TPR/R_{21}						TTR/R_{21}			
R_1	Γ_1	1.05	1.10	1.15	1.20	1.25	0.9	0.8	0.7	0.6	0.5
1	0						1	1	1	1	1
0.9905	0.20028	0.90288	0.81010	0.71896	0.62464	0.51015	0.98833	0.98567	0.98219	0.97752	0.97092
0.9805	0.28538	0.89257	0.79862	0.70561	0.60746	0.47281	0.97603	0.97039	0.96306	0.95311	0.93887
0.9705	0.34909	0.98167	0.78638	0.69114	0.58795		0.96362	0.95491	0.94346	0.92781	0.90502
0.9605	0.40170	0.87070	0.77396	0.67617	0.56634		0.95113	0.93917	0.92336	0.90145	0.86895
0.9505	0.44716	0.85967	0.76134	0.66059	0.54143		0.93854	0.92317	0.90267	0.87385	0.82996
0.9405	0.48747	0.84857	0.74849	0.64426	0.50987		0.92585	0.90668	0.88129	0.84469	0.78699
0.9305	0.52382	0.83740	0.73538	0.62697			0.91305	0.89026	0.85913	0.81358	0.73817
0.9205	0.55699	0.82614	0.72197	0.60840			0.90011	0.87325	0.83601	0.77991	0.67958
0.9105	0.58751	0.81479	0.70820	0.58802			0.88705	0.85583	0.81173	0.74273	0.59942
0.9005	0.61579	0.80333	0.69401	0.56485			0.87383	0.83790	0.78600	0.70030	
0.8905	0.64211	0.79175	0.67931	0.53659			0.86045	0.81938	0.75843	0.64889	
0.8805	0.66671	0.78005	0.66399	0.49256			0.84687	0.80017	0.72836	0.57630	
0.8705	0.68977	0.76821	0.64790				0.83309	0.78013	0.69473		
0.8605	0.71144	0.75620	0.63077				0.81706	0.75905	0.65547		
0.8505	0.73185	0.74400	0.61225				0.80745	0.73668	0.60533		
0.8405	0.75110	0.73158	0.59165				0.79012	0.71259	0.51176		
0.8305	0.76928	0.71890	0.56759				0.77512	0.68612			
0.8205	0.78645	0.70593	0.53602				0.75967	0.65612			
0.8105	0.80269	0.69259					0.74371	0.62004			
0.8005	0.81806	0.67880					0.72708	0.56982			
0.7905	0.83259	0.66447					0.70964				
0.7805	0.84634	0.64943					0.69113				
0.7705	0.85934	0.63344					0.67121				
0.7605	0.87162	0.61611					0.64924				
0.7505	0.88322	0.59678					0.62408				
0.7405	0.89416	0.57379					0.59285				
		0.54135									

Table 8.5 Generalized Isothermal Flow Table

| R_1 | Γ_1 | TPR/R_{21} ||||| K_{iso}/R_{21} ||||| Γ_1 | TTR/R_{21} ||||
		0.999	0.99	0.97	0.95	0.2	1.0	1.5	10.0	20.0		1.1	1.3	1.5	2.0
1	0	0	0	0	0	1	1	1	1	1	0	0.90909	0.76923	0.66667	0.5
0.9905	0.20028	0.84700	0.45640	0.28179	0.21971	0.98857	0.98080	0.94095	0.88852	0.77243	0.20543	0.89806	0.75610	0.65140	0.47909
0.9805	0.28538	0.90323	0.58587	0.38422	0.30432	0.97646	0.96011	0.87335	0.74916	0.35253	0.28894	0.88696	0.74270	0.63552	0.45587
0.9705	0.34909	0.91806	0.65640	0.45101	0.36223	0.96427	0.93886	0.79776	0.55944		0.35193	0.87578	0.72897	0.61890	0.42909
0.9605	0.40170	0.92108	0.70005	0.49924	0.40599	0.95198	0.91697	0.70943			0.40412	0.86452	0.71488	0.60135	0.39588
0.9405	0.48747	0.91452	0.74786	0.56406	0.46854	0.92709	0.87074				0.48938	0.84172	0.68531	0.56231	
0.9205	0.55699	0.90141	0.76924	0.60399	0.51059	0.90169	0.82001				0.55858	0.81845	0.65319	0.51365	
0.9005	0.61579	0.88560	0.77720	0.62914	0.53965	0.87563	0.76208				0.61715	0.79461	0.61703	0.42681	
0.8805	0.66671	0.86841	0.77738	0.64452	0.55978	0.84871	0.69028				0.66789	0.77000	0.57321		
0.8605	0.71144	0.85043	0.77262	0.65306	0.57332	0.82062	0.57273				0.71249	0.74437	0.50545		
0.8405	0.75110	0.83195	0.76454	0.65659	0.58184	0.79086					0.75203	0.71729			
0.8205	0.78645	0.81313	0.75412	0.65634	0.58642	0.75853					0.78729	0.68800			
0.8005	0.81806	0.79407	0.74197	0.65314	0.58783	0.72176					0.81880	0.65489			
0.7805	0.84634	0.77485	0.72853	0.64762	0.58667	0.67525					0.84701	0.61347			
0.7605	0.87162	0.75549	0.71408	0.64022	0.58337										
0.7405	0.89416	0.73608	0.69883	0.63127											

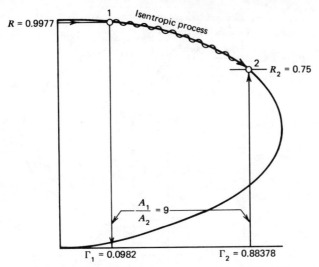

Figure 8.18 Isentropic process of Example 8.14.

Example 8.15: Fanno Flow. In the subsonic flow of air in a pipe of relative roughness, $e/D = 0.002$, find the exit static and total pressures if the pipe I.D. = 0.75 in., $T_t = 540°R$, $p_{t1} = 15$ psia, $G = 20$ lbm/ft^2 sec, and $L/D = 80$ (see Figure 8.19).

Solution: From Table 8.6 the viscosity of air at about 80°F is $\mu = 124.2 \times 10^{-7}$ lbm/ft sec; hence the Reynolds number is

$$R_D = \frac{GD}{\mu} = \frac{20 \times 0.75}{12 \times 124.2 \times 10^{-7}} = 1 \times 10^5.$$

For the given roughness, at this R_D, $f \simeq 0.025$, and the adiabatic loss coefficient

$$K_f = 0.025 \times 80 = 2.$$

The flow number

$$\alpha = \frac{20}{15 \times 144} \sqrt{\frac{53.35 \times 540}{32.174}} = 0.2771,$$

and $\Gamma_1 = 0.2771/0.68473 = 0.4047$. From Table 8.2 or 8.3, $R_1 \simeq 0.96$. From Table 8.3, at $R_1 = 0.96$, and under $K = 2$, we find $R_{21} = 0.86936$ and $TPR = 1.09461$. Also, $\Gamma_2 = \Gamma_1 \times TPR = 0.4047 \times 1.09461 = 0.4430$, and $R_2 = 0.9516$. It follows that $p_{t2} = p_{t1}/TPR = 13.7035$ psia and $p_2 = R_{21} \times p_{t1} = 13.0404$ psia.

336 Flow of Real Gases in Pipes

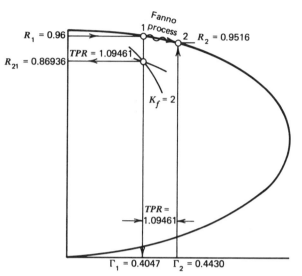

Figure 8.19 Fanno process Example 8.15.

Example 8.16: Rayleigh Flow. In the subsonic flow of air, heating, with T_{wall} varying linearly with length from 2000 to 3000°R, find all the pertinent conditions for a smooth 1 in. pipe, 36 in. long, at $G=10$ lbm/ft² sec, $R_1 = 0.931$, and $T_{t1} = 625°$R (see Figure 8.20).

Solution: (in One Step).

$$\bar{T}_w = \frac{T_{w1} + T_{w2}}{2} = \frac{2000 + 3000}{2} = 2500°\text{R},$$

$$\frac{\bar{T}_w}{T_{t1}} = \frac{2500}{625} = 4, \qquad \frac{T_1}{T_{t1}} = 0.98, \qquad T_1 = 152°\text{F}.$$

Also, $h'_c = f(G, D) = 55$ for an assumed $\bar{T} = 475°$F, and $h_c/h'_c = 0.546$. Thus

$$K_q = \frac{55 \times 0.546 \times 3 \times 12}{900 \times 10 \times 0.24} = 0.5, \qquad e^{K_q} = 1.65,$$

$$\frac{1}{TTR} = 4\left(1 - \frac{1}{1.65}\right) + \frac{1}{1.65} = 2.18, \qquad TTR = 0.4587, \qquad T_{t2} = 1362°\text{R},$$

$$\Gamma_1 = 0.52209, \qquad \left(\frac{T_{t1}}{T_{t*}}\right)_{\text{Ray}} = 0.386, \qquad \left(\frac{T_{t2}}{T_{t*}}\right)_{\text{Ray}} = \frac{0.386}{0.4587} = 0.842.$$

$$\Gamma_2 = 0.86007, \qquad R_2 = 0.76992, \qquad \frac{T_2}{T_{t2}} = 0.928,$$

$$T_2 = 1264°\text{R} = 805°\text{F}$$

Generalized Compressible Pipe Flow 337

Table 8.6 Some Properties of Air

	Air at Atmospheric Pressure		
T (°F)	ρ (lbm/ft^3)	μ (lbm/ft sec $\times 10^5$)	ν (ft^2/sec $\times 10^5$)
0	0.0862	1.086	12.6
20	0.0827	1.125	13.6
40	0.0794	1.159	14.6
60	0.0763	1.206	15.8
80	0.0735	1.242	16.9
100	0.0709	1.276	18.0
120	0.0684	1.293	18.9

Therefore

$$\bar{T} = \frac{152 + 804}{2} \simeq 475°F,$$

$$TPR = \frac{\Gamma_2(TTR)^{1/2}}{\Gamma_1} = 1.1148,$$

$$R_{21} = \frac{R_2}{TPR} = 0.69062.$$

Note that these values are consistent with those given in Table 8.4. Note

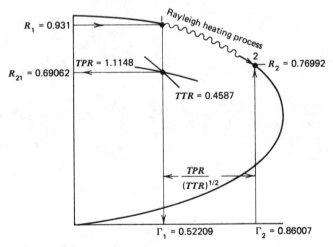

Figure 8.20 Rayleigh heating process of Example 8.16.

338 Flow of Real Gases in Pipes

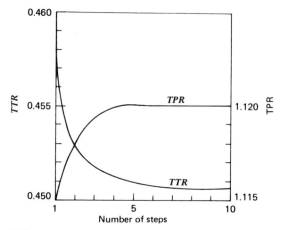

Figure 8.21 Effect of number of steps used in the Rayleigh solution of Example 8.16.

also that, for a heating process,

$$\frac{T_{t1}}{\overline{T}_w} < TTR < 1, \quad \text{that is, } 0.25 < 0.459 < 1.$$

Note further that we did only one step toward a solution. Two steps would lead to $TPR = 1.118$, and Figure 8.21 indicates the general effect of the number of steps on the Rayleigh flow solution, and indicates, in particular, that for this example five steps would be adequate.

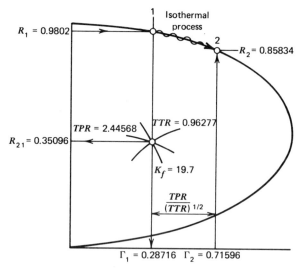

Figure 8.22 Isothermal process of Example 8.17.

Generalized Compressible Pipe Flow 339

Example 8.17: Isothermal Flow. In the subsonic flow of air in a buried pipe, find the earth temperature and other pertinent conditions if $D = 1$ in., $L = 800$ in., $G = 20$ lbm/sec ft^2, $p_{t1} = 20.5$ psia, $T_{t1} = 508°$R, and $e/D = 0.002$ (see Figure 8.22).

Solution:

$$\alpha = \frac{20}{20.5 \times 144} \sqrt{\frac{53.35 \times 508}{32.174}} = 0.19663,$$

$$\Gamma_1 = \frac{0.19663}{0.68473} = 0.28716, \quad R_1 = 0.9802,$$

$$\frac{T_1}{T_{t1}} = 0.9943, \quad T_1 = 45.1°F, \quad \mu = 117 \times 10^{-7} \text{ lbm/ft sec},$$

$$h'_c = f(G, D) = 100, \quad \frac{h_c}{h'_c} = 0.472, \quad h_c = 47.2,$$

$$K_q = \frac{47.2 \times 800}{900 \times 20 \times 0.24} = 8.74074, \quad e^{K_q} = 6.25252 \times 10^3,$$

$$R_D = \frac{GD}{\mu} = \frac{20 \times \frac{1}{12}}{117 \times 10^{-7}} = 1.42 \times 10^5, \quad f = 0.0246,$$

$$K_f = f\frac{L}{D} = 0.0246 \times \frac{800}{1} = 19.7.$$

From Table 8.5, or by iteration according to Section 8.5.2, at R_1 and K_{iso}, we read $R_{21} = 0.35096$ and $TPR = 2.44568$. Hence

$$R_2 = \frac{R_{21}}{TPR} = 0.85834 \quad \text{and} \quad \Gamma_2 = 0.71596.$$

Thus

$$TTR = \left(\frac{TPR \times \Gamma_1}{\Gamma_2}\right)^2 = 0.96277$$

and, finally,

$$T_w = \frac{T_{t1}(1/TTR - 1/e^{K_q})}{1 - (1/e^{K_q})} = 527.648°R = 67.648°F.$$

This pipe wall temperature must be maintained to achieve isothermal flow under the given conditions.

340 Flow of Real Gases in Pipes

Example 8.18. Redo Example 3.1, using the generalized approach, given $p_1 = 20$ psia, $p_{t1} = 20.043784$ psia, and $A_1/A_2 = 4$. The isentropic static pressure drop is required.

Solution:

$$R_1 = \frac{p_1}{p_t} = 0.997816.$$

By Table 8.2 or (8.64), $\Gamma_1 = 0.09627$. By (8.65), $\Gamma_2 = 0.38508$ and $R_2 = 0.963835$. But

$$p_1 - p_2 = p_{t1}(R_1 - R_2) = 0.681 \text{ psi},$$

which checks the result of Example 3.1 very closely.

REFERENCES

1. R. P. Benedict, "Some comparisons between compressible and incompressible treatments of compressible fluids," *Trans. ASME, J. Basic Eng.*, September 1964, p. 527.
2. R. P. Benedict and N. A. Carlucci, "Flow with losses," *Trans. ASME, J. Eng. Power*, January 1965, p. 37.
3. A. H. Shapiro, *The Dynamics and Thermodynamics of Compressible Fluid Flow*, Vol. 1, Ronald Press, 1953, p. 167.
4. J. K. Vennard and R. L. Street, *Elementary Fluid Mechanics* 5th ed, John Wiley, 1975, p. 413.
5. R. M. Rotty, *Introduction to Gas Dynamics*, John Wiley, 1962, p. 110.
6. R. P. Benedict, "On the determination and combination of loss coefficients for compressible fluid flows," *Trans. ASME, J. Eng. Power*, January 1966, p. 67.
7. R. P. Benedict, J. S. Wyler, J. A. Dudek, and A. R. Gleed, "Generalized flow across an abrupt enlargement," *Trans. ASME, J. Eng. Power*, Vol. 98, No. 3, July 1976, p. 377.
8. A. R. Shouman and J. L. Massey, Jr., "Stagnation pressure losses of compressible fluids through abrupt area changes neglecting friction at the walls," *ASME Pap.* 68-WA/FE-46.
9. R. P. Benedict, A. R. Gleed, and R. D. Schulte, "Air and water studies on a diffuser-modified flow nozzle," *Trans. ASME, J. Fluids Eng.*, June 1973, p. 169.
10. W. B. Hall and E. M. Orme, "Flow of a compressible fluid through a sudden enlargement in a pipe," *Proc. Inst. Mech. Eng.*, Vol. 169, No. 49, 1955.

11. L. Crocco, "One-dimensional treatment of steady gas dynamics," Section B of *Fundamentals of Gas Dynamics*, H. W. Emmons, Ed., Vol. III, *High Speed Aerodynamics and Jet Propulsion*, Princeton University Press, 1958, p. 292.
12. D. W. Roberts and J. P. Johnston, "Development of a new internal flow aeroacoustic facility," *Rep*. PD-18, Thermosciences Division, Stanford University, September 1974.
13. R. P. Benedict and W. G. Steltz, "A generalized approach to one-dimensional gas dynamics," *Trans. ASME, J. Eng. Power*, June 1962, p. 49; see also *Handbook of Generalized Gas Dynamics*, Plenum Press, 1966.
14. R. P. Benedict and N. A. Carlucci, "Flow with losses," *Trans. ASME, J. Eng. Power*, January 1965, p. 37; see also, *Handbook of Specific Losses in Flow Systems*, Plenum Press, 1966.
15. R. P. Benedict, "Some generalizations in compressible flow characteristics," *Trans. ASME, J. Eng. Power*, April 1973, p. 65.

NOMENCLATURE

Roman

a, b	coefficients
A	area
AR	area ratio
c_p	specific heat capacity
D	pipe diameter
E	error function
f	friction factor
F	loss, function
G	flow rate per unit area
g_c	gravitational constant
h	enthalpy
h_c	convective film coefficient
h_{loss}	head loss
K	coefficient
K_f	loss coefficient
K_q	heat transfer coefficient
K.E.	kinetic energy
L	length
$\dot{m}$	mass flow rate

- M Mach number
- p pressure
- P pressure function
- Q heat
- R static/total pressure ratio
- $\bar{R}$ specific gas constant
- T absolute temperature
- V volumetric average velocity
- W work
- x length coordinate

Greek

- α flow number
- Δ finite difference
- γ specific heat ratio
- Γ compressible flow function
- ρ fluid density
- φ area ratio

Subscripts

- $0, 1, 2, N$ axial locations
- adi   adiabatic
- iso   isothermal
- s static
- t total
- w wall
- $*$ critical

9

Flow of Liquid-Vapor Mixtures in Pipes

...when the pressure of the blood-columns in the outgoing arteries mounts above that in the relaxing ventricles, the valves that had swung open, bang shut... —*Gustav Eckstein*

9.1 GENERAL REMARKS

Whenever water flows at or near the saturated liquid condition, there is always the possibility of a *two-phase flashing flow* of water and steam somewhere in the piping system. This means that some portion of the flowing water may evaporate (i.e., flash to steam) upon expansion to a lower pressure.

The actual flashing process has been shown to lie somewhere between a constant enthalpy and a constant entropy process. Thus there will be some increase in entropy and at the same time some decrease in enthalpy in the actual flashing process. Both of these limiting processes have been analyzed in the literature [1–3], and the use of either has been shown to yield approximately the same maximum flow rate. For simplicity we will consider here that the vapor portion of the flashing mixture follows an *isentropic* process such as is shown in Figure 9.1.

As a model for analysis, let us consider a typical low pressure piping system, such as a drain from a turbine heater (A) to a condenser (B), as shown in Figure 9.2. Naturally, pressure drop, pipe sizing calculations, and flow rate determinations will be more difficult for such a piping system if flashing occurs. We determine whether a flashing situation obtains by

344 Flow of Liquid-Vapor Mixtures in Pipes

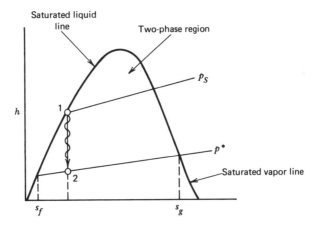

Figure 9.1 Mollier diagram for an isentropic flashing process.

comparing the system back pressure (p_B), at the piping exit, with the saturation pressure (p_S) corresponding to the water temperature at inlet.

9.2 LIQUID FLOW SITUATION

If $p_B > p_S$, there will be *liquid* flow throughout the piping system, and the usual methods of incompressible flow analysis apply. Specifically, by applying the energy balance of (7.4) between stations 1 and B, we obtain

$$G^2 = \frac{2g_c \rho_A (p_1 - p_B)144 + 2g\rho_A^2(Z_1 - Z_2)}{1 + K_o}, \tag{9.1}$$

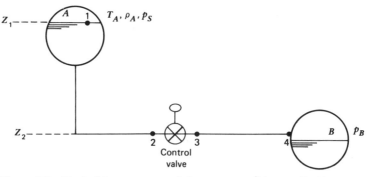

Figure 9.2 Typical low pressure piping system subject to flashing flow.

where G = flow rate per area (lbm/ft^2 sec),
  ρ = water density (lbm/ft^3),
  p = absolute pressure (lbf/in.^{2}),
  Z = elevation (ft),
  K_o = overall loss coefficient of system, a pure number.

9.3 FLASHING FLOW SITUATION

Conversely, if $p_B < p_S$, we must expect two-phase flashing flow somewhere in the piping system.

When the flashing situation obtains, a unique critical pressure (p^*) becomes important. This critical pressure is a function of the given G and p_S, and signifies the pressure at which the flow just goes sonic at the piping exit. Thus the considerably increased specific volume of the vapor over that of the liquid may actually limit the flow rate that can be passed by the piping system.

Example 9.1. Water is collected before the last stage of a certain turbine where the pressure is 1.8 psia and the specific volume of the water is 0.016 ft^3/lbm. If this water expands through a drain hole to a condenser pressure of 0.5 psia, what is the ratio of the volumes of the fluid after and before the expansion, using the 1967 ASME steam tables?

Solution: A constant entropy process between 1.8 psia and 0.5 psia yields a quality (x) of the mixture of 0.039 [where $x = (S_1 - S_{f2})/S_{fg2}$ according to the notation of Figure 9.1], which indicates that 3.9% of the water flashes into steam. But steam at 0.5 psia has a specific volume of 641.5 ft^3/lbm, and 3.9% of this is 25 ft^3/lbm. Since the water volume is negligible by comparison, we note that the after and before volumes stand in the ratio $25/0.016 \simeq 1600$.

Clearly, the flow that can be passed by a given piping system may be limited since the volume of the fluid can easily increase some 1600 times. Whether this *choked* flow situation is actually encountered depends on the relation between the critical pressure (p^*) and the system back pressure (p_B).

1. If $p_B > p^*$, the exit flow is subsonic, and the piping pressure at exit (p_4) equals the back pressure. In such cases the critical state is imaginary and lies beyond the piping exit.

346 Flow of Liquid-Vapor Mixtures in Pipes

2. If $p_B < p^*$, the flow is choked, and the piping pressure at exit equals the critical pressure. In such cases the critical state is real and is encountered at the piping exit.

Thus, in a horizontal pipe, for example, the static pressure decreases with losses until it reaches the saturation pressure (p_S), at which point flashing occurs. After flashing, the pipe pressure continues to decrease from p_S, because of losses, until it reaches either p_B (for a subsonic exit flow) or p^* (for a sonic exit flow), whichever pressure is greater.

9.3.1 Maximum Flashing Flow

Since, in the presence of flashing, the exit pressure cannot be less than p^*, it follows that the maximum flow per unit area (G_o) occurs when $p^* = p_S$. Under this condition, liquid flow will just persist throughout the piping, no matter how low is the back pressure.

The pressure drop that attends this maximum flow (G_o) is $p_1 - p_S$. If the flow is less than G_o, $p^* < p_S$, which means that the lower flow is accompanied by more pressure drop (namely, $p_1 - p^*$) whenever $p_B < p_S$.

A thermodynamic expression for G_o has been given [4] as

$$G_o = \left[\frac{144 \alpha g_c}{(d\beta/dp) + v_f(d\alpha/dp) - 144/J} \right]^{1/2}_{p_S}, \qquad (9.2)$$

Table 9.1 Maximum Flashing Flow Rate per Unit Area for Water, as a Function of Saturation Pressure at Inlet

p_S (psia)	G_o (lbm/ft² sec)
1	20.759
2	38.482
3	55.088
4	70.979
5	86.341
10	158.047
20	287.363
30	406.246
40	518.481
50	625.844
100	1116.237
250	2361.329

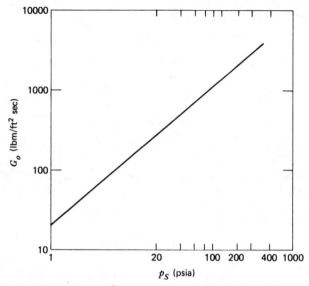

Figure 9.3 Maximum flashing flow rate per unit area for water, as a function of saturation pressure.

where $\alpha = h_{fg}/v_{fg}$, $\beta = h_f - \alpha v_f$, J is the mechanical equivalent of heat, and the gradients in α and β are evaluated by finite differences about p_S. Equation 9.2 has been solved for G_o as a function of p_S, and is given in Table 9.1 and graphed in Figure 9.3. An empirical expression for G_o, developed from the curve of Figure 9.3, is

$$G_o = 21.19 p_S^{0.861}. \tag{9.3}$$

This is good to about $\pm 3\%$. Note that G_o represents the maximum flow rate per unit area under the flashing flow situation, and can serve as a useful reference quantity for two-phase flow.

9.3.2 Pipe Critical Pressure

The critical pressure that is consistent with the given inlet saturation pressure and the corresponding flow rate per unit area can be determined by constructing a graph relating these parameters, somewhat as follows [2]:

1. A p^* is selected.
2. According to Figure 9.1, when $p_1 = p_S$ and $p_2 = p^*$, the quality can be

348 Flow of Liquid-Vapor Mixtures in Pipes

obtained from

$$x^* = \frac{S_s - S_f^*}{S_{fg}^*}.\tag{9.4}$$

3. From x^* the density at this critical state can be obtained from

$$\rho^* = \frac{1}{v_f^* + x^* v_{fg}^*}.\tag{9.5}$$

4. By similar calculations x and ρ are determined at two other pressures, usually one on either side of the specified p^*. This allows determination of the critical velocity according to

$$V^* = \sqrt{g_c(\Delta p / \Delta \rho)_s}.\tag{9.6}$$

Example 9.2. At $p^* = 10$ psia and $p_S = 100$ psia, find G, using the 1967 ASME steam tables.

Solution: By (9.4):

$$x^* = \frac{0.4743 - 0.2836}{1.5043} = 0.12677.$$

By (9.5):

$$\rho^* = \frac{1}{0.016592 + 0.12677 \times 38.404} 0.204705 \text{ lbm/ft}^3.$$

Similarly,

$$x_{10.5 \text{ psia}} = 0.125 \quad \text{and} \quad \rho_{10.5} = 0.217262,$$

and

$$x_{9.5 \text{ psia}} = 0.128563 \quad \text{and} \quad \rho_{9.5} = 0.192427.$$

Hence

$$\Delta p = 10.5 - 9.5 = 1 \text{ psi},$$
$$\Delta \rho = 0.21726 - 0.19243 = 0.02483,$$
$$V^* = \sqrt{\frac{32.174 \times 144 \times 1}{0.02483}} = 431.962 \text{ ft/sec},$$

and

$$G = \rho^* V^* = 0.204705 \times 431.962 = 88.425 \text{ lbm/ft}^2 \text{ sec.}$$

In Figure 9.4 p^* is given in terms of G and p_S; one point on these curves is at the intersection of $p_S = 100$ psia and $G = 88.425$ lbm/ft² sec, where we find $p^* = 10$ psia. Note that, for increased accuracy, we have constructed Figure 9.4 using $\Delta p = 0.01$ psi in (9.6) rather than the $\Delta p = 1$ psi of Example 9.2.

Thus we can obtain the critical pressure (p^*) at the piping exit graphically by entering Figure 9.4 with the given flow rate per unit area (G) and the given saturation pressure (p_S).

In an attempt to generalize Figure 9.4, we have made use of the dimensionless quantities p^*/p_S and G/G_o. The result is Figure 9.5, which

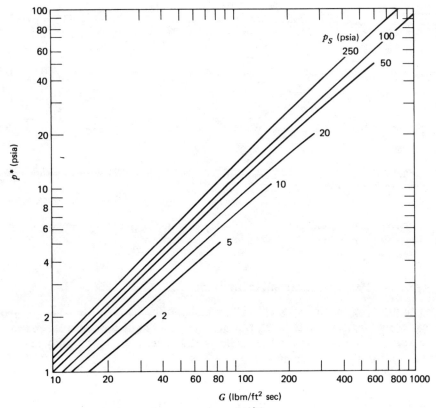

Figure 9.4 Flow rate per unit area versus critical pressure, with parameters of saturation pressure.

350 Flow of Liquid-Vapor Mixtures in Pipes

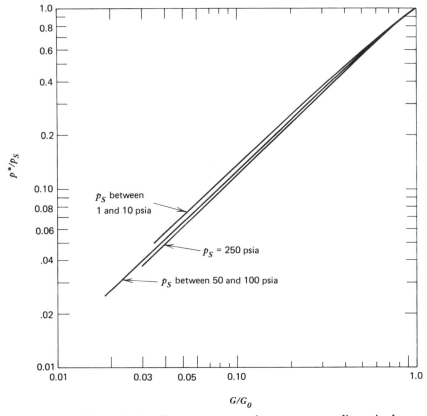

Figure 9.5 Dimensionless flow rate per unit area versus dimensionless critical pressure.

indicates an essential collapse of the p_S parameter to a single curve for the range of $1 \leqslant p_S \leqslant 10$.

9.3.3 Pseudo-Isentropic Exponent

A pseudo-isentropic exponent (γ^*), which can be used in the conventional gas dynamics equations to approximate certain characteristics of a flashing mixture, has been defined as follows [2]. The velocity of (9.6) can be written for an ideal gas as

$$V^* = \sqrt{\frac{\gamma^* g_c p^*}{\rho^*}} = \frac{G}{\rho^*}, \qquad (9.7)$$

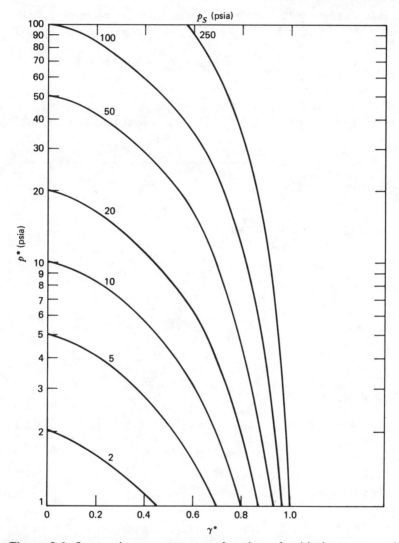

Figure 9.6 Isentropic exponent as a function of critical pressure, with parameters of saturation pressure.

and solving for γ^* yields

$$\gamma^* = \frac{G^2}{144 g_c p^* \rho^*} \qquad (9.8)$$

for conventional units.

Example 9.3. At $p^* = 10$ psia and $p_S = 100$ psia, find the pseudo-isentropic exponent, γ^*.

352 Flow of Liquid-Vapor Mixtures in Pipes

Solution: According to Example 9.2, the flow rate per area that is consistent with the given p^* and p_S is 88.425 lbm/ft² sec. Hence, by (9.8),

$$\gamma^* = \frac{88.425^2}{144 \times 32.174 \times 10 \times 0.204705} = 0.82443.$$

A graph that relates this exponent to the critical pressure, for parameters of p_S, is given in Figure 9.6, where one point on these curves is at the intersection of $p^* = 10$ psia and $p_S = 100$ psia, where we find $\gamma^* = 0.82443$.

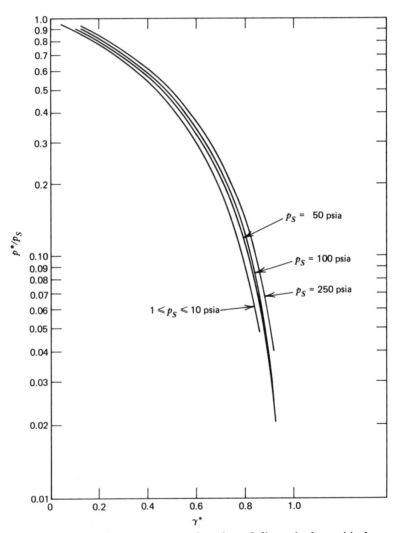

Figure 9.7 Isentropic exponent as a function of dimensionless critical pressure.

In an attempt to generalize Figure 9.6, we have made use of the dimensionless quantity p^*/p_S. The result is Figure 9.7, which indicates an essential collapse of the p_S parameter to a single curve for the range of $1 \leq p_S \leq 10$.

9.4 THE CONTROL VALVE

Often the piping system will include a control valve, as shown in Figure 9.2. It is good practice to avoid flashing before the control valve; that is, p_2 of Figure 9.2 should not be less than p_S if the valve is to remain in control. If this is a problem, an increased static head $(Z_1 - Z_2)$ can be used to ensure control, as can be seen by applying the Bernoulli equation between 1 and 2, that is,

$$\left(\frac{p_1}{\rho_A} + \frac{g}{g_c} Z_1\right) - \frac{\Delta p_i}{\rho_A} = \left(\frac{p_S}{\rho_A} + \frac{V_2^2}{2g_c} + \frac{g}{g_c} Z_2\right),$$

where $\Delta p_i/\rho_A$ represents the head loss between 1 and 2. Solving for ΔZ, we have

$$\frac{g}{g_c}(Z_1 - Z_2) = \frac{1}{\rho_A}[\Delta p_i - (p_1 - p_S)] + \frac{V_2^2}{2g_c}. \tag{9.9}$$

Thus the static head $g/g_c(Z_1 - Z_2)$ preceding the control valve must exceed $1/\rho_A[\Delta p_i - (p_1 - p_S)]$ if flashing is to be avoided before the control valve. Note that Δp_i here represents the pressure loss up to the valve inlet.

Even if it is assured that the fluid entering the valve is liquid, by maintaining a sufficient static head, by subcooling, or by both means, there is still another condition that will cause the valve to lose control. The pressure downstream of the valve, at 3 (in Figure 9.2), should not be allowed to rise above the valve critical pressure (p_{cv}), where p_{cv} is also called the valve vena contracta pressure. In other words, for the valve to remain in control, p_{cv} must exceed p_3. To check this condition, it is necessary to be able to calculate both p_{cv} and p_3.

9.4.1 Valve Critical Pressure

If the fluid entering the control valve at 2 is saturated liquid, critical flow will exist at the valve orifice. An approximate liquid critical pressure ratio factor (F_F) has been given [5] for water as

$$F_F = 0.96 - 0.28\sqrt{\frac{p_S}{3206}} \tag{9.10}$$

354 Flow of Liquid-Vapor Mixtures in Pipes

to be used in the relation

$$p_{cv} = F_F \times p_S. \qquad (9.11)$$

Example 9.4. At $T_A = 199.89°F$, find the valve critical pressure.

Solution: By steam tables: $p_S = 11.5$ psia.
 By (9.10): $F_F = 0.943$.
 By (9.11): $p_{cv} = 10.84$ psia.

This result indicates that the pipe downstream of the control valve must be sized so that p_3 is less than 10.84 psia.

9.5 DETERMINATION OF PRESSURE DOWNSTREAM OF VALVE

The pressure (p_3) just downstream of the valve can be determined as indicated by the following method, dependent greatly on whether or not the flow is choked at the piping exit.

9.5.1 Choked Flow

For choked flashing flow at the piping exit (i.e., when $p_4 = p^* > p_B$), the actual piping loss coefficient between any point x and the exit point 4 (i.e., $K_{x,4}$) equals the critical loss coefficient ($K_{x,*}$) between these same points. For two-phase flashing flow, the pseudo-isentropic exponent (γ^*) of (9.8) and Figures 9.6 and 9.7 is used in the conventional gas dynamics equation [see (8.18)]:

$$K_{x,*} = \frac{1}{\gamma^*}\left(\frac{1}{M_x^2} - 1\right) + \left(\frac{\gamma^* + 1}{2\gamma^*}\right)\ln\left[\frac{(\gamma^* + 1)M_x^2}{2 + (\gamma^* - 1)M_x^2}\right]. \qquad (9.12)$$

For a straight horizontal pipe between x and 4, this same loss coefficient is given by

$$K_{x,*} = f\frac{L_{x,4}}{D}. \qquad (9.13)$$

Thus, for given values of $L_{x,4}$, D, and f [as by the Colebrook equation of (6.19)], we can obtain $K_{x,*}$ via (9.13).

Determination of Pressure Downstream of Valve

To provide a graphical solution for M_x, we can construct a suitable general graph by first selecting a value for M_x. Then, for given values of γ^*, we can solve for $K_{x,*}$ by (9.12).

Example 9.5. At $M_3 = 0.5$ and $\gamma^* = 0.2$, find $K_{3,*}$.

Solution: By (9.12):

$$K_{3,*} = \frac{1}{0.2}\left(\frac{1}{0.5^2} - 1\right) + \left(\frac{0.2+1}{2\times 0.2}\right)\ln\left[\frac{(0.2+1)0.5^2}{2+(0.2-1)0.5^2}\right]$$

$$= 15 + 3\ln\left(\frac{1.2\times 0.25}{2 - 0.8\times 0.25}\right) = 9.62.$$

In Figure 9.8, $K_{x,*}$ is given in terms of M_x and γ^*, where one point on these curves is at the intersection of $M_3 = 0.5$ and $\gamma^* = 0.2$, where we find $K_{3,*} = 9.62$.

To obtain p_3, the conventional gas dynamics relation [6] between the pressure at any point and the critical pressure is used with γ^*, that is,

$$\frac{p_x}{p^*} = \frac{1}{M_x}\sqrt{\frac{\gamma^*+1}{2+(\gamma^*-1)M_x^2}}. \tag{9.14}$$

Example 9.6. Find p_3 for the conditions of Example 9.5 if $p^* = 10$ psia.

Solution: By (9.14):

$$\frac{p_3}{p^*} = \frac{1}{0.5}\sqrt{\frac{0.2+1}{2+(0.2-1)0.25}} = 1.633,$$

and $p_3 = 1.633 \times 10 = 16.33$ psia.

Equation 9.14 is graphed in Figure 9.9, the parameter being γ^*.

9.5.2 Subsonic Flow

For subsonic flow at the piping exit (i.e., when $p_4 = p_B > p^*$), the actual loss coefficient between any point x and the exit point (i.e., $K_{x,4}$) is less than the critical loss coefficient of (9.12) by the amount $K_{4,*}$. In other words, according to Figure 8.5,

$$K_{x,*} = K_{x,4} + K_{4,*}, \tag{9.15}$$

where $K_{x,*}$ and $K_{4,*}$ are given by (9.12), while $K_{x,4}$ is given by (9.13).

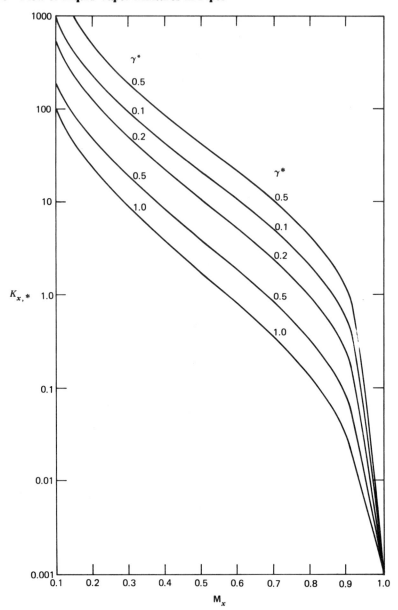

Figure 9.8 Mach number at any point in piping as a function of loss coefficient, with parameters of the pseudo-isentropic exponent.

Determination of Pressure Downstream of Valve 357

Example 9.7. Find p_3 if $\gamma^* = 0.36$, $K_{3,4} = 2.52$, $p^* = 6.2$ psia, and $p_B = 8$ psia.

Solution: Since $p^* < p_B$, we have subsonic flow at exit.
By (9.14): At $p_4/p^* = 8/6.2 = 1.291$, $M_4 = 0.69$.
By (9.12): At $\gamma^* = 0.36$ and $M_4 = 0.69$, $K_{4,*} = 1.25$.
By (9.15): $K_{3,*} = 2.52 + 1.25 = 3.77$.
By (9.12): At $K_{3,*} = 3.77$ and $\gamma^* = 0.36$, $M_3 = 0.55$.
By (9.14): At $M_3 = 0.55$ and $\gamma^* = 0.36$, $p_3/p^* = 1.55$.
Thus

$$p_3 = 1.55 \times 6.2 = 9.61 \text{ psia.}$$

Example 9.8. Find whether the valve controls in a situation like the one in Figure 9.2 if $T_A = 162.24°F$, $\dot{m} = 6480$ lbm/hr, $p_1 = p_S$, $f = 0.015$, $L_{3,4} = 31.1$ ft, $D = 4$ in., $g = g_c$, $Z_1 - Z_2 = 8$ ft, $\Delta p_i = 3$ psi, and p_B varies from 1 to 2 psia.

Solution:

1. By steam tables, $p_S = 5$ psia and $\rho_A = 1/0.016407 = 60.95$ lbm/ft^3.
2. Since $p_B < p_S$, there will be flashing in this system.
3. $G = \dot{m}/A$, $A = \pi D^2/4 = 0.0872665$ ft^2, and $G = 6480/(3600 \times 0.0872665) = 20.626$ lbm/ft^2 sec.
4. By Figure 9.3: $G_o = 86$ lbm/ft^2 sec. [By (9.2) we get $G_o = 86.341$, while by (9.3) we have 84.712, the latter being off about 1.9%.]
5. $G/G_o = 20.626/86 = 0.24$.
6. By Figure 9.5: $p^*/p_S = 0.3$, and $p^* = 1.5$ psia.
7. By Figure 9.7: $\gamma^* = 0.6$.
8. We consider the control valve: p_2 must be less than p_S, for control. By (9.9):

$$8 > \tfrac{1}{61}(3 \times 144) = 7.08.$$

We conclude that there will be liquid flow up to the control valve inlet.
By (9.10) and (9.11): $p_{cv} = 4.745$ psia.

9. To determine p_3, we must first see whether the flow is subsonic or choked at 4.

At $p_B = 1$ psia, $p_B < p^*$, and the flow is choked.
At $p_B = 2$ psia, $p_B > p^*$, and the flow is subsonic.

Both of these cases must be considered.

358 Flow of Liquid-Vapor Mixtures in Pipes

10. *For choked flow*: $p_4 = p^* = 1.5$ psia.
    By (9.13): $K_{3,4} = K_{3,*} = 0.015 \times 31.1/(4/12) = 1.4$.
    By Figure 9.8: At $K_{3,*} = 1.4$ and $\gamma^* = 0.6$, $M_3 = 0.6$.
    By Figure 9.9: At $M_3 = 0.6$ and $\gamma^* = 0.6$, $p_3/p^* = 1.55$, and $p_3 = 1.55 \times 1.5 = 2.325$ psia.
    Since $p_3 < p_{cv}$, we conclude that the valve will control in this choked flow condition.

11. *For subsonic flow*: $p_4 = p_B = 2$ psia.
    By (9.13): $K_{3,4} = 1.4$.
    $p_4/p^* = 2/1.5 = 1.333$ psia.
    By Figure 9.9: At $p_4/p^* = 1.333$ and $\gamma^* = 0.6$, $M_4 = 0.71$.
    By Figure 9.8: At $M_4 = 0.71$ and $\gamma^* = 0.6$, $K_{4,*} = 0.6$.
    By (9.15): $K_{3,*} = 1.4 + 0.6 = 2.0$.
    By Figure 9.8: At $K_{3,*} = 2$ and $\gamma^* = 0.6$, $M_3 = 0.56$.

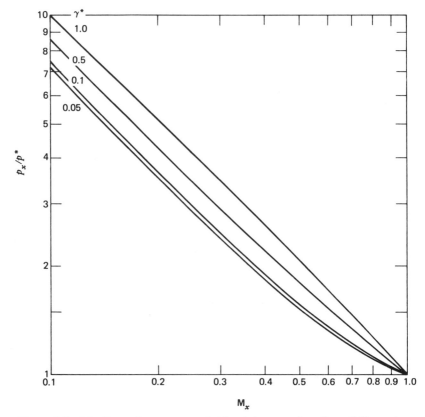

Figure 9.9 Mach number at any point in piping as a function of dimensionless critical pressure, with parameters of the pseudo-isentropic exponent.

By Figure 9.9: At $M_3 = 0.56$ and $\gamma^* = 0.6$, $p_3/p^* = 1.65$ and $p_3 = 1.65 \times 1.5 = 2.475$ psia.

Since $p_3 < p_{cv}$, we conclude that the valve will control in this subsonic flow condition.

9.6 PIPE SIZING FOR FLASHING FLOW

Note that the 4 in. pipe in Example 9.8 was sufficiently large to keep p_3 less than p_{cv}, and hence to keep the control valve in control. One use of the two-phase flow analysis of this chapter is to size the downstream pipe, *before installation*, to accomplish just this.

Example 9.9. Suppose that a 2 in. pipe were proposed for the downstream pipe of Example 9.8. Find whether this pipe size would be adequate to keep the valve in control.

Solution: As in Example 9.8, $p_S = 5$ psia, $G_o = 86$ lbm/ft² sec, and $p_{cv} = 4.745$ psia.

1. $G = \dot{m}/A = 6480/(3600 \times 0.0218166) = 82.506$ lbm/ft² sec.
2. $G/G_o = 0.959$.
3. By Figure 9.5: $p^*/p_S = 0.96$, and $p^* = 4.8$ psia.
4. By Figure 9.7: $\gamma^* = 0.05$.
5. Since p^* is greater than either p_B, the flow will be choked.
6. By (9.13): $K_{3,4} = (0.015 \times 31.1)/(2/12) = 2.8$.
7. By Figure 9.8: At $K_{3,*} = 2.8$ and $\gamma^* = 0.05$, $M_3 = 0.815$.
8. By Figure 9.9: At $M_3 = 0.815$ and $\gamma^* = 0.05$, $p_3/p^* = 1.08$ and $p_3 = 1.08 \times 4.8 = 5.184$ psia.
9. Since $p_3 > p_{cv}$, we conclude that the valve *will not* control for this 2 in. downstream choked pipe situation.

This example indicates that we would be required to go to a larger diameter downstream pipe. As it turns out, either a 3 or a 4 in. pipe would be satisfactory.

9.7 FLOW RATE DETERMINATION FOR FLASHING FLOW

Another use of this two-phase flashing flow analysis is to determine the flow rate that can be passed by a given piping system. A method for doing

360 Flow of Liquid-Vapor Mixtures in Pipes

this can be outlined quite simply as follows (see also [4]):

1. Determine the overall loss coefficient (K_o) of the system.
2. Assume $G = G_{\text{liquid}}$ of (9.1), and obtain $K_{1,S}$, which is the loss coefficient from inlet to flashing point.
3. Determine the percentage of the system loss that is in water via the relation

$$\% \text{ Water loss} = \frac{K_{1,S} \times 100}{K_o}. \tag{9.16}$$

4. Determine G_0 via (9.2) or Figure 9.3.
5. From G/G_0, obtain p^*/p_S, γ^*, M_S, and $K_{3,4}$ via Figures 9.5, 9.7, and 9.8.
6. From $K_{S,4}$, determine the percentage of the system loss that is in the two-phase region via the relation

$$\% \text{ Two-phase loss} = \frac{K_{S,4} \times 100}{K_o}. \tag{9.17}$$

7. Add (9.16) and (9.17) and decide:
   A. If sum = 100%, the correct flow was assumed.
   B. If sum < 100%, a smaller flow must be assumed.
   C. If sum > 100%, a larger flow must be assumed.
8. By successive guesses of G, obtain a consistent solution.

Example 9.10. How much flow can be passed by a system if $T_A = 401°F$, $p_1 = 424$ psia, $p_B = 220$ psia, $K_o = 22.7$,[†] and $D = 3$ in.?

Solution:

Try 1

1. $p_S = 250$ psia, and $\rho_A = 1/0.018655 = 53.605$ lbm/ft³.
2. $p_B < p_S$, so that two-phase flashing flow is expected in the piping system.
3. Assume that G by (9.1) is 2067.7 lbm/ft² sec.
4. Also by (9.1): $K_{1,S} = 2g_c\rho_A(p_1-p_S)144/G^2 - 1 = 19.215$.
5. By (9.16); % Water loss = $(19.215 \times 100)/22.7 = 84.6\%$.
6. By (9.2): $G_0 = 2363$, and $G/G_o = 0.875$.

[†]Some references suggest the use of a different K_o for the two-phase region than for the liquid region, but this does not seem warranted in the approximate methods presented here.

7. By Figure 9.5: $p^*/p_S = 0.9$ and $p^* = 225$ psia.
8. Conclude that, since $p^* > p_B$, the pipe is choked at exit and $p_4 = p^* = 225$ psia.
9. By Figure 9.7: $\gamma^* = 0.17$.
   By Figure 9.9: At $p_S/p^* = 1.111$, $M_S = 0.8$.
   By Figure 9.8: $K_{S,*} = K_{S,4} = 1.1$.
   By (9.17): % Two-phase loss $= (1.1 \times 100)/22.7 = 4.8\%$.
   Sum % losses $= 84.6 + 4.8 = 89.4\%$, which shows that the flow rate is too high. Therefore a smaller flow must be tried.

Try 2

1. Assume that $G = 1965$ lbm/ft² sec. This leads to $G/G_o = 0.832$, $p^*/p_S = 0.856$, $p_S/p^* = 1.168$, $M_S = 0.75$, $K_{1,S} = 21.383$, and $K_{S,*} = 1.3$.
2.
$$\text{\% Water loss} = (21.383 \times 100)/22.7 = 94.2\%.$$

$$\text{\% Two-phase loss} = (1.3 \times 100)/22.7 \quad = \underline{5.7\%}.$$

$$\text{Sum} \quad = 99.9\%.$$

3. Conclude that this is close enough to 100%, so that $\dot{m} = GA = 1965 \times 0.049087 = 96.46$ lbm/sec.

REFERENCES

1. W. F. Allen, Jr., "Flow of a flashing mixture of water and steam through pipes and valves," *Trans. ASME*, April 1951, p. 257.
2. G. S. Liao and J. K. Larson, "Analytical approach for determination of steam/water flow capability in power plant drain systems," *ASME Pap.* 76-WA/PWR-4, December 1976.
3. S. Levy, "Prediction of two-phase critical flow rate," *Trans. ASME, J. Heat Transfer*, February 1965, p. 53.
4. M. Sajben, "Adiabatic flow of flashing liquids in pipes," *Trans. ASME, J. Basic Eng.*, December 1961, p. 619.
5. H. Boger, "Sizing control valves for flashing service," *Instrum. Control Syst.*, January 1970, p. 85.
6. A. H. Shapiro, *The Dynamics and Thermodynamics of Compressible Fluid Flow*, Vol. 1, Ronald Press, 1953, p. 168.

NOMENCLATURE

Roman

- A area
- D pipe diameter
- F_F valve factor
- g acceleration of gravity
- G mass flow rate per area
- h enthalpy
- K loss coefficient
- $\dot{m}$ mass flow rate
- M Mach number
- p absolute pressure
- S entropy
- v specific volume
- V directed velocity
- x steam quality
- Z elevation

Greek

- $\alpha = h_{fg}/v_{fg}$
- $\beta = h_f - \alpha v_f$
- γ isentropic exponent
- Δ finite difference
- ρ fluid density

Subscripts

- $1, 2, x$ axial positions
- B back
- cv valve critical
- f saturated liquid
- g saturated vapor
- i inlet
- o overall, maximum
- s at constant entropy
- S saturation at inlet temperature

Superscript

- $*$ critical

10
Loss Characteristics of Piping Components

> ...on the blood's return trip no more force is wasted than on the way out, just enough left at the last to spill the blood into the right auricle, zero pressure, or nearly... —*Gustav Eckstein*

10.1 GENERAL REMARKS

In this chapter we get down to brass tacks. Actual loss characteristics of many common piping components are dealt with here, including straight viscous pipes, curved pipes and stepped elbows, branching tees, elements showing gradual area changes such as reducers and diffusers, those exhibiting abrupt area changes such as contractions and expansions, and differential pressure producing fluid meters such as orifices, nozzles, and venturis. Finally, we consider losses across valves and screens.

All in all, this is what pipe flow analysis is about. Previously, we were given typical loss coefficients so that we could solve various problem types. Now, we become specific and discuss, and present the basis for, many of the loss coefficients in use throughout industry today.

10.2 VISCOUS PIPES

The most common loss element in a piping system is a straight run of the pipe itself. Pipe loss is characterized in part by the friction factor (f), as described by the Colebrook equation of (6.19). Other factors involved in the pipe loss coefficient are the pipe length (L) and the pipe diameter (D). These three factors combine according to (7.8) to yield

$$K_{\text{viscous pipe}} = f\frac{L}{D}. \tag{7.8}$$

The experimental and theoretical work which is the basis of f was presented at great length in Chapter 6 and need not be discussed further here.

Applications involving $f = f(R_D, e/D)$ for various Reynolds numbers and pipes of various relative roughnesses also were given in Chapters 7 and 8.

Whenever the length of the pipe is small compared to its diameter [more specifically, for $L/(DR_D) < 10^{-1}$, according to Section 6.13], friction factors other than those defined by the fully developed *turbulent* flow friction factor of Colebrook may be applicable. For example, for developing *laminar* boundary layers the mean apparent laminar friction factor of (6.40) can be used. For developing turbulent boundary layers, the mean apparent turbulent friction factor of (6.41) is suggested.

10.3 ELBOWS

The pipe bend, also called an elbow, is a very common loss element in many piping systems. Much experimental work has been reported on the elbow, especially the 90° elbow, and empirical predictions of losses in elbows have been presented in the form of graphs, equations, and nomographs (see, e.g., [1]–[7]).

The two most common variables reported involving pipe bends are the relative radius (R/r), where R represents the radius of curvature of the centerline of the elbow, and r represents the inside radius of the elbow; and the pipe Reynolds number (R_D). Other variables to be considered are the relative roughness (e/D) of the inner wall of the elbow, and the entry and exit conditions of the elbow.

Many results are given in terms of smooth elbows, where "smooth" indicates that the elbow is flanged or otherwise joined into the piping system so that a smooth internal surface is presented to the flow. Another type of elbow commonly used is the so-called screw elbow, wherein there is an abrupt step-up at inlet and an abrupt step-down at exit (see Figure 10.1). Naturally, we should expect more loss in the screw elbow than in the smooth one. Very few data are available for compressible fluid flow through elbows at any significant Mach numbers, and it is conventional to consider the flow to be incompressible in most elbow loss work.

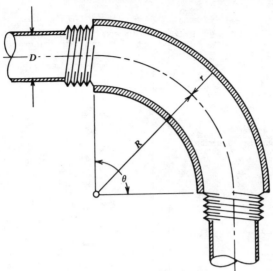

Figure 10.1 Screw-type elbow.

Of the many references available on elbow losses, the work of Ito [1, 2] is essentially definitive because of his careful and systematic experiments, because he presented empirical equations that well represent his experimental results, and because in the main his work agrees with that of many other experimenters.

Basically, about 50 pipe diameters of length downstream of an elbow are required before the elbow loss is fully realized in a piping system. Furthermore, it is difficult to measure pressures accurately in the near vicinity of the elbow. Hence it is usual practice to sample pressures at a number of points *around* the pipe and *along* the inlet and exit pipes of the elbow under investigation, and to establish from these the normal slope of the viscous pipe loss (see Figure 10.2), and ultimately the elbow loss alone.

It follows from this discussion that, when fittings and other piping loss elements are placed close to each other, such elements influence those downstream in an undetermined manner. This means that predictions of losses of elements placed in series cannot be too reliable, and that only tests conducted under nearly identical setups can yield meaningful results in such cases.

Also, in connection with this discussion, the loss coefficients published and described here *do not* include the straight pipe viscous losses before and after the elbow. These must be added to the basic elbow loss according to the procedure described in Section 7.4.2.

366 Loss Characteristics of Piping Components

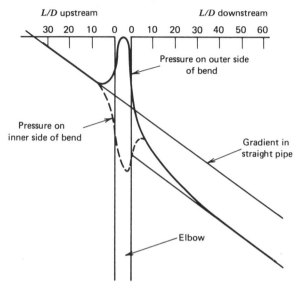

Figure 10.2 Pressure distribution along pipeline containing a 90° elbow (after Ito [1]).

Ito [1] tested his smooth elbows with exit pipes at least 50 pipe diameters in length, but his results can be applied successfully for elbows mounted with about 20 diameters of straight pipe upstream and about 30 pipe diameters downstream of the elbow. The water Reynolds numbers on which Ito's results are based varied from about 10^4 to 10^6; however, it is believed that the same results apply at higher Reynolds numbers.

Specific Loss in Smooth Elbows. Graphically, the loss coefficients for smooth elbows are presented in Figure 10.3. It can be seen that the more the flow must turn, the greater will be the loss coefficient. The curves in Figure 10.3 are based on the empirical equations given next.

For $R_D(r/R)^2 < 91$, Ito gives:

$$K_{\text{smooth elbow}} = 0.00873 \, \alpha f \theta \frac{R}{r}, \tag{10.1}$$

where θ is the bend angle in degrees, f is the friction factor, given by Ito for fully developed turbulent flow in smooth curved pipes as

$$f = \left(\frac{r}{R}\right)^{1/2} \left\{ 0.029 + \frac{0.304}{\left[R_D(r/R)^2\right]^{0.25}} \right\}, \tag{10.2}$$

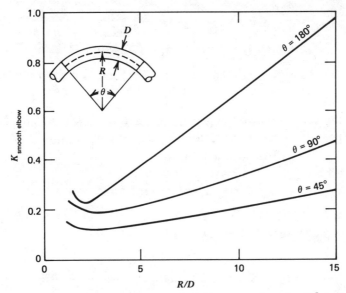

Figure 10.3 Loss coefficient for smooth elbows at $R_D = 2 \times 10^5$ (after Ito [1]).

and α is an empirical factor given by Ito as

$$\alpha_{\theta=45°} = 1 + 14.2\left(\frac{R}{r}\right)^{-1.47}, \tag{10.3}$$

$$\alpha_{\theta=90°} = 0.95 + 17.2\left(\frac{R}{r}\right)^{-1.96}. \tag{10.4}$$

For $R_D(r/R)^2 > 91$, Ito gives

$$K_{\text{smooth elbow}} = 0.00241 \alpha \theta \left(\frac{R}{r}\right)^{0.84} R_D^{-0.17}. \tag{10.5}$$

Note that, if $R_D(r/R)^2 < 0.034$, the friction factor of (10.2) coincides with the Colebrook friction factor of (6.19).

Example 10.1. Find the loss coefficients of a 45° and a 90° elbow if the pipe Reynolds number is 2×10^5 and the radius ratio $R/r = 4$.

Solution: $R_D(r/R)^2 = 2 \times 10^5 \times (\frac{1}{4})^2 = 12{,}500$, which is greater than 91. Therefore (10.5) applies.
 By (10.3): $\alpha_{45°} = 1 + 14.2(4)^{-1.47} = 2.8504$.
 By (10.4): $\alpha_{90°} = 0.95 + 17.2(4)^{-1.96} = 2.0863$.

368 Loss Characteristics of Piping Components

By (10.5):

$$K_{45°} = 0.00241 \times 2.8504 \times 45(4)^{0.84}(2 \times 10^5)^{-0.17} = 0.124,$$

$$K_{90°} = 0.00241 \times 2.0863 \times 90(4)^{0.84}(2 \times 10^5)^{-0.17} = 0.182.$$

Specific Loss in Screw Elbows. Since the screw elbow involves inlet and exit steps in the flow, an extra term made up of the sum of K_{up} and K_{down} must be added to $K_{smooth\ elbow}$, that is,

$$K_{screw\ elbow} = K_{smooth\ elbow} + K_{up} + K_{down}. \tag{10.6}$$

The abrupt enlargement loss, called here K_{up}, was discussed in Section 7.4.3 and described by (7.13) as

$$K_{up} = (1 - \beta^2)^2, \tag{10.7}$$

where β is the ratio of the inlet pipe diameter to the elbow diameter. According to the notation of the Figure 10.1, we have

$$\beta = \frac{D}{2r}. \tag{10.8}$$

The abrupt contraction loss [8], to be discussed in Section 10.8, is called here K_{down} and can be obtained from Figure 10.4, again in terms of the β of (10.8).

Ito [1] performed tests on screw elbows also, and the empirical curves of Figure 10.5 are examples of his results.

Example 10.2. Show that the loss coefficient of a screw elbow can be well approximated by the sum of the step loss coefficients at inlet and exit and the smooth elbow loss coefficient. Use Ito's experimental values: $R = 28.4$ mm, $r = 22.7$ mm, $D = 34.9$ mm, $\theta = 90°$, and $R_D = 2 \times 10^5$.

Solution: $R_D(r/R)^2 = 2 \times 10^5 \times (22.7/28.4)^2 = 127{,}775$, which is greater than 91. Therefore (10.5) applies.

By (10.4): $\alpha_{90°} = 0.95 + 17.2(1.25)^{-1.96} = 12.057$.
By (10.5): $K_{smooth\ elbow} = 0.00241 \times 12.057 \times 90(1.25)^{0.84}(2 \times 10^5)^{-0.17} = 0.396$.
By (10.8): $\beta = 34.9/(2 \times 22.7) = 0.769$.
By (10.7): $K_{up} = (1 - 0.769^2)^2 = 0.167$.
By Figure 10.4: $K_{down} \simeq 0.2$.

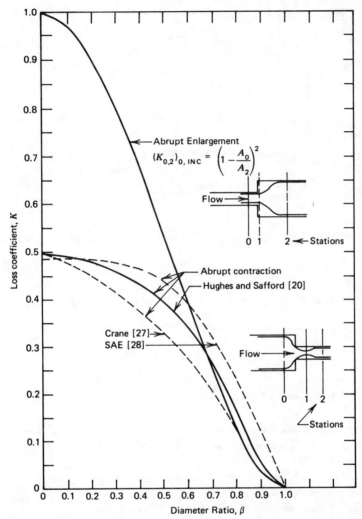

Figure 10.4 Constant density loss coefficients for abrupt area changes (after Benedict et al. [8]).

By (10.6):

$$K_{\text{screw elbow}} = 0.396 + 0.167 + 0.2 = 0.763,$$

which is seen to check very closely the results of Figure 10.5.

Equivalent Length Approach. Often, it is convenient to express the effect of a loss element in terms of an equivalent length of straight pipe. Note that equivalent straight pipe, as used here, means that such a pipe will have

370 Loss Characteristics of Piping Components

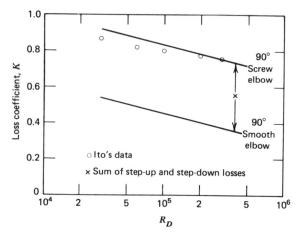

Figure 10.5 Theoretical and experimental characteristics of a 90° screw elbow.

the same dissipation (i.e., the same total pressure drop) as the elbow it is to replace. From (7.8) it follows that

$$\left(\frac{L}{D}\right)_{\text{elbow equivalent}} = \frac{K_{\text{elbow}}}{f_{\text{straight pipe}}}, \quad (10.9)$$

where f is from the Colebrook equation of (6.19) at the given pipe R_D.

Example 10.3. Find the equivalent length of straight pipe of a smooth 90° elbow of $R/r = 20$ operating at $R_D = 2 \times 10^5$.

Solution: $R_D(r/R)^2 = 2 \times 10^5 (\frac{1}{20})^2 = 500$, which is greater than 91. Therefore (10.5) applies.
  By (10.4): $\alpha_{90°} = 0.95 + 17.2(20)^{-1.96} = 0.998$.
  By (10.5): $K_{\text{smooth elbow}} = 0.00241 \times 0.998 \times 90(20)^{0.84}(2 \times 10^5)^{-0.17} = 0.336$.
  By (6.19): $f_{\text{smooth}} = 0.01564$ (also by Table 6.1 and Figure 6.7).
  By (10.9):

$$\left(\frac{L}{D}\right)_{\text{elbow equivalent}} = \frac{0.336}{0.01564} = 21.48,$$

which indicates that it would take a length of straight smooth pipe of about 21.5 diameters to yield the equivalent loss of the 90° smooth elbow.

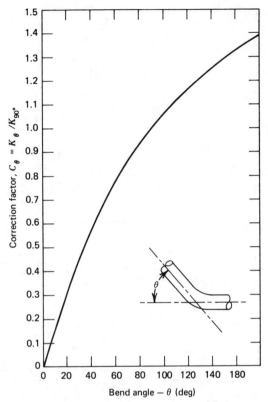

Figure 10.6 Correction factor for bends other than 90° (after SAE [6]).

Correction Curve Approach. The SAE [6] approaches elbow loss in a slightly different manner. A correction term is applied to the basic 90° smooth elbow result to get the effect of bend angle (θ), that is,

$$(C_\theta)_{\text{elbow}} = \frac{K_\theta}{K_{90°}}, \qquad (10.10)$$

where C_{elbow} is shown in Figure 10.6.

Example 10.4. Find the loss coefficient of a 45° smooth elbow, and compare it with Ito's formulation.

Solution: Using the conditions and results of Example 10.1, we have

$$(K_{90°})_{\text{Ito}} = 0.182 \quad \text{and} \quad (K_{45°})_{\text{Ito}} = 0.124.$$

Table 10.1 Nominal Loss Coefficients for Elbows

Loss Coefficient	Smooth	Screw
$K_{45°}$	0.1	0.4
$K_{90°}$	0.2	0.8

By Figure 10.6: $(C_{45°})_{\text{elbow}} = 0.63$.
By (10.10):

$$(K_{45°})_{\text{SAE}} = 0.63 \times 0.182 = 0.115,$$

which agrees fairly well with the Ito result.

Nominal Loss Coefficients for Elbows. In Table 10.1 we summarize nominal loss coefficients for elbows, both smooth and screw types, based on $R_D \simeq 2 \times 10^5$ and $R/r \simeq 4$.

10.4 TEES

Branches are often encountered in piping systems. Thus the flow may separate, splitting from one pipe at inlet to two or more pipes at outlet. These outlet pipes may go off at 90° to the inlet pipe, or they may leave at any angle (α). Similarly, the flow may converge where several inlet branches feed one outlet pipe. These feeders may be located at 90° to the outlet pipe, or they may enter at any angles (β, γ). Some of these arrangement possibilities are illustrated in Figure 10.7.

Various studies on tees are available in the literature. Of these, only two; believed to be typical, have been analyzed [9, 10]. These results are summarized in Figure 10.8 for 90° separating and joining flow tees.

As with elbows and friction factors, most of the work reported in the literature is for incompressible flow or for gas flows at low Mach numbers, so that compressibility effects are necessarily small.

The following equations are offered for branch losses. These are based on the work of Vazsonyi [9], because his results are in good agreement with most of the available data.

For Separating Flows.

$$(K_{0,1})_{\text{tee}} = \lambda_1 + (2\lambda_2 - \lambda_1)\left(\frac{V_1}{V_0}\right)^2 - 2\lambda_2\left(\frac{V_1}{V_0}\right)\cos\alpha', \qquad (10.11)$$

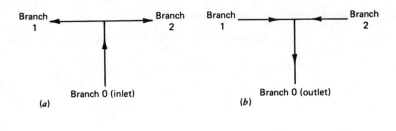

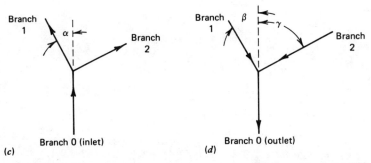

Figure 10.7 Some examples of tee arrangements in piping systems. (*a*) 90° separating flow tee. (*b*) 90° joining flow tee. (*c*) General separating flow tee. (*d*) General joining flow tee.

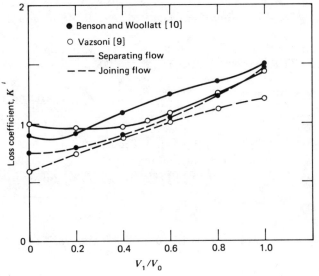

Figure 10.8 Loss coefficients for 90° tees for separating and joining flows (after Vazsonyi [9] and Benson and Woolatt [10]).

373

374 Loss Characteristics of Piping Components

where K is based on the K.E. of the combined flow in branch 0, and

$$\lambda_1 = 0.0712\alpha^{0.7041} + 0.37 \quad \text{for} \quad \alpha < 22.5°,$$
$$\lambda_1 = 1.0 \quad \text{for} \quad \alpha \geqslant 22.5°; \tag{10.12}$$

$$\lambda_2 = 0.0592\alpha^{0.7029} + 0.37 \quad \text{for} \quad \alpha < 22.5°,$$
$$\lambda_2 = 0.9 \quad \text{for} \quad \alpha \geqslant 22.5°; \tag{10.13}$$

and

$$\alpha' = 1.41\alpha - 0.00594\alpha^2. \tag{10.14}$$

The factors λ_1 and λ_2 are plotted in Figure 10.9, and α' is given in Figure 10.10.

For Joining Flows.

$$(K_{1,0})_{\text{tee}} = \lambda_3 \left(\frac{V_1}{V_0}\right)^2 + 1 - 2\left[\left(\frac{V_i}{V_0}\right)\left(\frac{Q_1}{Q_0}\right)\cos\beta' + \left(\frac{V_2}{V_0}\right)\left(\frac{Q_2}{Q_0}\right)\cos\alpha'\right], \tag{10.15}$$

where K again is based on the K.E. of the combined flow in branch 0, Q is the volumetric flow rate ($=AV$), λ_3 is defined by Figure 10.11, and β' and γ' are defined by the same line described by (10.14), as shown in Figure 10.10.

Data from tees indicate that there is no variation of the tee loss coefficient with Reynolds number for $R_D > 1000$.

Example 10.5. Find the loss coefficient for a 90° separating flow tee if $V_1/V_0 = 0.5$.

Solution: By (10.12): $\lambda_1 = 1.0$ (also from Figure 10.9).
By (10.13): $\lambda_2 = 0.9$ (also from Figure 10.9).
By (10.14): $\alpha' = 1.41(90) - 0.00594(90)^2 = 78.786°$ (also from Figure 10.10).
By (10.11):

$$(K_{0,1})_{\text{tee}} = 1 + (1.8 - 1)(0.5)^2 - 2 \times 0.9 \times 0.5 \cos 78.786 = 1.025,$$

which agrees with Figure 10.8.

Example 10.6. Find the loss coefficient for a 90° joining flow tee if $\beta = \gamma = 90°$, $Q_1/Q_0 = Q_2/Q_0 = 0.5$, and $V_1/V_0 = V_2/V_0 = 0.5$.

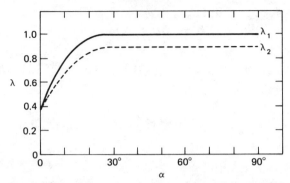

Figure 10.9 Separating tee factor (after Vazsonyi [9]).

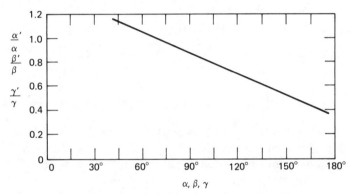

Figure 10.10 Various factors for use in separating and joining flow tees (after Vazsonyi [9]).

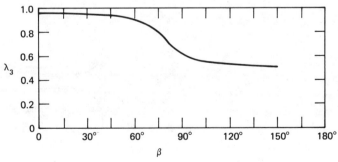

Figure 10.11 Joining tee factor (after Vazsonyi [9]).

375

Solution: From Figure 10.11: $\lambda_3 = 0.61$.
From Figure 10.10: $\gamma' = \beta' = 78.786°$.
By (10.15):

$$(K_{1,0})_{tee} = 0.61(0.5)^2 + 1 - 2(2 \times 0.5 \times 0.5 \cos 78.786) = 0.96,$$

which agrees with Figure 10.8.

Example 10.7. Find the loss coefficient for a joining flow tee if $\beta = 30°$, $\gamma = 45°$, $Q_1/Q_0 = Q_2/Q_0 = 0.5$, $V_1/V_0 = 0.4$, and $V_2/V_0 = 0.6$.

Solution: From Figure 10.11: $\lambda_3 = 0.95$.
From (10.14): $\beta' = 1.41(30) - 0.00594(30)^2 = 36.954°$ and $\gamma' = 1.41(45) - 0.00594(45)^2 = 51.4215°$, both of which can be obtained from Figure 10.10.
By (10.15):

$$(K_{1,0})_{tee} = 0.95(0.4)^2 + 1 - 2(0.4 \times 0.5 \cos 36.954 + 0.6 \times 0.5 \cos 51.4215)$$
$$= 0.46.$$

10.5 REDUCERS

When the piping flow area is *reduced* gradually, as shown in Figure 10.12, the number of velocity heads lost is very small indeed. The number most commonly quoted to represented reducers and/or well-rounded inlet loss elements is

$$K_{reducer} = 0.05, \tag{10.16}$$

based on the flow area (or velocity) of the smaller piping section.

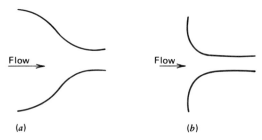

Figure 10.12 Loss elements with gradual area reduction. (*a*) Reducer. (*b*) Well-rounded inlet.

10.6 DIFFUSERS

Whenever it is necessary to *increase* the flow area gradually, as shown in Figure 10.13, the conical diffuser is widely used. Much of the work on the diffuser has been reported in terms of diffuser efficiency; however, we will also present this information in terms of a loss coefficient.

Diffuser Efficiency. Following the notation of Figure 10.13, we can give the efficiency of a diffuser as

$$\eta_{\text{diffuser}} = \frac{p_2 - p_1}{p_2' - p_1}, \tag{10.17}$$

provided that the pressures at stations 1 and 2 are measured where uniform velocities prevail. The numerator represents the actual static pressure recovery across the diffuser. The denominator represents the ideal pressure recovery, predicated on the zero loss assumption, that is, $p_{t2}' = p_{t1}$, or

$$p_2' + \frac{\rho V_2^2}{2g_c} = p_1 + \frac{\rho V_1^2}{2g_c}. \tag{10.18}$$

It follows from (10.18) and the basic continuity expression of (2.53) that

$$\frac{p_2' - p_1}{\rho V^2 / 2g_c} = 1 - \left(\frac{A_1}{A_2}\right)^2 = 1 - \beta^4,$$

where $\beta = D_1 / D_2$. Thus (10.17) can be written as

$$\eta_{\text{diffuser}} = \frac{p_2 - p_1}{(\rho V_1^2 / 2g_c)(1 - \beta^4)}. \tag{10.19}$$

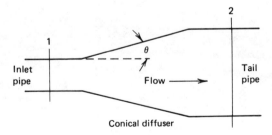

Figure 10.13 Loss element with gradual area increase.

In 1938, Patterson [11] gave the classical study on diffusers employed with tail pipes, such as shown in Figure 10.13, where one-dimensional velocity distributions at 1 and 2 are good assumptions. This work is summarized in Figure 10.14, where for reasonably small expansion angles (like $4° < \theta < 7°$) the diffuser efficiency is seen to be about 90%. This value also is borne out by the more recent work of Hilbrath et al. [12] in 1973. Sovran and Klomp [13], in 1967, gave a massive compilation of recent data on diffusers, discharging into a plenum, however, so that the one-dimensional pipe flow assumption at station 2 is not satisfied. Hence these results are not readily applied in the case of pipe flow.

Diffuser Loss Coefficient. The loss coefficient of a diffuser can be defined [14] as

$$K_{\text{diffuser}} = (1 - \eta)_{\text{diffuser}}(1 - \beta^4), \tag{10.20}$$

which follows directly from (10.19), from the basic incompressible definition

$$K = \frac{\Delta p_t}{(\rho V_1^2 / 2g_c)},$$

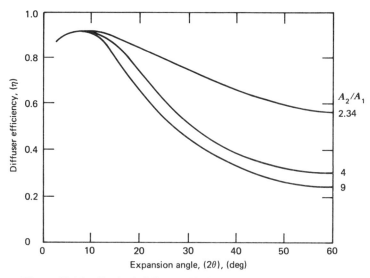

Figure 10.14 Conical diffuser efficiencies (after Patterson [11]).

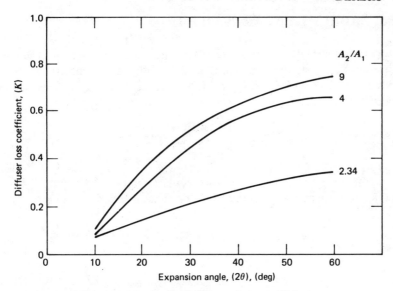

Figure 10.15 Conical diffuser loss coefficients.

and from the fact that

$$\Delta p_t = -(p_2 - p_1) + \frac{\rho}{2g_c}(V_1^2 - V_2^2).$$

Combining (10.20) with the diffuser efficiencies of Figure 10.14, we obtain the diffuser loss coefficients of Figure 10.15.

Example 10.8. Find the loss coefficient and the total pressure ratio for a conical diffuser mounted in a piping system, where the inlet diameter is 2 in., the exit diameter is 4 in., and the inlet pressure ratio is 0.9.

Solution: Taking $\eta_{\text{diffuser}} \simeq 0.9$ and $\beta = 0.5$, we have, by (10.20),

$$K_{\text{diffuser}} = (1 - 0.9)(1 - 0.5^4) = 0.09375.$$

By (7.14):

$$\frac{p_{t2}}{p_{t1}} = 1 - (1 - R_1)K_1$$

$$= 1 - 0.1 \times 0.09375 = 0.99.$$

380 Loss Characteristics of Piping Components

10.7 ABRUPT ENLARGEMENTS

Abrupt changes in flow areas are often encountered when working with ordinary piping systems. These have already been dealt with extensively in this study. In Section 2.6.3 we considered the *ideal* incompressible pressure recovery across an abrupt enlargement, as typified by (2.54), namely

$$\left(\frac{p_2-p_1}{\rho}\right)_{ideal} = \frac{V_1^2}{2g_c}(1-\beta^4). \tag{10.21}$$

In Section 7.4.3 we derived various equations (i.e., 7.13, 7.14, 7.19) that characterized an abrupt enlargement for incompressible flow *with losses*, namely,

$$K_{inc} = (1-\beta^2)^2, \tag{10.22}$$

$$\left(\frac{p_{t2}}{p_{t1}}\right)_{inc} = 1 - (1-R_1)K_{inc}, \tag{10.23}$$

and

$$\left(\frac{p_2}{p_1}\right)_{inc} = 1 + \left(\frac{1-R_1}{R_1}\right)(2\beta^2 - 2\beta^4). \tag{10.24}$$

In Section 8.4.5 we derived various equations (8.50, 8.51, 8.52, 8.53) characterizing abrupt enlargements for subsonic and supercritical flows *with losses*, namely,

$$\frac{M_2[2+(\gamma-1)M_2^2]^{1/2}}{1+\gamma M_2^2} = \frac{M_1[2+(\gamma-1)M_1^2]^{1/2}}{1+\gamma M_1^2 + (p_1'/p_1)[(1-\varphi)/\varphi]}, \tag{10.25}$$

$$\left(\frac{p_1'}{p_1}\right)_{subsonic} = 1, \tag{10.26}$$

$$\left(\frac{p_1'}{p_1}\right)_{supersonic} = \left(\frac{\varphi}{1-\varphi}\right)(1+\gamma)^{1/2}\left\{\frac{1+\gamma M_2^2}{M_2[2+(\gamma-1)M_2^2]^{1/2}} - (1+\gamma)^{1/2}\right\}, \tag{10.27}$$

$$\frac{p_{t2}}{p_{t1}} = \varphi\left(\frac{M_1}{M_2}\right)\left[\frac{2+(\gamma-1)M_2^2}{2+(\gamma-1)M_1^2}\right]^{(\gamma+1)/[2(\gamma-1)]}, \tag{10.28}$$

and

$$\frac{p_1}{p_2} = \frac{1}{\varphi}\left(\frac{M_2}{M_1}\right)\left[\frac{2+(\gamma-1)M_2^2}{2+(\gamma-1)M_1^2}\right]^{1/2}. \qquad (10.29)$$

Experimental results for flows across abrupt enlargements were given in Figures 7.2 and 7.3 for the incompressible cases, and in Figure 8.3 for compressible flows. Numerical results, in graphical form, were presented for the compressible case in Figure 8.4, and for incompressible flows in Figure 10.4. All of this work is based primarily on References 8 and 15. In this section we simply call attention to this material.

Example 10.9. Compare the total pressure loss and the static pressure recovery for an abrupt enlargement of $\beta = 0.5$ for incompressible flow and for the subsonic flow of air if $R_1 = 0.9$.

Solution: For $\beta = 0.5$, $\varphi = \beta^2 = 0.25$.

Incompressible Results
By (10.22): $K_{\text{inc}} = (1 - 0.25)^2 = 0.5625$.
By (10.23): $(p_{t2}/p_{t1})_{\text{inc}} = 1 - (1 - 0.9)0.5625 = 0.94375$.
Thus

$$(TPR)_{\text{inc}} = \frac{p_{t1}}{p_{t2}} = 1.0596.$$

By (10.24):

$$\left(\frac{p_2}{p_1}\right)_{\text{inc}} = 1 + \left(\frac{0.1}{0.9}\right)(2 \times 0.25 - 2 \times 0.0625) = 1.0417.$$

Compressible Results
By Table 3.6 or (3.74): $\Big\}$ $M_1 = 0.39095$ at $R_1 = 0.9$.
By Table 8.2 or (8.19):
By (10.25): $M_2 = 0.0953$ (an iterative solution).
By (10.28):

$$\frac{p_{t2}}{p_{t1}} = \frac{0.25 \times 0.39090}{0.0953}\left(\frac{2+0.4 \times 0.0953^2}{2+0.4 \times 0.39090^2}\right)^{2.4/0.8} = 0.942112.$$

Thus

$$(TPR)_{\text{comp}} = \frac{p_{t1}}{p_{t2}} = 1.0614.$$

382 Loss Characteristics of Piping Components

Note that this same result can be obtained by the generalized flow function of (8.66), that is, by

$$(TPR)\varphi\Gamma_1 = \Gamma_2,$$

where
$$\Gamma_1 = f(M_1) = 0.61715,$$
$$\Gamma_2 = f(M_2) = 0.16379 \text{ by Table 8.2}.$$

Thus

$$(TPR)_{comp} = \frac{0.16379}{0.61715 \times 0.25} = 1.0616.$$

By (10.29):

$$\frac{p_1}{p_2} = \frac{0.0953}{0.25 \times 0.39090} \left(\frac{2 + 0.4 \times 0.0953^2}{2 + 0.4 \times 0.39090^2} \right)^{1/2} = 0.96136.$$

Thus

$$(SPR)_{comp} = \frac{p_2}{p_1} = 1.0402.$$

Summary of Results of Example 10.9. Ideally, there is no loss in total pressure across an abrupt enlargement.

In real liquid flow the total pressure is reduced by 5.96%. In real gas flow the total pressure is reduced by 6.16%.

Similarly, the static pressure rise across an abrupt enlargement is 4.17% for real liquids and 4.02% for real gases.

Example 10.10. Find the total pressure loss across an abrupt enlargement if $\beta = 0.5$, $R_1 = 0.52828$, and $R_2 = p_2/p_{t2} = 0.9$.

Solution: For $R_1 = 0.52828$, M_1 is 1, and $\Gamma_1 = 1$, by Table 8.2. This means we have supercritical flow across the abrupt enlargement, with $M_2 = 0.39090$ at $R_2 = 0.9$, and $\Gamma_2 = 0.61715$.

By (8.66):

$$TPR = \frac{\Gamma_2}{\Gamma_1 \varphi} = \frac{0.61715}{1 \times 0.25} = 2.4686.$$

or

$$\frac{1}{TPR} = 0.4051, \quad \text{in agreement with Figure 8.3.}$$

Note the large unrecoverable total pressure loss in the supercritical case, as compared with the subsonic case of Example 10.9.

10.8 ABRUPT CONTRACTIONS

While abrupt enlargements have been dealt with at some length in this study, we have neglected the inverse situation. Often the flow area of a piping system must be *reduced* abruptly. A flow model for such a situation is shown in Figure 10.16.

The empirical equation usually given [16–19] to describe the maximum head loss involved when a constant density fluid flows across an abrupt decrease in the piping area is

$$(h_{\text{loss}})_{0,2} = 0.5 \frac{V_2^2}{2g_c}. \tag{10.30}$$

Although the analysis of this piping element appears quite similar to that of an abrupt enlargement, it is not possible to obtain an explicit analytic expression for the loss coefficient of an abrupt contraction such as (10.22)

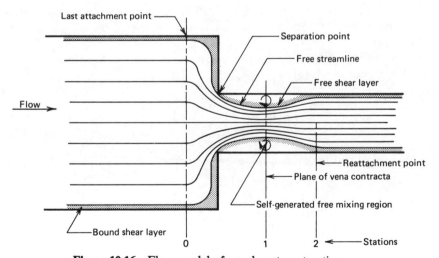

Figure 10.16 Flow model of an abrupt contraction.

384 Loss Characteristics of Piping Components

for the abrupt enlargement. However, analytic expressions based on the use of one or two experimentally determined coefficients can be given.

In Figure 10.16 there is an inlet loss from 0 to 1 which arises from the change in velocity (and/or the change in velocity distribution) involved in the acceleration. Then the boundaries of the fluid enlarge abruptly from the vena contracta area at 1 to the exit area at 2. Thus there is an abrupt enlargement loss from 1 to 2.

One expression for the overall loss coefficient for an abrupt contraction, *from a plenum inlet*, which is simply the sum of the inlet and enlargement losses, has been given by Hughes and Safford [20] as

$$(K_{0,2})_{2,\text{inc}} = \frac{1}{C_v^2 C_c^2} - \frac{2}{C_c} + 1, \tag{10.31}$$

where $(K_{0,2})_2$ is read as the loss coefficient (K) from 0 to 2 based on the velocity at 2; C_v is the velocity coefficient, defined as

$$C_v \equiv \frac{V}{V'}, \tag{10.32}$$

and ratios the actual to the ideal velocity, as at the vena contracta; and C_c is the contraction coefficient, defined as

$$C_c \equiv \frac{A_1}{A_2}. \tag{10.33}$$

Note that successful application of (10.31) hinges on a knowledge of the two coefficients, C_v and C_c, and is seriously restricted by the plenum inlet assumption.

Another expression for the overall loss coefficient of an abrupt contraction has been given [8] as

$$(K_{0,2})_{2,\text{inc}} = \left(\frac{1}{C_D^2} - 1\right)(1 - \beta^4). \tag{10.34}$$

This is obtained by writing the energy equation of (7.5), applied between 0 and 2, as

$$\frac{p_0 - p_2}{\rho} = \frac{V_2^2 - V_0^2}{2g_c} + (K_{0,2})_2 \frac{V_2^2}{2g_c} = \frac{(V_2')^2 - (V_0')^2}{2g_c}, \tag{10.35}$$

where the primes signify ideal conditions. The discharge coefficient (C_D) in

(10.34) is defined, for the same pressure drop, as

$$C_D = \frac{Q_{\text{actual}}}{Q_{\text{ideal}}} = \frac{A_2 V_2}{A_2' V_2'}; \qquad (10.36)$$

and, since the pipe flows full at 2, we have in terms of (10.32)

$$C_D = \frac{V_2}{V_2'} = C_{v2}, \qquad (10.37)$$

whereas by continuity we have

$$\frac{A_2}{A_0} = \beta^2 = \frac{V_0}{V_2} = \frac{V_0'}{V_2'}. \qquad (10.38)$$

Equation 10.34 is recommended for the constant density, abrupt contraction loss coefficient because it depends on a knowledge of only one empirical coefficient, namely, the experimentally determinable C_D, and because it is free of the plenum flow restriction of (10.31).

Typical velocity coefficients, as measured by Weisbach [21] in water tests, are given in Table 10.2, and indicate a flow dependence. Nevertheless, several authors [20–22] use a mean value of 0.975 for C_v, while others [16, 18, 23] neglect inlet losses entirely when forming the abrupt contraction loss coefficients of (10.31).

Typical contraction coefficients, as measured by Weisbach [21, 24] in free efflux water tests, but suggested for use in short pipes also, are given in Table 10.3 and in Figure 10.17. These experimental values essentially were

Table 10.2 Typical Velocity Coefficients[a]

Head (m):	0.02	0.5	3.5	17.0	103.0
C_v:	0.959	0.967	0.975	0.994	0.994

[a] After Weisbach [21].

Table 10.3 Typical Contraction Coefficients[a]

	\multicolumn{10}{c}{$(D_1/D_2)^2$}
	0.1	0.2	0.3	0.4	0.5	0.6	0.7	0.8	0.9	1.0
$(C_c)_{\text{Weisbach}}$	0.624	0.632	0.643	0.659	0.681	0.712	0.755	0.813	0.892	1.0
$(C_c)_{\text{Freeman}}$	0.632	0.644	0.659	0.676	0.696	0.717	0.744	0.784	0.890	1.0

[a] After Weisbach [21, 24] and Freeman [22].

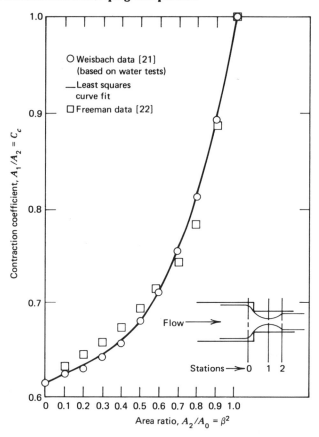

Figure 10.17 Several free discharge contraction coefficients.

confirmed analytically by von Mises [25] for two-dimensional orifice flow, and Kirchhoff [26] gave a theoretical minimum contraction coefficient (for $\beta = 0$) for a perfect liquid passing through a long narrow slit or a circular opening as

$$C_c = \frac{\pi}{\pi + 2} = 0.611. \tag{10.39}$$

Because of their wide acceptance and closeness to our flow model, we have expressed the Weisbach data on contractions by the least squares equation

$$C_c = 0.61375 + 0.13318\beta^2 - 0.26095\beta^4 + 0.51146\beta^6. \tag{10.40}$$

However, a different set of contraction coefficients has been measured by Freeman [20, 22] in free efflux water tests of square ring nozzles, and these also are given in Table 10.3 and Figure 10.17.

Although C_c is undoubtedly Reynolds number dependent, most authors [16, 18, 20, 23] use the flow-independent Weisbach or Freeman data given in Table 10.3 and Figure 10.17 to form the abrupt contraction loss coefficients of (10.31).

If interest is confined to the turbulent flow regime, it may be that both C_v and C_c are indeed flow independent and, hence, functions of β only. On this basis, and for this practical flow situation, one might generate abrupt contraction loss coefficients via (10.31) by using Weisbach's C_c and a C_v (in lieu of better information) varying linearly with β from 0.98 to 1. Such a formulation does indicate a maximum loss coefficient at $\beta = 0$ of 0.516, which is in fair agreement with (10.30). But note that the velocity and contraction coefficients were determined by Weisbach and Freeman for free discharges from orifices or nozzles and not for internal discharges in piping systems such as are under discussion here.

The mean discharge coefficient is given as 0.815 by Weisbach [20], based on his water tests in short pipes. According to (10.34), this yields a maximum loss coefficient at $\beta = 0$ of 0.506, a value that not only is in agreement with, but actually is the basis of, (10.34).

The constant density total pressure ratio is as significant in considerations of the abrupt contraction as it is for the abrupt enlargement, and for the same reasons. For the step-down TPR, (7.14) is applied to yield

$$\frac{p_{t2}}{p_{t0}} = \left(\frac{1}{TPR}\right)_{\text{inc, step-down}} = \frac{1}{1 + (1 - R_2)(K_{0,2})_2}, \quad (10.41)$$

where R_2 is the exit pressure ratio (p_2/p_{t2}). Equation 10.41 is the contraction counterpart of the enlargement equation, 10.23.

Note that the abrupt contraction loss concerns the flow within the boundaries extending from the last station in the larger pipe, where attachment of the fluid stream is assured, that is, where the minimum static pressure exists, to the downstream location in the smaller pipe, where reattachment of the fluid stream with the pipe wall has occurred, that is, where the maximum static pressure in that pipe exists.

Various conventional constant density loss coefficients have been summarized graphically in Figure 10.4. Included are the loss coefficients based on (10.31), along with those suggested by Crane [27] and the SAE [28].

In Figure 10.18, experimental results obtained in both air and water tests are presented for abrupt contractions. Differences between compressible and constant density loss coefficients (based on exit velocities) can be seen, although K_{comp} approaches K_{inc} as the flow approaches zero. For convenience the air and water curves of Figure 10.18 have been extrapolated to $R_2 = 1$, and these results have been expressed by the least squares equation

$$(K_{0,2})_{2,\text{ inc, step-down}} = 0.57806 + 0.39543\beta - 4.53854\beta^2$$
$$+ 14.24265\beta^3 - 19.2214\beta^4 + 8.54038\beta^5. \quad (10.42)$$

Solutions to (10.42) for various β's are given for abrupt contractions in Table 10.4

It is clear from Figure 10.18 that compressible loss coefficients across abrupt contractions differ significantly from, but approach (as the flow nears zero), the relevant constant density loss coefficients. This means that the incompressible loss coefficients of (10.42) should not be used for compressible flows except at the lowest flows. However, the total pressure loss (expressed as the ratio, TPR) across an abrupt contraction, for the

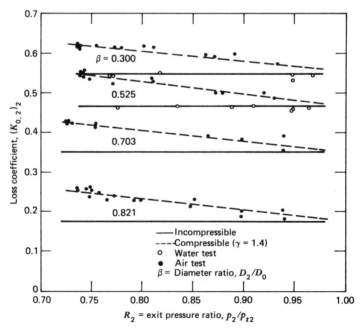

Figure 10.18 Comparison of abrupt contraction loss coefficients for compressible and constant density flows (after Benedict et al. [8]).

Table 10.4 Typical Loss Coefficients for Abrupt Contractions [Based on (10.42)]

β	$K_{\text{step-down}}$
0	0.58
0.1	0.58
0.2	0.56
0.3	0.54
0.4	0.52
0.5	0.49
0.6	0.43
0.7	0.34
0.8	0.21
0.9	0.07
1.0	0

same pertinent static/total pressure ratio (R_2), essentially is conserved in the sense of being approximately the same for all fluids [8]. This means that the simplistic constant density relation of (10.41) can be used to predict the drop in total pressure for any fluid flowing over an abrupt contraction in a piping system.

Example 10.11. Compare the loss coefficients across an abrupt contraction as obtained by Crane, the SAE, and (10.31), (10.34), and (10.42), for $\beta = 0.5$.

Solution: By Figure 10.4:

$$K_{\text{SAE}} = 0.445, \quad K_{\text{Crane}} = 0.330.$$

By (10.31): At $C_c = 0.64$, from Figure 10.17 and Table 10.3,
$K_{10.31} = 0.366$ if $C_v = 0.99$, as in text,
$K_{10.31} = 0.443$ if $C_v = 0.975$, from Table 10.2.
By (10.34): At $C_D = 0.815$, as in text,
$K_{10.34} = 0.474$.
By (10.42):

$$K_{10.42} = 0.487.$$

Example 10.12. Find the total pressure loss across an abrupt contraction if $\beta = 0.5$ and $R_2 = 0.9$.

390 Loss Characteristics of Piping Components

Solution: By example 10.11, $(K_{0,2})_2 \simeq 0.48$ at $\beta = 0.5$.
By (10.41):

$$\frac{p_{t2}}{p_{t0}} = \frac{1}{1 + 0.1 \times 0.48} = 0.954,$$

which indicates that the total pressure drops across this abrupt contraction by about 4.6%.

10.9 INLETS AND EXITS

The fluid experiences a certain pressure drop when entering a piping system from a plenum. This is called the inlet loss, and it takes on several values, depending on the inlet piping configuration (see Figure 10.19). We express this loss in terms of the dimensionless loss coefficient, where usual

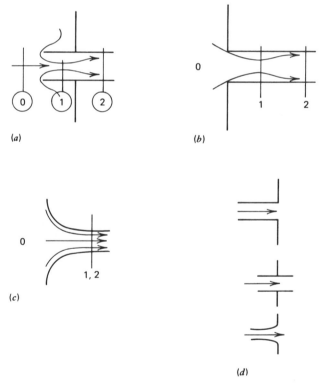

Figure 10.19 Various inlet and exit piping configurations. (*a*) Borda inlet. (*b*) Flush inlet. (*c*) Well-rounded inlet. (*d*) Various exits.

means are employed to convert the loss coefficient to equivalent pressure drop. Note that all the inlet losses are special cases of the abrupt contraction loss of Section 10.8.

Borda Tube Inlet. A Borda inlet (see Figure 2.23) is an orifice with a short pipe extending before and after the opening. The objective of this reentrant tube arrangement is to ensure *zero wall force* on the issuing jet. Following the notation of Figure 10.19a, we can derive the *constant density* loss coefficient of the Borda inlet by combining a momentum and energy balance between stations 0 and 1.

Momentum:

$$(p_0 - p_1)A_2 = \left(\frac{\rho V_1 A_1}{g_c}\right) V_1.$$

Energy:

$$\frac{p_0}{\rho} = \frac{p_1}{\rho} + \frac{V_1^2}{2g_c}.$$

Combined:

$$\frac{\rho A_1 V_1^2}{A_2 g_c} = \frac{\rho V_1^2}{2g_c}$$

or, by (10.33),

$$(C_c)_{\text{Borda, inc}} = \frac{A_1}{A_2} = \frac{1}{2}. \tag{10.43}$$

Equation 10.43 also follows directly from (2.51), the incompressible force coefficient (fi), when f_{Borda} is set to zero.

Three assumptions are inherent in (10.43):

1. Flow is from a plenum where $V_0 = 0$.
2. The pressure at 1 is assumed to act over A_2.
3. There are no wall shear stresses between 0 and 1, that is, there is no loss between 0 and 1, and hence $C_{v1} = 1$.

The loss coefficient for a Borda inlet is obtained from (10.31), at $C_{v1} = 1$ and $C_c = 0.5$, to yield

$$(K_{0,2})_{2, \text{ Borda, inc}} = \frac{1}{0.5^2} - \frac{2}{0.5} + 1 = 1. \tag{10.44}$$

392 Loss Characteristics of Piping Components

The *compressible* Borda contraction coefficient for *subsonic* flow can be given as

$$(C_c)_{\text{Borda, subsonic}} = \frac{1-R_b}{[2\gamma/(\gamma-1)]R_b^{1/\gamma}[1-R_b^{(\gamma-1)/\gamma}]}, \qquad (10.45)$$

where $R_b = p_2/p_t$. This follows directly from the generalized contraction coefficient of (3.118), evaluated at the Borda conditions of $\beta = 0$ and $f = 0$.

When R_1, at the vena contracta, reaches the critical pressure ratio (R^*) of (3.125) and (8.62), we obtain from (3.118) the Borda contraction coefficient for *choked* flow at inlet as

$$(C_c)_{\text{Borda, choked}} = \frac{1-R_b}{R^*(\gamma+1) - R_b}. \qquad (10.46)$$

Example 10.13. Evaluate the contraction coefficient of a Borda inlet at $R_b = 1$, 0.9, and 0.2 for air at $\gamma = 1.4$.

Solution:

At $R_b = 1$. Equation 10.45 cannot be evaluated by direct means since it yields the indeterminate form, $C_c = 0/0$. However, by L'Hôpital's rule:

$$\underset{R_b \to 1}{\text{Limit}} (C_c)_{\text{Borda, subsonic}} = \frac{\lim \dfrac{d}{dR_b}(1-R_b)}{\lim \dfrac{d}{dR_b}\left\{\left(\dfrac{2\gamma}{\gamma-1}\right) R_b^{1/\gamma}\left[1 - R_b^{(\gamma-1)/\gamma}\right]\right\}}$$

$$= \frac{0-1}{\left(\dfrac{2\gamma}{\gamma-1}\right)\left[\dfrac{1}{\gamma} R_b^{(1-\gamma)/\gamma} - 1\right]}.$$

When this expression is evaluated at $R_b = 1$ and $\gamma = 1.4$, we obtain

$$(C_c)_{\text{Borda, } R_b \to 1} = \frac{-1}{(2.8/0.4)[(1/1.4)-1]} = \frac{1}{2}.$$

It is satisfying that the limiting contraction of the compressible formulation of (10.45) reduces to the constant density result of (10.43), as it should.

At $R_b = 0.9$: Equation 10.45 yields

$$(C_c)_{\text{Borda, subsonic}} = \frac{1-0.9}{(2.8/0.4) \times 0.9^{1/1.4}(1-0.9^{0.4/1.4})} = 0.5194.$$

At $R_b = 0.2$: Equation 10.46 yields

$$(C_c)_{\text{Borda, choked}} = \frac{1-0.2}{0.52828 \times 2.4 - 0.2} = 0.74915.$$

Flush Inlet (see Figure 10.19b). This is Weisbach's short pipe model with $\beta = 0$, where $C_D = 0.815$ and (10.34) applies. Thus

$$(K_{0,2})_{2,\text{ flush inlet}} = \frac{1}{0.815^2} - 1 = 0.5. \qquad (10.47)$$

Well-Rounded Inlet (see Figures 10.12 and 10.19c). Empirically, the loss in a well-rounded inlet pipe is extremely small. Because of the gradual fluid acceleration into the pipe, there is essentially no contraction, and likewise the velocity coefficient is essentially 1. Thus, as for reducers, (10.16) applies, and we have

$$(K_{0,2})_{2,\text{ well-rounded inlet}} = 0.05. \qquad (10.48)$$

Exits. Similarly to the inlet loss, the fluid experiences another pressure loss when going *from* a piping system to a plenum. This is called the exit loss, and, in terms of the dimensionless loss coefficient, it takes on only one value, that is, exit loss is simply a special case of the abrupt enlargement loss of Section 10.7, where $\beta = 0$. Thus, according to Figure 10.19d and (10.22), we have

$$(K_{0,1})_{0,\text{ exit}} = 1, \qquad (10.49)$$

where (10.49) applies regardless of whether the pipe protrudes into the exit plenum, is well rounded at exit, or is flush. Equation 10.49 indicates that one velocity head is lost when the fluid leaves any piping system.

10.10 DIFFERENTIAL PRESSURE TYPE FLUID METERS

Much has been written concerning the differential pressure type fluid meter. This includes work on the nozzle, the venturi, and the orifice meter. Here we limit our interest to the flow and loss characteristics of each of these fluid meters.

The most elementary indicator of the unrecoverable loss across a fluid meter is the drop in total pressure $(p_{t1} - p_{t4})$ (see Figure 10.20). In this figure, 1 represents the inlet section; 2, the geometric throat section; 3, the fluid jet throat section; and 4, the downstream location where the expanding jet first reattaches to the pipe wall (i.e., where the maximum static pressure is attained).

394 Loss Characteristics of Piping Components

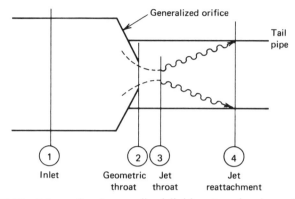

Figure 10.20 Schematic of generalized fluid meter, showing axial locations.

The total pressure drop is often expressed in terms of a loss coefficient, as in (7.15) and (8.38). According to the notation of Figure 10.20, we have

$$p_{t1} - p_{t4} = (K_{1,4})_2 \left(\frac{\rho V_2^2}{2g_c} \right), \tag{10.50}$$

where $(K_{1,4})_2$ signifies the loss coefficient from 1 to 4, based on the velocity at 2.

Limiting this discussion to constant density flow, we note that such loss coefficients are additive, provided that they are based on the same reference area. Hence

$$(K_{1,4})_2 = \left[(K_{1,3})_3 + (K_{3,4})_3 \right] \left(\frac{A_2}{A_3} \right)^2. \tag{10.51}$$

To express this loss coefficient in terms of the dimensionless geometry, discharge coefficient, and contraction coefficent parameters, it is necessary to first write expressions for the conservation of mass and of energy.

The *actual* flow rate can be expressed as

$$\dot{m} = \rho A_1 V_1 = \rho A_2 V_2 = \rho A_3 V_3. \tag{10.52}$$

Similarly, a *reference* flow rate, based on the same maximum static pressure drop across the fluid meter $(p_1 - p_3)$ (where p_3 should be measured at the vena contracta of an orifice or in the throat of a nozzle), and the same geometric areas, A_1, A_2, A_4, but in the absence of flow losses and jet contractions, can be expressed as

$$\dot{m}'' = \rho A_1 V_1'' = \rho A_2 V_2'', \tag{10.53}$$

where the double prime signifies the no-loss/no-contraction condition. Since the fluid meter discharge coefficient is defined as

$$C_D = \frac{\dot{m}}{\dot{m}''} = \frac{A_3 V_3}{A_2 V_2''}, \qquad (10.54)$$

and the fluid meter contraction coefficient is defined as

$$C_c = \frac{A_3}{A_2}, \qquad (10.55)$$

it follows that

$$C_D = C_c \frac{V_3}{V_2''}. \qquad (10.56)$$

Just as two expressions were given for continuity, namely, (10.52) and (10.53), so from actual and ideal energy balances we have

$$\frac{p_1}{\rho} + \frac{V_1^2}{2g_c} = \frac{p_3}{\rho} + \frac{V_3^2}{2g_c} + (K_{1,3})_3 \frac{V_3^2}{2g_c} \qquad (10.57)$$

and

$$\frac{p_1}{\rho} + \frac{V_1''^2}{2g_c} = \frac{p_3}{\rho} + \frac{V_2''^2}{2g_c}. \qquad (10.58)$$

From these four equations there results

$$V_2''^2 (1 - \beta_{2,1}^4) = V_3^2 \left[1 + (K_{1,3})_3 - \beta_{2,1}^4 C_c^2 \right], \qquad (10.59)$$

where $\beta_{2,1} = D_2/D_1$, or, in terms of (10.56),

$$\frac{C_D^2}{1 - \beta_{2,1}^4} = \frac{C_c^2}{1 + (K_{1,3})_3 - \beta_{2,1}^4 C_c^2}. \qquad (10.60)$$

Equation 10.60 can be rearranged to yield

$$(K_{1,3})_3 = \frac{C_c^2 (1 - \beta_{2,1}^4)}{C_D^2} - (1 - \beta_{2,1}^4 C_c^2). \qquad (10.61)$$

The loss coefficient from stations 3 to 4 takes two forms, depending on whether an uncontrolled expansion takes place from 3 to 4 or a diffuser is provided to guide the expansion. For orifices and nozzles, from 3 to 4, there results the familiar Carnot-Borda relation of (7.13), that is,

$$(K_{3,4})_{3,\text{ orifice and nozzle}} = \left(1 - \frac{A_3}{A_4}\right)^2 = (1 - \beta_{2,4}^2 C_c)^2. \qquad (10.62)$$

For venturis, which have a diffusing section after the throat, there results (10.20), that is,

$$(K_{3,4})_{3,\text{ venturi}} = (1 - \beta_{2,4}^4)(1 - \eta), \qquad (10.63)$$

where η represents the diffuser efficiency of (10.19) and Figure 10.14.

Combining (10.61) and (10.62) or (10.63), according to (10.51), yields the required generalized loss coefficients for fluid meters [29], namely,

$$(K_{1,4})_{2,\text{ orifices and nozzles}} = \left[\left(\frac{1 - \beta_{2,1}^4}{C_D^2}\right) - \frac{1}{C_c^2} + \beta_{2,1}^4\right]_{1,3}$$

$$+ \left(\frac{1}{C_c^2} - 2\frac{\beta_{2,4}^2}{C_c} + \beta_{2,4}^4\right)_{3,4} \qquad (10.64)$$

and

$$(K_{1,4})_{2,\text{ venturi}} = \left[\left(\frac{1 - \beta_{2,1}^4}{C_D^2}\right) - 1 + \beta_{2,1}^4\right]_{1,3}$$

$$+ \left[(1 - \beta_{2,4}^4)(1 - \eta)\right]_{3,4}. \qquad (10.65)$$

Applications. For the usual orifice installation in a pipe, that is, when $\beta_{2,1} = \beta_{2,4} = \beta$, (10.64) becomes

$$(K_{1,4})_{2,\text{ pipe orifice}} = \left(\frac{1 - \beta^4}{C_D^2}\right) - 2\beta^2\left(\frac{1}{C_c} - \beta^2\right). \qquad (10.66)$$

For the inlet plenum installation, that is, when $\beta_{2,1} = 0$, (10.64) becomes

$$(K_{1,4})_{2,\text{ inlet plenum orifice}} = \frac{1}{C_D^2} - 2\frac{\beta_{2,4}^2}{C_c} + \beta_{2,4}^4. \qquad (10.67)$$

For the usual nozzle installation in a pipe, that is, when $C_c = 1$, (10.64) becomes

$$(K_{1,4})_{2,\text{ pipe nozzle}} = \left(\frac{1-\beta^4}{C_D^2}\right) - 2\beta^2(1-\beta^2). \tag{10.68}$$

And, for the inlet plenum nozzle installation, (10.64) becomes

$$(K_{1,4})_{2,\text{ inlet plenum nozzle}} = \frac{1}{C_D^2} - 2\beta_{2,4}^2 + \beta_{2,4}^4. \tag{10.69}$$

For the usual venturi installation in a pipe, (10.65) becomes

$$(K_{1,4})_{2,\text{ pipe venturi}} = (1-\beta^4)\left(\frac{1}{C_D^2} - \eta\right). \tag{10.70}$$

And, for the inlet plenum venturi, (10.65) becomes

$$(K_{1,4})_{2,\text{ inlet plenum venturi}} = \left(\frac{1}{C_D^2} - 1\right) + (1-\beta_{2,4}^4)(1-\eta). \tag{10.71}$$

ASME Loss Parameter. The report of the ASME Research Committee on Fluid Meters (4th, 5th, and 6th editions) [30] expresses the loss in total pressure in terms of the maximum static pressure drop across the fluid meter. The resulting dimensionless ratio:

$$P_{\text{ASME}} = \frac{p_{t1} - p_{t4}}{p_1 - p_2} \tag{10.72}$$

has been published for the case when $D_1 = D_4$, in terms of experimentally determined curves (to be presented shortly).

This parameter P_{ASME}, of (10.72), has been related [29] to the generalized loss coefficient through the fluid meter discharge coefficient as

$$P_{\text{ASME}} = \frac{(K_{1,4})_2 C_D^2}{1 - \beta_{2,1}^4}. \tag{10.73}$$

It remains to determine C_D and C_c for orifices, nozzles, and venturis, before we can obtain the loss coefficients of (10.66) through (10.71) and the ASME loss parameter of (10.73).

Discharge Coefficient of a Nozzle. There are several methods for obtaining the discharge coefficient of a nozzle. The basic approach consists of obtaining the *actual flow rate*, by a catching-weighing-timing procedure using a liquid, and comparing it with the *ideal flow rate*, as deduced from pressure drop measurements taken across the nozzle, according to (10.36). This experimental method is discussed at some length in Reference 31. The ASME Research Committee on Fluid Meters has summarized in graphical form the results of thousands of such tests. An empirical equation, which best represents these graphical results, has been given [32] in the form of a cubic as

$$C_D = 0.19436 + 0.152884(\ln R_d) - 0.0097785(\ln R_d)^2 + 0.00020903(\ln R_d)^3, \tag{10.74}$$

where R_d is the Reynolds number based on the throat diameter.

Another approach toward determining the nozzle C_D is by a friction factor-kinetic energy coefficient formulation. Such an approach has been suggested by a number of workers in this area [33–35]. A generalized formulation of this type has been given [36] as

$$C_D = \frac{C_c(1-\beta^4)^{1/2}}{(K_{1,2} + \alpha_2 - \alpha_1 C_c^2 \beta^4)^{1/2}}, \tag{10.75}$$

where α represents the kinetic energy coefficient of (5.65). Equation 10.75 can be applied to any differential pressure type fluid meter. Thus, for a throat tap nozzle, for example, (10.75) has been expressed [36] as

$$C_{DL} = \left(\frac{1-\beta^4}{1-\beta^4 + 9.7156 R_d^{-0.5} - 0.4505 \beta^{3.8} R_d^{-0.2}} \right)^{1/2} \tag{10.76}$$

for the laminar boundary layer regime, and as

$$C_{DT} = \left(\frac{1-\beta^4}{1-\beta^4 + (0.17 - 0.4505 \beta^{3.8}) R_d^{-0.2}} \right)^{1/2} \tag{10.77}$$

for the turbulent regime.

The boundary layer approach, which was outlined in Chapter 4, and is detailed in Chapter 14, provides still another method for determining the C_D of flow nozzles (see e.g., [37]–[41]). This work has been summarized [42] for throat tap nozzles by the relation

$$C_D = \left(1 - 2\frac{\delta_2^*}{R_2} \right) \left[\frac{1-\beta^4}{1 - \beta^4 \left(\frac{1 - 2\delta_2^*/R_2}{1 - 2\delta_1^*/R_1} \right)^2} \right]^{1/2} \tag{10.78}$$

Differential Pressure Type Fluid Meters 399

where δ_1^* and δ_2^* represent the boundary layer displacement thicknesses of (4.13) at the nozzle inlet and throat respectively.

The numerical results of (10.74), (10.76), (10.77), and (10.78) are compared, graphically, in Figure 10.21, along with the ASME nominal value of $C_D = 0.977$ for pipe wall taps at $R_d = 10^5$. Comparisons between throat and pipe wall tap nozzle discharge coefficients are given in [43].

All of the material given for nozzle C_D's thus far in this section concerns the flow of an *incompressible* fluid. When one must deal with *compressible* fluid metering, the effects of compressibility in both the subsonic and supercritical flow regimes must be considered. Representative works dealing with these rather complex perturbations on the relatively simple constant density results can be found in the following references [44-46].

Discharge Coefficient of an Orifice. As with the nozzle, there are several methods for defining the discharge coefficient of an orifice. Such coefficients have been tabulated in detail only for the sharp-edged orifice.

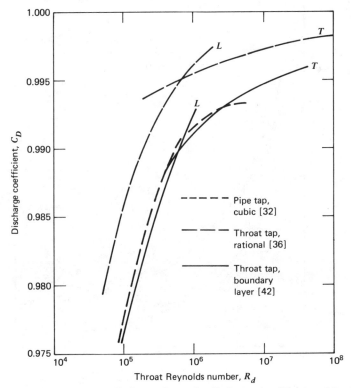

Figure 10.21 Various predictions of nozzle discharge coefficients ($\beta = 0.5$). L = laminar; T = turbulent.

400 Loss Characteristics of Piping Components

Rounded orifices, quadrant orifices, and other orifice edge shapes are not dealt with here. For more detail on the effect of orifice edge sharpness, however, the recent literature (e.g., [47]–[49]) should be consulted.

The ASME Research Committee on Fluid Meters has summarized the results of thousands of tests on sharp-edged orifices, in tabular form and by complex empirical equations. These will be given here only in representative graphs, as in Figure 10.22.

The generalized friction factor-kinetic energy coefficient formulation of (10.75) has been applied to the orifice (just as already given for the nozzle) and has been expressed [36] as

$$C_D = \left[\frac{1-\beta^4}{(1/C_c^2) - \beta^4 + 0.26 - 1.511(\beta - 0.35)^2 - 15R_d^{-0.5} - 0.4505\beta^{3.8}R_d^{-0.2}} \right]^{1/2}$$

(10.79)

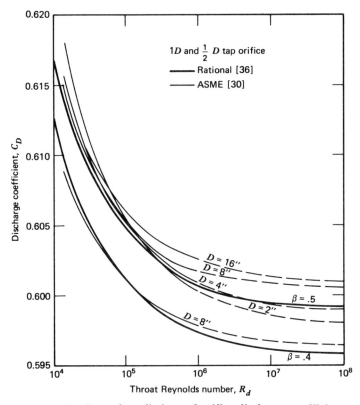

Figure 10.22 Several predictions of orifice discharge coefficients.

for pressure taps placed at one diameter upstream of the orifice face and for downstream taps placed either at the vena contracta location or at one-half pipe diameter downstream of the orifice face. In (10.79) the contraction coefficient of (10.40) is recommended, and representative values of C_c are given in Table 10.5.

The boundary layer approach to orifice discharge coefficients has not received much attention in the literature. A pioneer work on this subject is that by Hall [50].

The various approximations to orifice discharge coefficents are compared graphically in Figure 10.22, along with ASME nominal values of C_D, as given in Table 10.5.

Discharge Coefficient of a Venturi. The nominal ASME value for the discharge coefficient of a venturi at $R_d = 10^5$ is 0.985, as given in Table 10.5. The friction factor-kinetic energy formulations are identical with those of a throat tap nozzle, already given as (10.76) and (10.77). Much work has also been done on the venturi discharge coefficient from a boundary layer approach. Representative references include 38, 40, 42, and 51. However, as with the friction factor-kinetic energy formulations, the results are so similar in regard to C_D to those for nozzles that they need not be given here.

Comparisons of Fluid Meter Loss Coefficients. Given the various discharge and contraction coefficients, as required for (10.66) through (10.71) and as summarized in Table 10.5, we can now solve these equations for the various fluid meter loss coefficients. These are summarized in Table 10.5 and in Figure 10.23.

Furthermore, once these same factors are available for the nozzle, venturi, and orifice, we can easily evaluate (10.73) for the ASME loss parameter. This work is presented in Figure 10.24, where the excellent agreement between the derived P_{ASME}, based on the generalized fluid meter loss coefficient of (10.64) and (10.65), and the experimental loss parameters published by the ASME [30] justifies increased confidence in our loss formulations. A new study of loss characteristics of fluid meters [52] also indicates that our loss predictions are well confirmed by experiment.

Example 10.14. Determine the expected loss in total pressure across a nozzle, installed in a pipe such that $\beta = 0.6$, when the static pressure drop across the nozzle is 22.5 psi.

Solution: By Figure (10.24): At $\beta = 0.6$, we read $P_{\text{ASME}} = 0.54$.

Table 10.5 Summary of Solutions to Generalized FluidMeter Loss Equations at $R_d = 10^5$

| | C_D | | C_C | Loss Coefficient, $(K_{1,4})_2$ | | | | | |
| | | | | Orifice | | Nozzle | | Venturi | |
$\beta_{2,4}$	Nozzle	Venturi	Orifice	Orifice	Pipe (10.66)	Plenum (10.67)	Pipe (10.68)	Plenum (10.69)	Pipe (10.70)	Plenum (10.71)
0.2	0.977	0.985	0.599	0.620	2.657	2.660	0.969	0.969	0.130	0.130
0.4	0.977	0.985	0.602	0.630	2.232	2.296	0.752	0.753	0.127	0.128
0.6	0.977	0.985	0.613	0.655	1.476	1.755	0.451	0.457	0.114	0.188
0.8	0.977	0.985	0.618	0.730	0.612	1.132	0.158	0.177	0.077	0.090

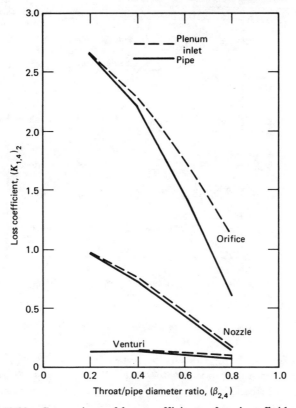

Figure 10.23 Comparison of loss coefficients of various fluid meters.

By (10.72):

$$\Delta p_t = P_{ASME} \times \Delta p = 0.54 \times 22.5 = 12.15 \, \text{psi}.$$

Incidentally, by (10.73) we find the loss coefficient to be

$$(K_{1,4})_2 = \frac{P_{ASME}(1-\beta^4)}{C_D^2} = \frac{0.54(1-0.6^4)}{0.977^2} = 0.49,$$

which compares favorably with the formulation of (10.68), using consistent values, that is,

$$(K_{1,4})_2 = \frac{1-0.6^4}{0.977^2} - 2 \times 0.6^2(1-0.6^2) = 0.45.$$

404 Loss Characteristics of Piping Components

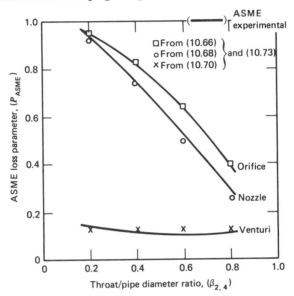

Figure 10.24 Comparison of experimental and derived values of the ASME loss parameter for fluid meters.

Methods for Reducing Fluid Meter Loss. It is well known, and can easily be verified by computing separately the two terms involved in (10.64) and (10.65), that most of the loss across a fluid meter is due to the unrecoverable loss in total pressure as the fluid diffuses from the throat to the downstream pipe.

One method used for reducing this portion of the overall fluid meter loss is to add a separable diffuser to a nozzle [14]. In effect, the nozzle approaches a venturi in respect to loss, while maintaining its unique and often well-established discharge coefficient. Figure 10.25 summarizes theoretical and experimental results for such a diffuser-modified nozzle. The great reduction in total pressure loss for the diffuser-modified nozzle, over the usual nozzle installation, is quite clear from this figure. It is further interesting, from a fluids viewpoint, that, as compressibility becomes important (i.e., as R_{throat} approaches R^*), the air flow pattern after the throat differs from the water flow pattern, and losses in air greatly exceed water losses. Finally, as a tie-in with Chapter 9, note that at sufficiently low pressures, approaching the vapor pressure in the throat (for the particular fluid temperature), the water flashes (vaporizes) and, in a manner exactly analogous to the air flow, losses are once more greatly amplified and the flow rate is again maximized at this cavitation point.

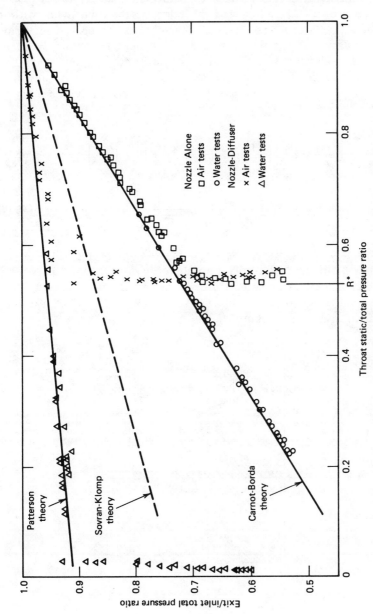

Figure 10.25 Loss characteristics of nozzle and diffuser-modified nozzle in air and water.

406 Loss Characteristics of Piping Components

A step diffuser also can be used to reduce fluid meter losses [53]. In practice, such a diffuser is realized by following the fluid meter with an intermediate pipe smaller in diameter than the final pipe. These various fluid meter arrangements under consideration here are shown in Figure 10.26. By methods already discussed at length in this chapter, it has been shown [53] that the total pressure loss across any nozzle arrangement involving any liner of diameter given by $\beta_{4,5} = D_4/D_5$ can be characterized by

$$\frac{p_{t5}}{p_{t1}} = 1 - \left[\left(\frac{1}{C_D^2} - 1 \right)(1 - \beta_{1,2}^4) + \left(1 - \frac{\beta_{1,2}^2}{\beta_{4,5}^2} \right)^2 \right.$$
$$\left. + \beta_{1,2}^4 \left(\frac{1}{\beta_{4,5}^2} - 1 \right)^2 \right] \frac{1 - R_1}{\beta_{1,2}^4}. \qquad (10.80)$$

The performances, in respect to loss, of the nozzle-step diffuser, the nozzle-conical diffuser, and of plain nozzles of various β's are compared in Figure 10.27. Clearly, one should use the largest β that is acceptable to minimize fluid meter losses or, alternatively, go to a downstream diffuser arrangement, as shown in Figure 10.26.

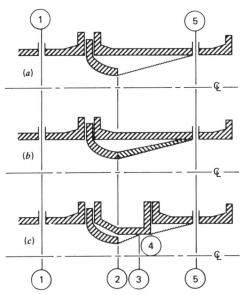

Figure 10.26 Various fluid meter arrangements to reduce loss. (a) = Typical nozzle. (b) = Nozzle-conical diffuser. (c) = Nozzle-step diffuser.

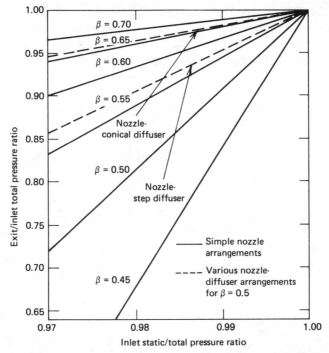

Figure 10.27 Loss curves for the design of various nozzle installations.

As a tie-in with the *equivalent length concept* of Example 10.3, note that the usual fluid meter arrangement involving a $\beta \leqslant 0.5$ nozzle is equivalent, in respect to loss, to about 500 ft of straight 1 ft diameter pipe, according to the following simplified version of (7.14):

$$\frac{p_{t5}}{p_{t1}} = 1 - \left(f\frac{L_{1,5}}{D}\right)(1 - R_1), \tag{10.81}$$

where the following values are taken as representative: $p_{t5}/p_{t1} \simeq 0.8$, $R_1 \simeq 0.98$, and $f = 0.02$.

10.11 VALVES

Many valve types are used in piping systems to do a variety of jobs. Some control the flow rate, others isolate portions of the piping, and still others check the flow, act as relief ports, meter the flow, and so on; all valves operate by introducing flow restrictions into the piping system. In this section our purpose is to set down typical loss characteristics of conventional valves.

408 Loss Characteristics of Piping Components

The flow of fluids through valves is difficult to predict by purely analytical means because of the complex geometries involved in valve bodies and plugs (see Figure 10.28). Thus all of the equations available in the literature involve *empirical* coefficients to handle these complexities.

Valve Flow Coefficient. One universal yardstick of valve capacity is the valve flow coefficient (C_V). This is defined as

$$C_V = GPM\sqrt{\frac{SG_x}{\Delta p_t}}, \qquad (10.82)$$

where *GPM* is the flow through the value (U.S. gal/min),
SG_x is the specific gravity of the flowing fluid
$= \rho_x / \rho_{H_2O \text{ at } 60°F}$, where $\rho_{H_2O \text{ at } 60°F} = 62.367 \text{ lbm/ft}^3$, and
Δp_t is the total pressure drop across the valve (psi).

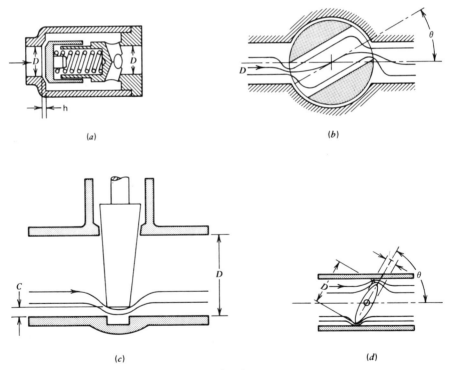

Figure 10.28 Schematics of several valve types. (*a*) Relief valve. (*b*) Ball valve. (*c*) Gate valve. (*d*) Butterfly valve.

Thus C_V indicates the water flow in gallons per minute that could be passed through the valve at a pressure drop of 1 psi across the valve when the valve is in the wide-open position [54]. The determination of C_V is made when *no flashing* flow is present and when the valve is installed in a *horizontal* section of the piping.

There does not seem to be a Reynolds number effect on C_V; however, test values of C_V are often found to be *lower* than factory quoted C_V's by about 10%. This is believed to be explanined by the absence of a sufficient number of straight pipe sections before and after the valve in the actual test situation, as compared with the factory arrangement. Note that a lower C_V indicates a greater pressure drop (Δp_t) across the valve.

Valve Loss Coefficient. The usual formulation for a loss coefficient can be given in terms of valve notation as

$$K_V = \frac{144 \Delta p_t}{\rho V^2 / 2g_c}, \qquad (10.83)$$

where Δp_t is in pounds per square inch and $\rho V^2/2g_c$ is in pounds force per square foot. This can be related quite simply to the valve flow coefficient of (10.82) as follows. Solving (10.82) and (10.83) for Δp_t, we have

$$\Delta p_t = \frac{K_V}{144} \frac{\rho V^2}{2g_c} = SG_x \left(\frac{GPM}{C_V} \right)^2$$

or

$$C_V = \sqrt{\frac{SG_x 2g_c}{\rho_x}} \left(\frac{GPM}{V} \right) \frac{12}{\sqrt{K_V}}. \qquad (10.84)$$

But

$$GPM \left(\frac{231 \text{ in.}^3}{\text{gal}} \times \frac{1 \text{ min}}{60 \text{ sec}} \times \frac{1 \text{ ft}^3}{1728 \text{ in.}^3} \right) = AV = \frac{\pi D^2 V}{4 \times 144}, \qquad (10.85)$$

where D is the pipe diameter in inches and V is the fluid velocity in feet per second.

Combining (10.84) and (10.85), we have

$$C_V = \frac{29.8395 D^2}{\sqrt{K_V}}. \qquad (10.86)$$

Example 10.15. Find the valve loss coefficient for a 4 in. valve if $C_V = 236$.

Solution: By (10.86):

$$K_V = \frac{29.84^2 D^4}{236^2} = 4.093 \simeq 4.1.$$

Example 10.16. If the measured total pressure drop across a 4 in. valve is 0.341 psi at a flow rate of 18.017 lbm/sec and a density of 62.323 lbm/ft³, find the loss coefficient and the flow coefficient for the valve.

Solution: $D = 4.025$ for the 4 in. valve.
By continuity:

$$\frac{\rho V^2}{2g_c} = \frac{\dot{m}^2}{2g_c \rho A^2}$$

$$= \frac{18.017^2}{2 \times 32.174 \times 62.323 \times \left[(\pi/4)(4.025/12)^2\right]^2}$$

$$= 10.3672 \text{ psf.}$$

By (10.83):

$$K_V = \frac{144 \times 0.341}{10.3672} = 4.736.$$

By (10.86):

$$C_V = \frac{29.84 \times 4.025^2}{\sqrt{4.736}} = 222.12.$$

And, as a check of consistency, by (10.85):

$$GPM = \frac{\dot{m}}{\rho} \left(\frac{60 \times 1728}{231} \right) = 129.753 \text{ gal/min,}$$

and, by (10.82):

$$C_V = 129.753 \sqrt{\frac{62.323}{62.367 \times 0.341}} = 222.12.$$

A good practical source for loss coefficients for many valve types at various opening positions is given by Dodge [55].

Table 10.6 summarizes typical loss coefficients for several valve types at the fully open position.

Gas Sizing for Control Valves. For compressible fluids, correction factors in addition to C_V are required to account for compressibility, gas specific heat, and so on. Buresh and Schuder [56, 57] have given a "universal" gas sizing equation which, for air, can be written as

$$Q = \sqrt{\frac{520}{T}} \, C_1 C_V p_1 \sin\left(\frac{3417}{C_1}\sqrt{\frac{\Delta p}{p_1}}\right), \qquad (10.87)$$

where Q is the volumetric flow rate in cubic feet per hour (of standard air at 14.7 psia and 60° F);

$$C_1 = \frac{C_g}{C_V}, \qquad (10.88)$$

which varies widely with valve internal geometry, but for the majority of wide-open valves $\simeq 32$ to 36; and C_g is an experimental coefficient obtained from critical pressure drop in air tests according to

$$C_g = \frac{Q^*}{p_1\sqrt{520/T}}. \qquad (10.89)$$

The quantity in parentheses in (10.87) is the valve angle in degrees. Above 90°, critical flow is indicated, and (10.87) reduces to (10.89). Equation 10.87 was designed to predict flow rates from nearly incompressible flows at very low pressure differentials to critical flows.

Table 10.6 Typical Valve Loss Coefficients for Fully Open Valves

Type of Valve	K_{valve}(open)	
	Screwed	Flanged
Globe	10	5
Angle	5	2
Check	3	—
Butterfly	0.5	—
Gate	0.2	0.1

412 Loss Characteristics of Piping Components

Example 10.17. Find the air flow than can be passed by a 4 in. butterfly valve at a 60° disk opening, with $t = 60°$ F, $p_1 = 44.3$ psia, $C_V = 255$, and $C_1 = 25$, at $\Delta p_1 = 10$ psi and at $\Delta p_2 = 30$ psi.

Solution: By (10.87):

$$Q_1 = \sqrt{\frac{520}{60+460}} \times 25 \times 255 \times 44.3 \sin\left(\frac{3417}{25}\sqrt{\frac{10}{44.3}}\right).$$

Here the angle is 64.9° and $Q_1 = 255{,}824$ ft^3/hr.

$$Q_2 = 1 \times 25 \times 255 \times 44.3 \sin\left(\frac{3417}{25}\sqrt{\frac{30}{44.3}}\right).$$

Here the angle is 112.5°; hence 90° is used, and

$$Q_2 = Q^* = 25 \times 255 \times 44.3 = 282{,}412 \text{ ft}^3/\text{hr}.$$

As a check of consistency, by (10.88):

$$C_g = C_1 C_V = 25 \times 255 = 6375,$$

while, by (10.89):

$$C_g = \frac{Q^*}{p_1\sqrt{520/T_1}} = \frac{282{,}412}{44.3} = 6375.$$

Other Considerations. Pipe reducers are sometimes added on either side of a valve to allow the use of smaller, less expensive valves. However, since pipe reducers introduce additional restrictions on the flow, they cause a decrease in valve capacity. These effects have been estimated in the literature (see, e.g., [58] or [59]), and such sources should be consulted for more details on this effect.

Kneisel [60] has suggested the use of an expansion factor for valves, such as is used in orifice work. His equation can be expressed in the form of (10.87), and has been shown to be in essential aggreement with it. The advantage of the Kneisel equation is that it is based on the extensive information already available for orifices.

10.12 SCREENS

The drag force across a screen can be given as

$$F_D = C_D A_{\text{blocked}} \left(\frac{\rho V^2}{2g_c} \right) = \Delta p A_{\text{total}},$$

where C_D is the drag coefficient of the screen, $\rho V^2/2g_c$ is the dynamic pressure of the flowing fluid, Δp is the pressure drop across the screen, and the areas are shown in Figure 10.29. From this, we can express the drag coefficient as

$$C_D = \left(\frac{\Delta p}{\rho V^2/2g_c} \right) \frac{A_{\text{total}}}{A_{\text{blocked}}}. \tag{10.90}$$

The quantity in parentheses in (10.90) is, of course, the conventional loss coefficient (K_{screen}). Thus C_D also can be expressed as

$$C_D = \frac{K_{\text{screen}}}{1-\gamma}, \tag{10.91}$$

where the screen porosity (γ), is defined as

$$\gamma = \frac{A_{\text{open}}}{A_{\text{total}}}. \tag{10.92}$$

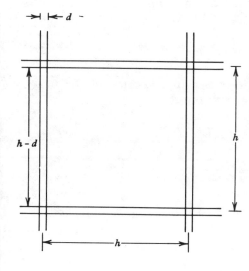

Figure 10.29 Various geometric relations for a square mesh screen.

$A_{\text{total}} = h^2, \quad A_{\text{open}} = (h-d)^2,$

$\gamma = \text{screen porosity} = \dfrac{A_{\text{open}}}{A_{\text{total}}},$

$A_{\text{blocked}} = A_{\text{total}} - A_{\text{open}}$
$= 2hd - d^2,$

$S = \text{screen solidity} = \dfrac{A_{\text{blocked}}}{A_{\text{total}}}.$

414 Loss Characteristics of Piping Components

This is related to the screen solidity (S), defined as

$$S = \frac{A_{\text{blocked}}}{A_{\text{total}}} \qquad (10.93)$$

by the elementary relation

$$S = 1 - \gamma. \qquad (10.94)$$

A third expression for C_D results from the experimental fact that at low Reynolds numbers the drag coefficient is inversely proportional to the screen Reynolds number ($R_d = \rho V d / \mu$), that is,

$$C_D = \frac{k}{R_d}, \qquad (10.95)$$

where k is the constant of proportionality and d is the screen wire diameter.

Equations 10.90, 10.91, and 10.95 lead to the nondimensional screen loss factor (λ):

$$\lambda = \frac{d \Delta p g_c}{\mu V} = \frac{k(1-\gamma)}{2}. \qquad (10.96)$$

The loss factor (λ) of (10.96) was defined and experimentally determined by Bernardi et al. [61], and the results were correlated by the least squares exponential regression:

$$\lambda = 599 e^{-7.01\gamma}. \qquad (10.97)$$

It has been found that the loss factor is a function, not of the screen Reynolds number, but only of the screen porosity.

Of course, the loss factor (λ) is simply related to the loss coefficient (K) via the expression

$$K_{\text{screen}} = \frac{2\lambda}{R_d}. \qquad (10.98)$$

Example 10.18. Find the loss coefficient of a screen of porosity 0.5 at a screen Reynolds number of 5.

Solution: By (10.97): $\lambda = 599 e^{-7.01(0.5)} = 17.998$.

By (10.98):

$$K_{\text{screen}} = \frac{2 \times 17.998}{5} = 7.2.$$

Another basic publication on screen loss coefficients is that of Cornell [62]. At low velocities, Cornell and Bernardi et al. are in essential agreement; Cornell also gives the screen loss coefficient for high velocity compressible flows, as shown in Figure 10.30.

Example 10.19. Find the loss coefficient of a screen of solidity 0.5 at an inlet Mach number of 0.25.

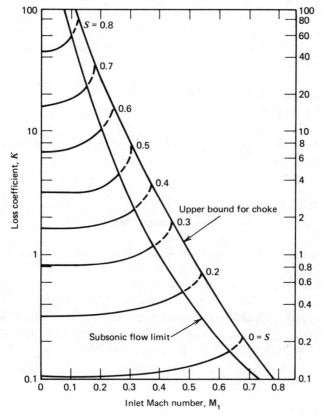

Figure 10.30 Screen loss coefficients for compressible flows (after Cornell [62]).

416 Loss Characteristics of Piping Components

Solution: By Figure 10.30:

$$K_{\text{screen}} = 4.$$

Although the compressible solutions of Figure 10.30 do not correspond exactly with the incompressible solutions of (10.97), the disparity is not large, and the variations with Mach number are clearly shown in Figure 10.30.

REFERENCES

1. H. Ito, "Pressure losses in smooth pipe bends," *Trans. ASME, J. Basic Eng.*, Vol. 82, 1960, p. 131.
2. H. Ito, "Friction factors for turbulent flow in curved pipes," *Trans. ASME, J. Basic Eng.*, Vol. 81, 1959, p. 123.
3. K. H. Beij, "Pressure losses for fluid flow in 90° pipe bends," *Natl. Bur. Stand. Publ.*, RP-1110, Vol. 21, July 1938.
4. R. J. S. Pigott, "Pressure losses in tubing, pipe and fittings," *Trans. ASME*, Vol. 72, 1950, p. 679.
5. R. J. S. Pigott, "Losses in pipe and fittings," *Trans. ASME*, Vol. 79, 1957, p. 1767.
6. *SAE Aero-Space Applied Thermodynamics Manual*, 1960, revised 1962, pp. A18-A20.
7. W. Sisson, "Equivalent length of straight duct for elbow," *Des. News*, February 21, 1972, p. 52.
8. R. P. Benedict, N. A. Carlucci, and S. D. Swetz, "Flow losses in abrupt enlargements and contractions," *Trans. ASME, J. Eng. Power*, January 1966, p. 73.
9. A. Vazsonyi, "Pressure loss in elbows and duct branches," *Trans. ASME*, April 1944, p. 177.
10. R. S. Benson and D. Woollatt, "Compressible flow loss coefficients at bends and T-junctions," *Engineer*, Vol. 222, January 28, 1966, p. 153.
11. G. N. Patterson, "Modern diffuser design," *Aircraft Eng.*, Vol. 10, 1938, p. 267.
12. H. Hilbrath, W. P. Dill, and W. A. Wacker, "The choking pressure ratio of a critical flow venturi," *ASME Pap.* 73-WA/FM-7, November 1973.
13. G. Sovran and E. D. Klomp, "Experimentally determined optimum geometries for rectilinear diffusers with rectangular, conical, or annular cross section," in *Fluid Mechanics of Internal Flow*, Ed., G. Sovran, Elsevier, 1967, p. 270.
14. R. P. Benedict, A. R. Gleed, and R. P. Schulte, "Air and water studies on a diffuser-modified flow nozzle," *Trans. ASME, J. Fluids Eng.*, June 1973, p. 169.

15. R. P. Benedict, J. S. Wyler, J. A. Dudek, and A. R. Gleed, "Generalized flow across an abrupt enlargement," *Trans. ASME, J. Eng Power*, Vol. 98, No. 3, July 1976, p. 329.
16. R. A. Dodge and M. J. Thompson, *Fluid Mechanics*, McGraw-Hill, 1937, p. 216.
17. H. Rouse, *Elementary Fluid Mechanics*, John Wiley, 1946, p. 265.
18. J. C. Hunsaker and B. G. Richtmire, *Engineering Applications of Fluid Mechanics*, McGraw-Hill, 1947, p. 152.
19. R. C. Binder, *Fluid Mechanics*, 4th ed., Prentice-Hall, 1962, p. 123.
20. H. J. Hughes and A. T. Safford, *Hydraulics*, Macmillan, 1912, p. 297.
21. J. Weisbach, *Mechanics of Engineering*, Van Nostrand, 1872, p. 821. Translated by E. B. Coxe.
22. J. R. Freeman, "The discharge of water through fire hose and nozzles," *Trans. ASCE*, Vol. 21, 1888, pp. 303–482.
23. M. P. O'Brien and G. H. Hickox, *Applied Fluid Mechanics*, McGraw-Hill, 1937, p. 210.
24. J. Weisbach, *Die Experimental-Hydraulik*, Engelhardt, 1855, p. 133.
25. R. von Mises, "Berechnung von Ausfluss-und Uberfallzahlen," *Z. VOI*, Vol. 61, 1917, p. 477. Translated as "Calculation of discharge- and overfall-numbers."
26. G. Kirchhoff, "Zur Theorie freier Flussigkeitsstrahlen," *Crelles J.*, Vol. 70, 1869, p. 289. Translated as "On the theory of free fluid jets."
27. "Flow of fluids," *Crane Co. Tech. Pap.* 410, 1957, p. A-26.
28. *SAE Aero-Space Applied Thermodynamics Manual*, Part A, "Incompressible flow," 1962, p. A-46.
29. R. P. Benedict, "Loss coefficients for fluid meters," *Trans. ASME, J. Fluids Eng.*, March 1977, p. 245.
30. H. S. Bean, Ed., *Fluid Meters—Their Theory and Application*, Report of ASME Research Committee on Fluid Meters, 6th ed., 1971.
31. R. P. Benedict, *Fundamentals of Temperature, Pressure, and Flow Measurements*, 2nd ed., John Wiley, 1977.
32. R. P. Benedict, "Most probable discharge coefficients for ASME flow nozzles," *Trans. ASME, J. Basic Eng.*, December 1966, p. 734.
33. W. S. Pardoe, "The effects of installation on the coefficient of venturi meters," *Trans. ASME*, Vol. 65, 1943, p. 337.
34. A. H. Shapiro and R. D. Smith, "Friction coefficients in the inlet length of smooth round tubes," *NACA TN* 1785, November 1948.
35. S. P. Hutton, "The prediction of venturi meter coefficients and their variation with roughness and age,"*Proc. Inst. Civil Eng.* Vol. 3, Part 3, 1954, p. 216.
36. R. P. Benedict and J. S. Wyler, "A generalized discharge coefficient for differential pressure type fluid meters," *Trans. ASME, J. Eng. Power*, October 1974, p. 440.

37. M. A. Rivas, Jr., and A. H. Shapiro, "On the theory of discharge coefficients for rounded-entrance flow meters and venturis," *Trans. ASME*, Vol. 78, April 1956, p. 489.
38. G. W. Hall, "Application of boundary-layer theory to explain some nozzle and venturi flow peculiarities," *Proc. Inst. Mech. Eng.*, Vol. 173, No. 36, 1959, p. 837.
39. K. C. Cotton and J. C. Westcott, "Throat tap nozzles used for accurate flow measurements," *Trans. ASME, J. Eng. Power*, October 1960, p. 247.
40. S. B. Au, "The prediction of axisymmetric turbulent boundary layer in conical nozzles," *J. Appl. Mech., Trans. ASME*, March 1974, p. 20.
41. R. P. Benedict and J. S. Wyler, "Analytical and experimental studies of ASME flow nozzles," *Trans. ASME, J. Fluids Eng.*, September 1978, p. 265.
42. R. P. Benedict, "Generalized fluid meter discharge coefficient based solely on boundary layer parameters," *Trans. ASME, J. Eng. Power*, October 1979, p. 572.
43. J. S. Wyler and R. P. Benedict, "Comparisons between throat and pipe wall tap nozzles," *Trans. ASME, J. Eng. Power*, October 1975, p. 569.
44. R. P. Benedict, "Generalized contraction coefficient of an orifice for subsonic and subcritical flows," *Trans. ASME, J. Basic Eng.*, June 1971, p. 99.
45. R. P. Benedict, "Generalized expansion factor of an orifice for subsonic and supercritical flows," *Trans. ASME, J. Basic Eng.*, June 1971, p. 121.
46. R. P. Benedict and R. D. Schulte, "A note on the critical pressure ratio across a fluid meter," *Trans. ASME, J. Fluids Eng.*, September 1973, p. 337.
47. F. R. Herning, "Studies concerning the problem of blunt edges of standard orifices and segment orifices," *Brennstoff-Waerme-Kraft*, Vol. 14, March 1962, p. 119.
48. K. A. Crockett and E. L. Upp, "The measurement and effects of edge sharpness on the flow coefficients of standard orifices," *Trans. ASME, J. Basic Eng.*, June 1973, p. 271.
49. R. P. Bendict, J. S. Wyler, and G. B. Brandt, "The effect of edge sharpness on the discharge coefficient of an orifice," *Trans ASME, J. Eng. Power*, October 1975, p. 576.
50. G. W. Hall, "Analytical determination of the discharge characteristics of cylindrical-tube orifices," *J. Mech. Eng. Sci.*, Vol. 5, No. 1, 1963, p. 91.
51. D. Lindley, "An experimental investigation of the flow in a classical venturimeter," *Proc. Inst. Mech. Eng.*, Vol. 184, Part 1, No. 8, 1969–1970, p. 133.
52. S. H. Alvi, K. Sridharan, and N. S. Lakshmana Rad, "Loss characteristics of orifices and nozzles," *Trans. ASME, J. Fluids Eng.*, September 1978, p. 299.
53. R. P. Benedict and A. R. Gleed, "Methods for reducing losses across fluid metering nozzles," *Trans. ASME, J. Fluids Eng.*, December 1976, p. 614.
54. H. W. Boger, "Recent trends in sizing control valves," *Instrum. Control Syst.*, 1969 Buyer' Guide, p. 16.
55. L. Dodge, "Fluid throttling devices," *Prod. Eng.*, March 30, 1964, p. 81.

56. J. F. Buresh and C. B. Schuder, "The development of a universal gas sizing equation for control valves," *IDA Trans.*, Vol. 3, No. 4, October 1964.
57. ISA Standard S39.3, "Control valve sizing equations for compressible fluids," April 1973.
58. A. Brodgesell, "Control valve sizing formulas," *Instrum. Technol.*, January 1972, p. 60.
59. H. D. Baumann, "Effect of pipe reducers on valve capacity," *Instrum. Control Syst.*, December 1968, p. 99.
60. O. Kneisel, "An empirical formula for air flow in a control valve," *Flow, Its Measurement and Control in Science and Industry*, Symposium, Pittsburgh, Pa., May 10–14, 1971.
61. R. T. Bernardi, J. H. Linehan, and L. H. Hamilton, "Low Reynolds number loss coefficient for fine mesh screens," *Trans. ASME, J. Fluids Eng.*, December 1976, p. 762.
62. W. G. Cornell, "Losses in flow normal to plane screens," *Trans. ASME*, May 1958, p. 791.

NOMENCLATURE

Roman

A area
C correction factor
C_c contraction coefficient
C_D discharge coefficient
   drag coefficient
C_g experimental valve coefficient
C_v velocity coefficient
C_V flow coefficient
D diameter
f friction factor
F force
g_c gravitational constant
GPM gallons per minute
h_{loss} head loss
k constant
K loss coefficient
L length
$\dot{m}$ mass flow rate

M   Mach number
p   pressure
P_{ASME} ASME loss parameter
Q   volumetric flow rate
r   inside elbow radius
R   elbow radius of curvature, static/total pressure ratio
R_D pipe Reynolds number
R_d screen Reynolds number
S   screen solidity
SG   specific gravity
t   empirical temperature
T   absolute temperature
TPR   total pressure ratio
V   fluid velocity

Greek

α empirical factor, angle
β diameter ratio, angle
γ specific heat ratio, screen porosity, angle
Γ generalized flow function
δ boundary layer thickness
Δ finite difference
λ tee factor, screen loss factor
η diffuser efficiency
μ fluid viscosity
ρ fluid density
φ area ratio

Subscripts

1,2   axial stations
t   total

11

Piping Networks

...a few moments ago, the single swift stream of the great artery, now the slower and slower streams of those smaller and smaller branches spreading out over the countryside, giving one a gay feeling, all that blood hurrying toward unnumbered and unnamed creeks and rills... —*Gustav Eckstein*

11.1 GENERAL REMARKS

A piping network is simply a number of piping components so connected that the flow to any component can come from several branches. Such piping systems can be extremely complicated, as, for example, the network of pipes used to distribute water throughout a city, or even the pipes involved in a single steam turbine plant. The number of piping junctions, the number of different piping loss elements, and the number of branches involved may be quite large, and the problem can be quite unwieldy.

In this chapter we discuss methods of analysis when multiple piping components are combined in various series and parallel arrangements. In Chapters 7 and 8 we considered piping components in series for both compressible and incompressible fluid flows. We also considered the method of combining losses of piping components arranged in series through the use of a common reference area. However, when more complex networks of pipes are encountered, additional methods of solution must be considered. For example, it is not always possible to solve a compressible flow network problem by the usual forward type solution, wherein one starts at the piping inlet and proceeds toward the exit by passing successively through each piping element. If the Mach number approaches unity, the flow may choke, in which case it may be necessary

422 Piping Networks

to obtain a solution by proceeding in the direction opposite to that of the flow. Once the network exceeds in complexity the simple series and parallel arrangements, special purpose computer solutions must be used. Such solutions will receive only scant attention here because of their highly specialized and highly complex nature. For more manageable networks, however, numerical examples will be given to illustrate some of the basic ideas involved in piping network solutions.

11.2 THE SERIES NETWORK

In a series network the overall piping consists of two or more piping elements connected so that the fluid flows through one element and then through another (see Figure 11.1). Since only one flow rate is involved, the solution is rather straightforward. For incompressible fluids we proceed as in Examples 7.1 through 7.7. To review briefly, we present one additional example.

Example 11.1. Water at 70°F flows from reservoir A to reservoir B, whose level is 20 ft below A, through a flush inlet and 1000 ft of 2 ft diameter pipe of absolute roughness (e_1) 0.005 ft, and then through 800 ft of 3 ft diameter pipe of roughness 0.001 ft (see Figure 11.1, but omit the elbows and valve for simplicity). Find the volumetric flow rate in cubic feet per second if $g = g_c$.

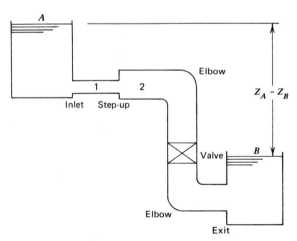

Figure 11.1 Piping components connected in series.

Solution: Assume that R_{D1}, the pipe Reynolds number in the 2 ft diameter pipe, is 10^5. Then, at $e_1/D_1 = 0.0025$, we read, from Figure 6.7, $f_1 = 0.026$. The consistent Reynolds number in pipe 2 is $R_{D2} = \frac{2}{3} R_{D1} = 6.7 \times 10^4$, that is,

$$\frac{R_{D2}}{R_{D1}} = \frac{V_2 D_2 / \nu}{V_1 D_1 / \nu} \quad \text{or} \quad R_{D2} = R_{D1}\left[\left(\frac{V_2}{V_1}\right)\left(\frac{D_1}{D_1}\right)\right]$$

$$= R_{D1}\left[\frac{D_1^2 D_2}{D_2^2 D_1}\right] = \frac{2}{3} R_{D1}.$$

At $e_2/D_2 = 0.00033$, we read $f_2 = 0.021$.

By (7.9):

$$(h_{\text{loss}})_{A,B} = \left[K_{\text{inlet}} + K_{\text{2 ft pipe}} + K_{\text{step-up}} + K_{\text{3 ft pipe}}\left(\frac{D_1}{D_2}\right)^4 + K_{\text{exit}}\left(\frac{D_1}{D_2}\right)^4\right]\frac{V_1^2}{2g_c},$$

based on the arbitrary choice of the 2 ft pipe for the reference area.

By (10.47):

$$K_{\text{inlet}} = 0.5.$$

By (7.8):

$$K_{\text{2 ft pipe}} = f_1 \frac{L_1}{D_1} = \frac{0.026 \times 1000}{2} = 13,$$

$$K_{\text{3 ft pipe}} = f_2 \frac{L_2}{D_2} = \frac{0.021 \times 800}{3} = 5.6.$$

By (10.22):

$$K_{\text{step-up}} = \left[1 - \left(\frac{D_1}{D_2}\right)^2\right]^2 = 0.3086.$$

By (10.49):

$$K_{\text{exit}} = 1.$$

Therefore:

$$(h_{\text{loss}})_{A,B} = (0.5 + 13 + 0.3086 + 5.6 \times 0.19753 + 0.19753)\frac{V_1^2}{2g_c}$$

or

$$(h_{\text{loss}})_{A,B} = 15.1123 \frac{V_1^2}{2g_c}.$$

But by (7.5):

$$(h_{\text{loss}})_{A,B} = \frac{g}{g_c}(Z_A - Z_B) = 20 \text{ ft-lbf/lbm},$$

since $p_A - p_B = 0$ and $V_A - V_B = 0$.

Equating the last two results, we have

$$V_1 = \left(\frac{20 \times 2 \times 32.174}{15.1123}\right)^{1/2} = 9.2282 \text{ ft/sec}.$$

It remains to check our original assumption for R_{D1}.

$$(R_{D1})_{\text{calc}} = \frac{V_1 D_1}{\nu} = \frac{9.2282 \times 2}{1.055 \times 10^{-5}} = 1.75 \times 10^6$$

rather than the assumed $R_{D1} = 10^5$.

For a second trial, we use $R_{D1} = 1.8 \times 10^6$ and $R_{D2} = 1.2 \times 10^6$, and find that $f_1 = 0.025$ and $f_2 = 0.016$. These new values of friction lead to $V_1 = 9.4703$ ft/sec and hence

$$Q = A_1 V_1 = 29.75 \text{ ft}^3/\text{sec}.$$

When the fluid is compressible, we proceed as above except that we must consider the Mach number and the possibility of choking the system somewhere. If we are dealing with subsonic flow throughout, the problem is complicated only because of the need to use compressible loss coefficients, as treated sufficiently in Chapters 8 and 10 and in Examples 8.8 through 8.18. If we must deal with high speed compressible fluid flows, we may need to resort to a *backward* type of solution, discussed in 11.5.

11.3 THE PARALLEL NETWORK

In a parallel network, two or more pipes branch out and then come together again downstream so that the flow divides among the branches (see Figure 11.2). The head losses are necessarily the same in every branch, and the individual branch flow rates are cumulative. Paralleling (also

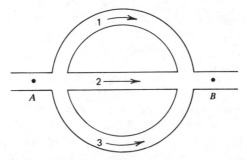

Figure 11.2 Piping branches connected in parallel.

called looping) is the usual method for arranging pipes to increase the capacity of the system.

It has long been noted that an analogy exists between the flow of a fluid and an electric current in both series and parallel circuits. Furthermore, in this same analogy, head loss in the fluid field is related to electric potential drop in the electric field (see Section 2.5.2 and Table 2.1). For complex piping networks, where advantage is often taken of this analogy by applying electric network analysis to fluid parameters, we can state mathematically [2, 3]:

1. The flow into each junction must equal the flow out of each junction. This is the continuity principle, wherein

$$Q_{total} = \Sigma Q = Q_1 + Q_2 + \cdots . \tag{11.1}$$

2. The algebraic sum of the pressure drops around each closed loop must be zero. Thus, for Figure 11.2, we can write

$$\Delta p_{AB} = \Delta p_{A1B} = \Delta p_{A2B} = \Delta p_{A3B}. \tag{11.2}$$

3. The Darcy-Weisbach equation, or its equivalent, must be satisfied for each branch. Thus

$$h_{loss} = f \frac{L}{D} \frac{V^2}{2g_c} = \frac{fL}{2g_c D} \left(\frac{16Q^2}{\pi^2 D^4} \right) = KQ^N, \tag{11.3}$$

which indicates that head loss is proportional to Q, the volumetric flow rate.

Simultaneous solution of (11.1), (11.2), and (11.3) leads to the flow division in the various piping branches, once the piping component loss characteristics are determined (see Chapter 10).

426 Piping Networks

Sometimes, in flow network analysis, the Hazen-Williams equation:

$$V = 1.318 C_1 R^{0.63} S^{0.54} \tag{11.4}$$

is used in place of the usual Darcy-Weisbach equation. In (11.4), C_1 is the Hazen-Williams *coefficient of relative roughness* of the pipe (see Table 11.1), R is the *hydraulic radius* of the pipe (ft), and S is the *slope* of the hydraulic gradient. Of course, (11.3) and (11.4) are related and entirely equivalent, as is shown by the following simplified development. Combining the definitions

$$R = \frac{\text{flow area}}{\text{perimeter}} = \frac{\pi D^2/4}{\pi D} = \frac{D}{4} \tag{11.5}$$

and

$$S = \frac{\text{head loss}}{\text{length}} = \frac{h_{\text{loss}}}{L} \tag{11.6}$$

with (11.4), we have

$$h_{\text{loss}} = \frac{2g_c(4^{0.63/0.54})}{1.318^{1/0.54}} \left(\frac{LV^2}{2g_c D}\right) \left(\frac{V^{1/0.54} D}{V^2 D^{0.63/0.54} C_1^{1/0.54}}\right).$$

Introducing the pipe Reynolds number ($R_D = VD/\nu$) results in

$$h_{\text{loss}} = \left(\frac{194.48 D^{-0.01852}}{C_1^{1.85185} \nu^{0.14815}}\right) \left(\frac{1}{R_D^{0.14815}}\right) \left(\frac{LV^2}{2g_c D}\right).$$

Table 11.1 Nominal Values of the Hazen-Williams Coefficient, $C_1{}^a$

Type of Pipe	Value
Extremely smooth and straight	140
New, smooth cast iron	130
Average cast iron, new riveted steel	110
Vitrified sewer	110
Cast iron, some years in service	100
Cast iron, in bad condition	80

a After Giles [3].

The Parallel Network 427

Neglecting the small effects of D and ν in the first parenthetical term, we see that the Hazen-Williams equation of (11.4) can be put into the form

$$h_{\text{loss}} = \varphi(C_1, R_D)\left(\frac{LV^2}{2g_c D}\right),$$

which can be recognized as the Darcy-Weisbach equation of (11.3) since the friction factor (f) also is a function of relative roughness and Reynolds number.

The advantage of the Hazen-Williams formula over the Darcy-Weisbach equation lies in the fact that C_1 depends only on relative roughness, and not on the flow rate, whereas f depends on both. This idea can best be illustrated by examples.

Example 11.2 [3]. Find by the Hazen-Williams formulation the flow distribution of water at 70°F in each horizontal branch of a parallel piping network of used cast iron pipe, as shown in Figure 11.2, if $p_A/\rho = 120$ ft-lbf/lbm, $p_B/\rho = 72$ ft-lbf/lbm, $D_A = D_B$, and the lengths and diameters are as follows: $L_1 = 12,000$ ft, $L_2 = 4000$ ft, $L_3 = 8000$ ft; $D_1 = 12$ in., $D_2 = 8$ in., $D_3 = 10$ in.

Solution: The three basic network equations, that is, (11.1), (11.2), and (11.3), apply.

By Bernoulli:

$$(h_{\text{loss}})_{A,B} = \frac{p_A}{\rho} - \frac{p_B}{\rho} = 120 - 72 = 48 \text{ ft-lbf/lbm}.$$

By (11.5):

$$R_1 = \frac{12}{4 \times 12} = 0.25 \text{ ft},$$

$$R_2 = \frac{8}{4 \times 12} = 0.1667 \text{ ft},$$

$$R_3 = \frac{10}{4 \times 12} = 0.2083 \text{ ft}.$$

By (11.6):

$$S_1 = \frac{48}{12,000} = 0.004,$$

$$S_2 = \frac{48}{4000} = 0.012,$$

$$S_3 = \frac{48}{8000} = 0.006.$$

428 Piping Networks

By Table 11.1, for cast iron pipes some years in service.

$$C_1 = 100.$$

By (11.4):

$$V_1 = 1.318 \times 100 \times 0.25^{0.63} \times 0.004^{0.54} = 2.79081 \text{ ft/sec},$$
$$V_2 = 1.318 \times 100 \times 0.1667^{0.63} \times 0.012^{0.54} = 3.9128 \text{ ft/sec},$$
$$V_3 = 1.318 \times 100 \times 0.2083^{0.63} \times 0.006^{0.54} = 3.0966 \text{ ft/sec}.$$

By continuity ($Q = AV = \pi D^2 V/4$):

$$Q_1 = 2.19072 \text{ ft}^3/\text{sec},$$
$$Q_2 = 1.36582 \text{ ft}^3/\text{sec},$$
$$Q_3 = 1.68866 \text{ ft}^3/\text{sec}.$$

Example 11.3. Show the relative complexity of the Darcy-Weisbach approach over the Hazen-Williams approach by recomputing the flow in branch 1 of Example 11.2.

Solution:
By Bernoulli:

$$h_{\text{loss}} = \frac{\Delta p}{\rho} = 48 \text{ ft-lbf/lbm}.$$

By Darcy-Weisbach:

$$\frac{\Delta p}{\rho} = f \frac{L}{D} \frac{V^2}{2g_c},$$

which, for $L_1 = 12{,}000$ ft and $D_1 = 12$ in., yields

$$48 = \frac{f(12{,}000) V^2}{1 \times 2 \times 32.174}. \tag{11.7}$$

Clearly, there are two unknowns in (11.7), namely f and V, and thus an iterative solution is required.

1. For a guess of R_D, the pipe Reynolds number, we can get f, once having decided on e/D, the relative roughness of the pipe.

2. By (11.7), we can now get V.
3. We can now calculate R_D by its definition and check our initial guess against this value. Thus:

$$V_1 = \sqrt{\frac{2 \times 32.174 \times 48}{12,000 f}} \quad . \tag{11.8}$$

By Figure 6.8, we find $(e/D)_{\text{new cast iron}} \simeq 0.0008$ for the 12 in. pipe of branch 1.
By (7.24) of Example 7.2, we find $(e/D)_{\text{used cast iron}} \simeq 0.0024$.
At an assumed R_D of 10^5, we obtain from (6.19), Table 6.2, or Figure 6.7, at $e/D = 0.0024$, $f \simeq 0.026$.
By (11.8):

$$V_1 = 3.146 \text{ ft/sec}$$

and

$$(R_D)_{\text{calc}} = \frac{VD}{\nu} = 2.982 \times 10^5$$

using $\nu = 1.055 \times 10^{-5}$ ft^2/sec from Table 7.1.
Since $(R_D)_{\text{calc}} \neq (R_D)_{\text{assumed}}$, we reguess $R_D = 3 \times 10^5$ and redo the calculations. This time, $f \simeq 0.025$, $V_1 = 3.209$, and $R_{DC} \simeq 3 \times 10^5$, so the iteration has led to a closed solution, and $Q_1 = A_1 V_1 = 2.52$ ft^3/sec.

It should be realized that the differences between the Hazen-Williams and the Darcy-Weisbach solutions are encountered mainly because of the uncertainty in e/D of the used cast iron pipe. Both the $C_1 = 100$ and the e/D correction of (7.24) should be recognized as purely nominal estimates that are in no way exact or compatible. On this basis the agreement between Examples 11.2 and 11.3 is good, and the objective of showing the relative complexity of these two approaches has been reached.

Example 11.4 [1]. Again referring to Figure 11.2, find the flow rate in each branch of the parallel piping network when 12 ft^3/sec of water flows at a density of 64.348 lbm/ft^3 and a kinematic viscosity of 3×10^{-5} ft^2/sec. The lengths, diameters, and absolute roughnesses for the pipes are as follows. $L_1 = 3000$ ft, $L_2 = 2000$ ft, $L_3 = 4000$ ft; $D_1 = 1$ ft, $D_2 = 0.6667$ ft, $D_3 = 1.3333$ ft; $e_1 = 0.001$ ft, $e_2 = 0.0001$ ft, $e_3 = 0.0008$ ft.

Solution: Assume that $R_{D1} = 10^5$, which yields $V_1 = 3$ ft/sec, $Q_1 = 2.3562$ ft^3/sec, $f_1 = 0.022$, and $(h_{\text{loss}})_{A,B} = 9.231$ ft-lbf/lbm.

430 Piping Networks

At this loss, assume that $f_2 = 0.020$ and get $V_2 = 3.1464$ ft/sec; this leads to $R_{D2} = 0.7 \times 10^5$, which checks the f_2 assumption closely. Hence $Q_2 = 1.0984$ ft^3/sec.

At this same loss, maintained across branch 3, and for an assumed $f_3 = 0.020$, find $V_3 = 3.1464$ ft/sec; this leads to $R_{D3} = 1.4 \times 10^5$, which checks the f_3 assumption closely. Hence $Q_3 = 4.3930$ ft^3/sec.

It remains to check continuity.

$$Q_T = Q_1 + Q_2 + Q_3 = 7.8476 \text{ ft}^3/\text{sec}$$

rather than the given flow rate of 12 ft^3/sec. This requires a change in all flow rates according to the following proportions:

$$Q_1 = \frac{2.3562}{7.8476} \times 12 = 3.603 \text{ ft}^3/\text{sec},$$

$$Q_2 = \frac{1.0984}{7.8476} \times 12 = 1.680 \text{ ft}^3/\text{sec},$$

$$Q_3 = \frac{4.3930}{7.8476} \times 12 = 6.7175 \text{ ft}^3/\text{sec}.$$

These changes satisfy continuity, but now the head loss requirement must be satisfied.

By similar calculations, using the *new* flow rates,

$V_1 = 4.5875$, $R_{D1} = 1.53 \times 10^5$, $f_1 = 0.0215$, $h_{\text{loss}} = 21.095$;

$V_2 = 4.8124$, $R_{D2} = 1.07 \times 10^5$, $f_2 = 0.0180$, $h_{\text{loss}} = 19.434$;

$V_3 = 4.8113$, $R_{D3} = 2.14 \times 10^5$, $f_3 = 0.0195$, $h_{\text{loss}} = 21.045$.

Further refinement does not seem necessary since both head loss and continuity requirements are approximately satisfied.

Evidently, the solution of any network problem must satisfy both *continuity* and *Bernoulli* principles throughout. Continuity states that the net flow rate into any piping junction must be zero. Bernoulli states that the net head loss around any loop of the network must be zero. Beginning with an assumed flow rate, one follows a trial and error procedure by satisfying the continuity and head loss equations to any degree of accuracy acceptable. One of the simplest formalized methods for accomplishing this iterative solution is due to Professor Hardy Cross [4]. This method has been so well documented in the recent literature (see e.g., [1]–[3] and [5]) that it need not be given here.

Networks that are more general in nature and involve compressible fluids are solved by many special piping programs and approaches that, for obvious reasons, cannot be given here. An entirely different approach, based on a multinode electric network applied to fluid piping networks, also has been documented (see, e.g., [6]–[8]).

11.4 THE BRANCHING NETWORK

In the branching network, two or more series combinations of piping components branch out, as in the parallel network; but, unlike the situation in the parallel arrangement, the branches do not come together again (see Figure 11.3). Theoretically, such networks can be solved by the

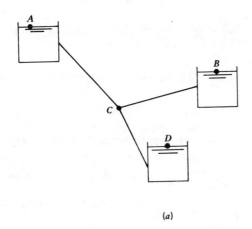

(a)

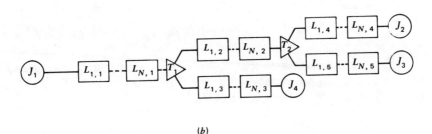

(b)

Figure 11.3 Several branching networks. (a) Three-reservoir network. (b) General branching network.

432 Piping Networks

methods already given. Practically, when the number of branches becomes large, special computational methods (beyond the scope of this introduction) usually are employed. To illustrate the basic approach, however, several simplified examples are given next.

Example 11.5 [3]. With reference to Figure 11.3a, the following facts are available:

Piping Section	L	D (ft)	C_1
AC	8000	2	100
BC	4000	1.333	120
CD	4000	1	100

with $Z_A = 212$ ft, $Z_B = 190$ ft, and $Z_D = 100$ ft. Find the flow distribution.

Solution: To simplify calculations, assume that $Z_C = Z_B = 190$ ft. Under these conditions there will be no flow into or out of B.
By (11.5): $R_{AC} = 0.5$, $R_{BC} = 0.333$, $R_{CD} = 0.25$.
By (11.6): $S_{AC} = (212 - 190)/8000 = 0.00275$, $S_{CD} = (190 - 100)/4000 = 0.0225$.
By (11.4): $V_{AC} = 3.528$ ft/sec, $V_{CD} = 7.904$ ft/sec.
By continuity: $Q_{AC} = 11.0835$ ft³/sec, $Q_{CD} = 6.2078$ ft³/sec.
We must conclude that Z_C was chosen improperly because $Q_{AC} > Q_{CD}$, whereas the flow into C must equal the flow out of C. To alleviate this situation, we assume Z_C to be *greater* then Z_B. Then some of the flow from A will head for B and some for D.
Arbitrarily, we assume that $Z_C = 200$ ft, and redo the calculations. This time we obtain:

$$S_{AC} = 0.0015, \quad S_{BC} = 0.0025, \quad S_{CD} = 0.025;$$
$$V_{AC} = 2.543 \text{ ft/sec}, \quad V_{BC} = 3.114 \text{ ft/sec}, \quad V_{CD} = 7.508 \text{ ft/sec};$$
$$Q_{AC} = 7.989 \text{ ft}^3/\text{sec}, \quad Q_{BC} = 4.346 \text{ ft}^3/\text{sec}, \quad Q_{CD} = 5.897 \text{ ft}^3/\text{sec}.$$

Again, we must conclude that Z_C was chosen improperly because now $Q_{AC} < Q_{BC} + Q_{CD}$. An elevation intermediate between 190 and 200 ft is next chosen. For $Z_C = 196.5$ ft we obtain:

$$S_{AC} = 0.00194, \quad S_{BC} = 0.001625, \quad S_{CD} = 0.024125;$$
$$V_{AC} = 2.922 \text{ ft/sec}, \quad V_{BC} = 2.468 \text{ ft/sec}, \quad V_{CD} = 7.365 \text{ ft/sec};$$
$$Q_{AC} = 9.180 \text{ ft}^3/\text{sec}, \quad Q_{BC} = 3.444 \text{ ft}^3/\text{sec}, \quad Q_{CD} = 5.784 \text{ ft}^3/\text{sec}.$$

Table 11.2 Results for Branching Network of Example 11.6

Piping Element	At Exit P_t (psia)	p (psia)	M	D (in.)	K at Minimum Area	
Inlet	42.889	42.000	0.173	2	0	Branch 1
Viscous pipe ($L=100$ in., $e/D=0.001$)	41.970	41.061	0.177	2	1.009	$\dot{m}_1 = 0.9061$ lbm/sec
Step-up	41.458	41.402	0.044	4	0.563	
Reducer	41.412	40.490	0.180	2	0.050	
Elbow ($90°$, $R/r=6$)	41.270	40.345	0.180	2	0.151	
Tee	41.270	40.924	0.110	2	0	Branch 2
Viscous pipe ($L=200$ in., $e/D=0.002$)	40.428	40.074	0.112	2	2.394	$\dot{m}_2 = 0.5586$ lbm/sec
Nozzle ($D_{\text{throat}}=1$ in.)	36.577	36.186	0.124	2	0.563	
Conical diffuser ($L=24$ in.)	36.521	36.497	0.031	4	0.142	
Step-down	36.394	36.000	0.125	2	0.131	
Tee	41.270	41.137	0.068	2	0	Branch 3
Step-up	41.195	41.187	0.017	4	0.563	
Step-down	41.152	41.019	0.068	2	0.131	
Constant K	39.790	39.652	0.071	2	10	$\dot{m}_3 = 0.3475$ lbm/sec
Orifice ($D_{\text{throat}}=1$ in.)	35.157	35.000	0.080	2	0.706	

434 Piping Networks

Since $Q_{AC} \simeq Q_{BC} + Q_{CD}$, we have obtained a solution.

Example 11.6. Air flows in an adiabatic branching network of the form of Figure 11.3b. The static pressures at J_1, T_2, and J_4 of this figure are given as 42, 36, and 35 psia, respectively. With the piping components and their geometries specified as shown in Table 11.2, find the flow in each branch, assuming the tee (T_1) loss to be negligible.

Solution: In a practical sense, this fairly complex network requires a computer solution, which will not be given here. However, the results, as shown in Table 11.2, are instructive for a number of reasons. For example, we see the continuous decrease in total pressure as we proceed through the various piping elements. We observe the rise and fall of static pressure, depending on the area of the piping element. And we see that continuity is satisfied after the iteration has been completed (i.e., $\dot{m}_1 = \dot{m}_2 + \dot{m}_3$).

11.5 FORWARD AND BACKWARD SOLUTIONS

As we have seen throughout this study, the flow through any real piping element (e.g., viscous pipe, orifice, elbow, diffuser) can be characterized by a loss in total pressure (see Figure 11.4). All of the solutions given thus far are based on an analysis proceeding in the *forward* direction only, that is, one begins with inlet conditions and obtains exit conditions, the analysis being in the direction of flow.

When several loss elements are connected in series and parallel arrangements, as in Figure 11.3b, it is clear that internal nodes, that is, junctions between the branches of the network, are the rule. All junction pressures cannot be specified, in general, at the start of a network analysis, so that iterative solutions also are the rule. During the course of iteration, various

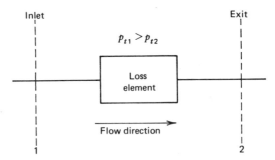

Figure 11.4 Single piping loss element.

Forward and Backward Solutions 435

trial values of flow rate are assumed, and some of these may lead to pressure ratios beyond the critical value [see (3.125) and (8.62)].

Thus, although the usual forward type analysis can handle incompressible and subsonic flows, it follows that supercritical values of flow rate and/or corresponding pressure ratios would necessarily be rejected. This is so because in the forward solution the assumed offending flow rate would be reduced in the calculation procedure, and the analysis would be started again. Therefore, although the forward approach is satisfactory for single-element analyses, or even for single-internal node analyses, it does not always work for more complex networks. The forward method does not always allow a complete calculation pass through all the piping elements in a given branch. This means that new values of pressure at all the interval nodes will not always be available for all sets of flow rates chosen. And, without these new pressures, the iteration process will not have the required information to proceed toward a solution.

However, when a branch of piping loss elements is considered from the *backward* direction, that is, in a direction opposite to that of the flow, it is possible always to obtain new values of total pressure at each interval node and thus to proceed with the iteration toward a solution.

This backward type of analysis always is possible because supercritical pressure ratios can be excluded in favor of critical pressure ratios at the exit of any loss element. This is accomplished by the temporary introduction of an additional unspecified loss element to account for flow losses from the just-critical exit of the element to the given downstream pressure condition (see Figure 11.5). Thus the backward method of analysis is preferred for piping elements arranged in a compressible flow network.

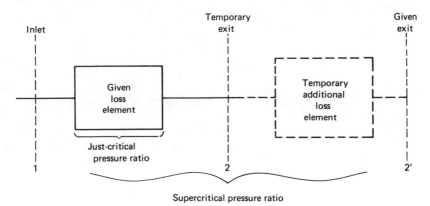

Figure 11.5 Means for handling supercritical pressure ratios.

436 Piping Networks

In the backward method, then, we proceed from initial values (guesses) of p_{t2} (the total pressure at exit) and $\dot{m}$ (the mass flow rate) to obtain, first, p_2 (the static pressure at exit). Then, through the use of the pertinent loss coefficient (K), which characterizes the loss element, we obtain p_{t1} and p_1 (the total and static pressures at inlet).

Backward Solution Method. In this method of solution, the flow rate, the exit total pressure, and the pertinent details of the loss element are given.

The quantity $(p_t\alpha)_2$, representing a flow per unit area term at exit, is formed via

$$(p_t\alpha)_2 = \frac{\dot{m}C}{A_2}, \tag{11.9}$$

where α is the point flow number defined by (3.68) and (8.8), and the constant C represents

$$C = \left(\frac{\bar{R}T_t}{g_c}\right)^{1/2}. \tag{11.10}$$

An iterative calculation is made to determine R_2 and p_2 by comparing (11.9) with the equivalent relation

$$(p_t\alpha)_{2,c} = p_2\left\{\left(\frac{2\gamma}{\gamma-1}\right)R_2^{2(1-\gamma)/\gamma}\left[1 - R_2^{(\gamma-1)/\gamma}\right]\right\}^{1/2}, \tag{11.11}$$

where the subscript c signifies a calculated value. The loss coefficient, as discussed in Chapters 8 and 10, is introduced next, and an iterative loop is required to obtain inlet conditions. Briefly, a hybrid flow number (α') is determined via

$$\alpha'_1 = \frac{\alpha_2}{p_{t1}}, \tag{11.12}$$

using an initial guess of $p_{t1} = p_{t2}$. Next, the inlet flow number is derived by

$$\alpha_1 = \alpha'_1\left(\frac{A'_2}{A_1}\right), \tag{11.13}$$

and the inlet pressure ratio is obtained from

$$\alpha_1 = \left\{\left(\frac{2\gamma}{\gamma-1}\right)R_1^{2/\gamma}\left[1 - R_1^{(\gamma-1)/\gamma}\right]\right\}^{1/2}, \tag{11.14}$$

which is the point flow number of (8.8) once more. The hybrid pressure ratio R_{21} is determined from (8.14), which yields, in turn,

$$p_{t1,c} = \frac{p_2}{R_{21}}. \tag{11.15}$$

This calculated value of p_{t1} is used as a better guess of p_{t1} to get a new value of α'. The $\alpha_1 - R_1 - R_{21} - p_{t1,c}$ process is repeated until $p_{t1} = p_{t1,c}$, at which time

$$p_1 = p_{t1} \times R_1. \tag{11.16}$$

Note that, when we speak of one quantity equaling another in an iterative solution, we really mean equality within a specified tolerance [9].

Specific Solution Method. Here we consider the specific method used to solve for flow losses through an individual loss element in the supercritical case. Among the 12 element types we have treated, we can discern three distinct groups. Group 1 includes viscous pipes, smooth elbows, and loss elements like valves and screens: all characterized by a flow area that is the same at inlet and exit. Group 2 includes reducers, diffusers, and venturis: characterized by a gradual change in flow area between inlet and exit. Group 3 includes step-ups, step-downs, nozzles, orifices, and screw elbows: characterized by an abrupt change in flow area and a step-up type of loss mechanism.

Consider the simplest of the flow elements, the *viscous pipe*, for example, Given p_{t2}, $\dot{m}$, e/D, L, and D, the problem is to recognize when conditions exist beyond the critical state. It is quite possible that the given values of p_{t2} and $\dot{m}$ will require an R_2 that is beyond the R^* of (8.62). This is noted when $(p_t\alpha)_2$ of (11.9) exceeds $(p_t\alpha)_{2,c}$ of (11.11). It means that the given values of p_{t2} and $\dot{m}$ are not compatible with subsonic flow or even with just-critical flow at the exit of the given element. We overcome this obstacle by introducing a temporary loss between the element exit and the given conditions (see Figure 11.5). In this situation we set $R_2 = R^*$, which is the limiting possibility for subsonic flow from inlet to exit. By also setting $(p_t\alpha)_{2,c}$ equal to $(p_t\alpha)_2$, we obtain p_2. And, finally, we have $p_{t2} = p_2/R^*$. Now, with known exit pressures that are compatible with subsonic flow in the given piping element, we proceed as described for the backward subsonic method of solution.

REFERENCES

1. V. L. Streeter and E. B. Wylie, *Fluid Mechanics*, 6th ed., McGraw-Hill, 1975, pp. 556–568.
2. J. K. Vennard and R. L. Street, *Elementary Fluid Mechanics*, 5th ed., John Wiley, 1975, pp. 424–439.

3. R. V. Giles, *Fluid Mechanics and Hydraulics*, 2nd ed., Schaum, 1962, pp. 115–129.
4. H. Cross, "Analysis of flow in networks of conduits and conductors," *Univ. Ill. Eng. Exp. Stn. Bull.* 286, November 1936.
5. P. T. Daniel, "The analysis of compressible and incompressible flow networks," *Trans. IChE*, Vol. 44, 1966, p. T77.
6. A. Lavi, G. Kusic, and C. F. Wood, "Computation of adiabatic flow in multibranch networks," *Westinghouse Res. Memo.* 70-7K4 = COMPA-M1, October 1970.
7. R. Schinzinger and R. R. Bizon, "Analysis of fluid flow circuits," *Westinghouse LRA Eng. Memo.* 67, June 1958.
8. T. Ito. H. Fukaya, Z. Okamoto, and H. Hoshino, "Analysis of hydraulic pipe line network using an electronic computer," *Mitsubishi Tech. Bull.* 46, 1967.
9. R. P. Benedict, "Uncertainties in iterative solutions," *Instrum. Control Syst.*, April 1969, p. 111.

NOMENCLATURE

Roman

A area
C a constant
C_1 Hazen-Williams constant
D pipe diameter
e absolute roughness
f friction factor
g_c gravitational constant
h_{loss} head loss
K loss coefficient
L length
$\dot{m}$ mass flow rate
p absolute pressure
Q volumetric flow rate
R hydraulic radius, point static/total pressure ratio
R_{21} Hybrid pressure ratio p_2/p_{t1}
$\bar{R}$ specific gas constant
R_D pipe Reynolds number

- S slope of hydraulic gradient
- T absolute temperature
- V volumetric average velocity
- Z elevation

Greek

- α flow number
- γ specific heat ratio
- Δ finite difference
- ν kinematic viscosity
- ρ fluid density

IV
THERMODYNAMIC MEASUREMENTS IN PIPES

Here we consider very briefly the measurement of temperature, pressure, and flow rate in piping systems. Only the most important equations and the largest corrections concerning these measurements are given. References to more complete works on these measurements are noted in the proper chapters. One chapter on special measurements in pipes is included to sample the interesting problems encountered when dealing with fluid flow in pipes.

12

Temperature Measurement in Pipes

> There are persons in the hills of India who never have had a thermometer under their tongues. We, most of us, began early, rolled one a bit, wished we could cheat, at least strongly hoped it would come out Fahrenheit 98.6°... —*Gustav Eckstein*

12.1 GENERAL REMARKS

In this chapter we give a brief introduction to the problem of measuring fluid temperature in a pipe or in piping components. We know that in temperature measurement we must sometimes distinguish between static and total conditions of the fluid, just as we did for pressure. We must consider such quantities as the stagnation factor, the adiabatic recovery factor, and the dynamic correction factor, to account for the temperature variations encountered in a moving fluid. Then there are the heat transfer effects on the temperature measurement. Various modes of heat transfer—radiation between the pipe walls, the fluid, and the probe; conduction between the pipe walls and the sensing tip; convection between the fluid and the probe—all play their parts. The choice of sensing instrument likewise contributes to the problem. Some probes stagnate the fluid, others yield an average temperature around the probe, still others are shielded from radiation and conduction, and some sensors simply look at the fluid nonintrusively, and hence experience neither heat transfer nor stagnation effects. Finally, the usual static calibration effects come into play concerning the indicated temperature versus the corrected temperature. These too are considered briefly in this chapter.

12.2 TEMPERATURE MEASUREMENT IN MOVING FLUIDS [1]

Whenever a fluid moves, the problems of temperature measurement are compounded. For *liquids*, we have seen by (7.29), that is, by

$$S_2 - S_1 = C \ln\left(\frac{T_2}{T_1}\right), \tag{12.1}$$

that the liquid temperature must remain constant in an isentropic flow process, independently of the fluid velocity (see Figure 7.7). However, for a liquid flowing with losses, such as we would expect in the flow of a real liquid in pipes, the liquid temperature must increase as the entropy increases according to (12.1) (see also Figure 7.8).

For *gases*, we know from (3.54), that is,

$$T_t = T + \frac{V^2}{2Jg_c c_p}, \tag{12.2}$$

that the gas temperature must decrease, even in an isentropic flow process, as the fluid velocity increases (see Figures 3.9 and 8.8). For gaseous flow with losses, such as we would expect in the flow of a real gas or vapor in a pipe, the gas temperature also decreases with velocity, but at a slower rate.

Idealized Relations. For an ideal (isentropic) liquid flow there is only one temperature of significance, and this can be called the liquid temperature (T_l). For an ideal gas flow there are three temperatures of significance:

1. Static temperature (T). This is the actual temperature of the gas at all times (whether in motion or at rest). The static temperature will be sensed by an adiabatic probe in thermal equilibrium and at rest with respect to the gas.
2. Dynamic temperature (Tv). This is the thermal equivalent of the directed kinetic energy of the gas continuum and is defined by

$$T_v = \frac{V^2}{2Jg_c c_p}. \tag{12.3}$$

3. Total temperature (T_t). This temperature is made up of the static temperature plus the dynamic temperature of the gas. The total temperature will be sensed by an idealized probe, at rest with respect to the piping boundaries, when it stagnates an idealized gas.

These three gas temperatures are related by (12.2).

Real Fluid Effects. There are many perturbation effects on the idealized concepts given above. For one thing, real fluid characteristics deviate from those of the idealized relations. Specifically, we never encounter real fluids of zero viscosity or of zero thermal conductivity, such as is assumed for idealized fluids. These real fluid effects influence temperature measurement as follows. The combined effects of fluid stagnation, fluid viscosity, and fluid thermal conductivity set up temperature and velocity gradients in the fluid boundary layers surrounding a probe. There is a consequent rise in temperature of the inner fluid layers as a result of viscous shear work on the fluid particles, and this impact conversion of directed kinetic energy to thermal effects is necessarily accompanied by a heat transfer through the fluid away from the adiabatic probe. The relative importance of these opposing effects, which upset the simple picture of temperature previously given, is indicated by the Prandtl number. The Prandtl number is the ratio of the fluid properties governing the transport of momentum by viscous effects (because of the velocity gradient), to the fluid properties governing the transport of heat by thermal diffusion (because of the temperature gradient), that is,

$$Pr = \frac{\text{kinematic viscosity}}{\text{thermal diffusivity}} = \frac{c_p \mu}{k}.$$

Note that, even if the Prandtl number is unity, we cannot discount the effects of conductivity and viscosity. Both are actually present. They simply counterbalance each other.

This phenomenon, wherein the fluid temperature changes in a constant enthalpy process, is known as the Joule-Thomson effect [2] (see Figure 12.1).

For liquids the Joule-Thomson coefficient is generally negative (this is true for water below 450°F). As the pressure drops isenthalpically, the temperature rises. This means that the liquid temperature will not remain constant in an isenthalpic change of state. Furthermore, for reasons already given concerning the interplay between viscous shear work and heat transfer in the liquid boundary layers surrounding a probe immersed in a real liquid, the probe will not realize the liquid temperature of (12.1). Even when the Prandtl number of the real liquid is unity, the adiabatic probe will not sense the idealized liquid temperature (T_l).

For gases a similar argument holds. The viscous shear forces present are synonomous with friction forces in the gas, and the temperature gradient set up in the boundary layers ensures local heat transfer to and from the

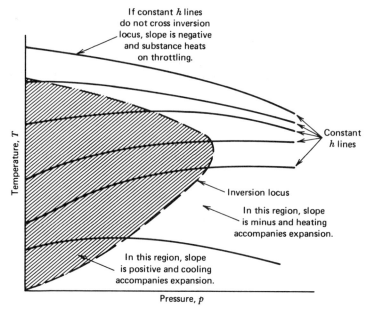

Figure 12.1 The Joule-Thomson effect.

probe. Hence boundary layer effects conflict with the isentropic assumption, and total temperature will not be realized by an idealized probe in a real gas, even if the Prandtl number is unity.

In summary: The fluid temperature will not be constant in an isenthalpic change of state of a real fluid, as indicated by the Joule-Thomson effect. Furthermore, even if the Prandtl number of the fluid is unity, an adiabatic probe will fail to indicate the gas total temperature or the liquid temperature.

The Recovery Factor. This brings us to a discussion of the recovery factor. Actually, there are several factors to be considered here, including the *stagnation recovery factor* (S), defined by

$$T_{\text{stagnation}} = T + ST_v, \tag{12.4}$$

the *flat plate recovery factor* (r), defined by

$$T_{\text{flat plate}} = T + rT_v, \tag{12.5}$$

and the *overall recovery factor* (R), defined by

$$T_{\text{adiabatic probe}} = T + RT_v. \tag{12.6}$$

The stagnation factor of (12.4) has been defined [1,3] in terms of thermodynamic quantities as

$$S = \frac{T}{v}\left(\frac{\partial v}{\partial T}\right)_p, \qquad (12.7)$$

and this factor is presented graphically in Figure 12.2 for air, water, and steam. Note that, on comparing (12.2), (12.3), and (12.4), it becomes clear

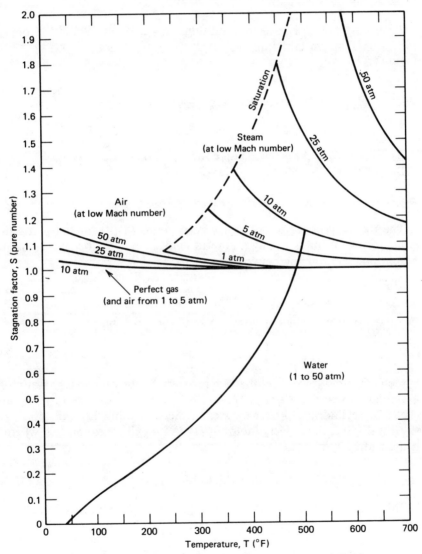

Figure 12.2 Stagnation recovery factors for several fluids.

448 Temperature Measurement in Pipes

that $T_{stagnation}$ does not equal T_{total} for real gases, nor does it equal T_{liquid} for real liquids.

Example 12.1. A probe is installed in a pipe so that a temperature sensor is located at an aerodynamic stagnation point on the probe. If the fluid is steam at $T=400°F$, $p=150$ psia, and $V=500$ ft/sec, find the difference between the stagnation temperature indicated by the probe and the thermodynamic total temperature of the steam.

Solution: At the given steam conditions, $c_p = 0.58$ Btu/lbm°R, from the 1967 ASME steam tables, p. 293.
By (12.2):

$$T_t = (400+460) + \frac{500^2}{2 \times 778 \times 32.174 \times 0.58}$$
$$= 860 + 8.6 = 868.6°R.$$

By Figure 12.2: At given steam conditions, $S = 1.28$.
By (12.4): $T_{stag} = 860 + 1.28 \times 8.6 = 871.0°R$. Thus

$$T_{stag} - T_t = 2.4°F.$$

The recovery factor of (12.5) has been defined [1,4] for flat plates in a laminar flow of fluids having a Prandtl number between 0.5 and 2, and for Mach numbers ranging between 0 and 10, as

$$r_{laminar} = Pr^{1/2}. \tag{12.8}$$

For flat plates in turbulent flow, the similar expression is

$$r_{turbulent} = Pr^{1/3}. \tag{12.9}$$

Some authorities [5] suggest that the recovery factor for a cylindrical probe is similar to that for a flat plate, and hence suggest the use of (12.8) and (12.9) for cylindrical probes as well. Others, notably El Agib et al. [6], suggest the overall recovery factor of (12.6) for cylindrical probes, where R is defined by

$$R = \bar{P}S + (1-\bar{P})r. \tag{12.10}$$

In this equation $\bar{P}$ represents the average value of the pressure coefficient around a circular cylindrical surface, which can be taken roughly as

$$\bar{P} = -0.7. \tag{12.11}$$

Temperature Measurement in Moving Fluids 449

Example 12.2. An adiabatic cylindrical temperature probe is installed perpendicularly to a moving air stream in a pipe. The air is at a temperature of 300°F and a velocity of 500 ft/sec, and has a Prandtl number of 0.686 and a specific heat of 0.24 Btu/lbm°R. If the boundary layer on the probe is turbulent, find the difference between the temperature indicated by the probe and the thermodynamic total temperature of the air.

Solution: By (12.2):

$$T_t = (300+460) + \frac{500^2}{2 \times 778 \times 32.174 \times 0.24}$$
$$= 760 + 20.8 = 780.8°R.$$

By Figure 12.2: At the given air condition, $S=1$.
By (12.9): $r_{turbulent} = 0.686^{1/3} = 0.882$.
By (12.10) and (12.11): $R = -0.7(1) + (1+0.7) \times 0.882 = 0.8$.
By (12.6):

$$T_{adiabatic\ probe} = 760 + 0.8 \times 20.8 = 776.6°R.$$

Thus

$$T_t - T_{adiabatic\ probe} = 4.2°F.$$

The Dynamic Correction Factor. A real probe, immersed in a real gas, tends to radiate to its surroundings. Also, there is a tendency for a conductive heat transfer along the probe stem. These two effects are just balanced by a convective heat transfer between the probe and the fluid. Also, real probes do not necessarily stagnate a moving fluid effectively. Thus it is no more realistic to consider an idealized probe than to consider an idealized fluid.

A *dynamic correction factor* (K), defined by

$$T_{real\ probe} = T + KT_v, \tag{12.12}$$

is used to correct the performance of a diabatic probe that attempts to stagnate a real moving fluid, where $T_{real\ probe}$ is the equilibrium temperature sensed by a stationary real probe.

The dynamic correction factor of (12.12) approaches the overall recovery factor of (12.6) as the fluid velocity increases, and the probe temperature is dominated by convective heat transfer (see Figure 12.3). However, at the lower fluid velocities, K can take on values that differ

450 Temperature Measurement in Pipes

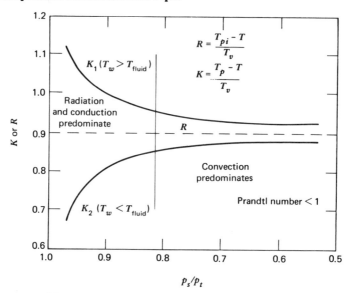

Figure 12.3 General dynamic correction factors.

greatly from R, depending on impact effects, viscosity and thermal conductivity effects, and heat transfer effects. It follows from (12.2) and (12.12) that

$$T_t - T_{\text{real probe}} = (1-K)T_v. \qquad (12.13)$$

Example 12.3. A temperature probe installed in a pipe indicates a temperature difference of 75°F from the theoretical total temperature of 1500°F when the air velocity is 150 ft/sec and the pipe wall temperature is 1000°F. What is the dynamic correction factor for this probe under these conditions?

Solution: [using $(c_p)_{\text{air}} = 0.24$ Btu/lbm°R]. By (12.3):

$$T_v = \frac{150^2}{2 \times 778 \times 32.174 \times 0.24} = 1.87°\text{R}.$$

By (12.13):

$$T_t - T_{\text{real probe}} = (1-K)(1.87) = 75°.$$

Hence

$$K = 1 - \frac{75}{1.87} = -39.$$

Note that in agreement with Figure 12.3, at low velocities and with $T_{\text{wall}} < T_{\text{fluid}}$, K can take on very large negative values.

12.3 HEAT TRANSFER EFFECTS ON TEMPERATURE MEASUREMENT

Under the combined influence of convective, radiative, and conductive heat transfer, the temperature distribution in a probe installed in a pipe in which a fluid flows has been given [1] by the following second-order, first-degree, nonlinear differential equation:

$$\frac{d^2T}{dx^2} + a_1(x)\frac{dT_x}{dx} - a_2(x,y)T_x = (-)a_2 a_3(x,y), \quad (12.14)$$

where

$$a_1(x) = \frac{dA_k}{A_k \, dx}, \quad (12.15)$$

$$a_2(x,y) = \frac{dA_c(h_r + h_c)}{kA_k \, dx}, \quad (12.16)$$

$$a_3(x,y) = \frac{h_c T_{\text{adi}} + h_r T_w}{h_c + h_r}. \quad (12.17)$$

Some of the terms used in these equations are clarified by Figure 12.4. The subscripts k, c, and r signify, respectively, conduction, convection, and radiation. The subscript w pertains to the pipe wall, and the subscript adi indicates the adiabatic condition. Of the remaining symbols, A signifies area; h, fluid film heat transfer coefficient; and k, thermal conductivity coefficient. Figure 12.5 ties in the temperature distribution in the fluid surrounding the thermometer well with the boundary layer concept.

Several approaches toward a solution of (12.14) are apparent.

Tip Solution. If all conduction effects are neglected, (12.14) becomes

$$T_{\text{tip}} = a_3 = \frac{h_r T_w + h_c T_{\text{adi}}}{h_r + h_c}. \quad (12.18)$$

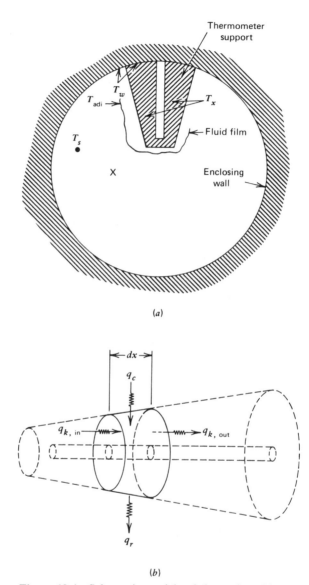

Figure 12.4 Schematic models of thermal problem.

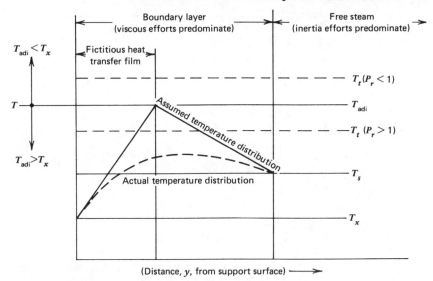

Figure 12.5 Temperature distribution in fluid surrounding a thermometer well.

This solution leads to a sensor tip temperature that is usually too high since any conduction away from the probe tip tends to reduce T_{tip}.

Stepwise Solution. More general solutions to (12.14), based on dividing the sensor into a number of elements (perpendicular to the heat conduction), and employing an iteration scheme to arrive at answers to specific installations, are summarized graphically in Figure 12.6.

Heat Transfer Coefficients. The coefficients required to work with (12.14) and (12.18) are approximated for easy reference in Figures 12.7 through 12.12. The use of the various heat transfer coefficients in the tip and stepwise solutions is best illustrated by examples.

Example 12.4. Find the adiabatic temperature of a gas flowing in a pipe if a nonconducting probe indicates 250°F when the pipe wall is at 200°F and the convective heat transfer coefficient is 20 Btu/hr ft² °F. Assume that the effective emissivity (ε') of the installation is unity.

Solution: Since no conduction is involved, the tip solution of (12.18) applies. Solving for the adiabatic gas temperature, we have

$$T_{\text{adi}} = T_{\text{tip}} + (T_{\text{tip}} - T_w)\left(\frac{h_r}{h_c}\right). \tag{12.19}$$

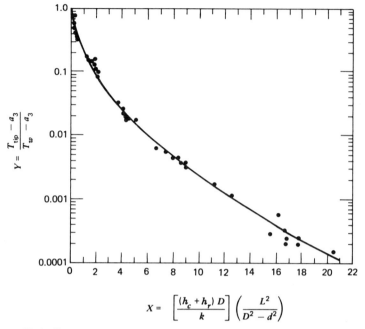

$$X = \left[\frac{(h_c + h_r)D}{k}\right]\left(\frac{L^2}{D^2 - d^2}\right)$$

Figure 12.6 Summary curve of 20 step-linearized solutions. Limits of variables $0 < \epsilon_{well} < 1$; $0 < \epsilon_{fluid} < 1$; $1 < k < 20$; $1 < h_c < 1000$; 2 in. $< L <$ 12 in.; $\frac{1}{16}$in. $< D < 2$ in.; $0 < d < \frac{1}{4}$in.; $70°F < T < 1200°F$.

By Figure 12.12: At $T_{tip} = 250°F$ and $T_w = 200°F$, we obtain $h_r/\epsilon' = h_r = 2.3$.

By (12.19):

$$T_{adi} = 250 + (250 - 200)\left(\frac{2.3}{20}\right) = 255.7°F.$$

Example 12.5. Air flows at a rate of 1 lbm/sec at a pressure of 2 atms in an uninsulated 6 in. pipe. The indicated well tip temperature is 200°F, and the pipe wall temperature is 180°F. The cylindrical well has a 3 in. immersion, $\frac{1}{2}$ in. O.D., and $\frac{1}{8}$ in. I.D.; $k = 20$ Btu/hr ft°F, $\epsilon_{well} = 0.9$, $\epsilon_{gas} = 0$, and μ_{air} at $200°F = 1.45 \times 10^{-5}$ lbm/ft sec. Find the adiabatic gas temperature.

Solution: The flow rate per unit area (G) is:

$$G = \frac{\dot{m}}{A} = \frac{1}{\pi\left(\frac{1}{2}\right)^2/4} = 5.093 \text{ lbm/sec ft}^2.$$

Heat Transfer Effects on Temperature Measurement 455

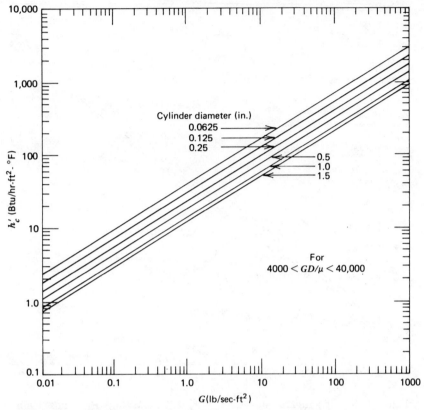

Figure 12.7 Generalized convective film coefficients, low R_D, across cylinders ($Nu = 0.193\,Re^{0.618}Pr^{0.31}$).

The Reynolds number, based on the pipe diameter, is

$$R_D = \frac{GD}{\mu} = \frac{5.093 \times 0.5}{1.45 \times 10^{-5}} = 1.75 \times 10^5.$$

By Figure 12.8: $h_c' = 42$ Btu/hr ft^2°F.
By Figure 12.9: $h_c/h_c' = 0.53$.
Thus:

$$h_c = 42 \times 0.53 = 22.26 \text{ Btu/hr ft}^2\text{°F}.$$

By Figure 12.12: $h_r/\varepsilon' = 1.9$.
Thus:

$$h_r = \varepsilon_{\text{well}} \times \frac{h_r}{\varepsilon'} = 0.9 \times 1.9 = 1.71 \text{ Btu/hr ft}^2\text{°F}.$$

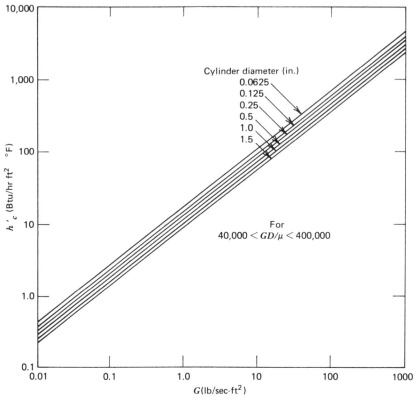

Figure 12.8 Generalized convective film coefficients, high R_D, across cylinders ($Nu = 0.02653 Re^{0.805} Pr^{0.31}$).

By Figure 12.6:

$$X = \left(\frac{(h_c + h_r)\,\overline{\text{O.D.}}}{k} \right) \left[\frac{L^2}{\overline{\text{O.D.}}^2 - \overline{\text{I.D.}}^2} \right] \quad (12.20)$$

$$= \left(\frac{22.26 + 1.71}{20} \times \frac{1}{24} \right) \left[\frac{3^2}{\left(\frac{1}{2}\right)^2 - \left(\frac{1}{8}\right)^2} \right] = 1.9176.$$

Note that, in the parenthetical term, O.D. is required in feet, whereas in the second term all dimensions can be in inches. By Figure 12.6: $Y \simeq 0.1$.
From the definitions:

$$Y = \frac{T_{\text{tip}} - a_3}{T_w - a_3} \quad (12.21)$$

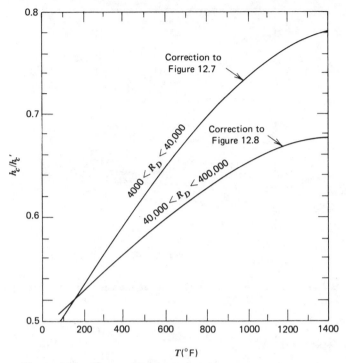

Figure 12.9 Correction curves for convective films in air.

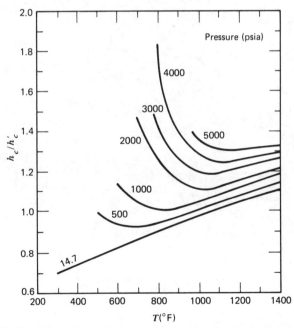

Figure 12.10 Correction curves for convective films in steam at low R_D.

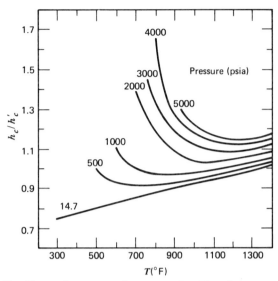

Figure 12.11 Correction curves for convective films in steam at high R_D.

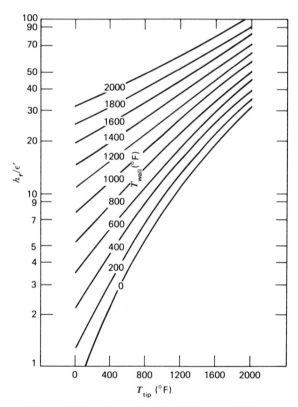

Figure 12.12 Generalized radiation coefficients.

and a_3 from (12.17), we obtain

$$T_{adi} = \left(\frac{h_c + h_r}{h_c}\right)\left(\frac{T_{tip} - YT_w}{1-Y}\right) - \left(\frac{h_r}{h_c}\right)T_w. \quad (12.22)$$

Thus:

$$T_{adi} = \left(\frac{22.26 + 1.71}{22.26}\right)\left[\frac{200 + 460 - 0.1(180 + 460)}{1 - 0.1}\right] - \frac{1.71}{22.26}(640),$$

or

$$T_{adi} = 663.93°R = 203.93°F.$$

If the emissivity of the fluid cannot be considered unity, the effective emissivity can be approximated by

$$\varepsilon' = \varepsilon_{well}(1 - \varepsilon_{fluid}). \quad (12.23)$$

If X of (12.20) is greater than 20, we set $Y=0$ and solve (12.22) for T_{adi} directly.

12.4 TEMPERATURE MEASUREMENT SYSTEMS

Many systems are used to measure temperature. Only the more common ones used in pipe flow work are discussed here. These include the liquid-in-glass thermometer (LIG), the resistance temperature detector (RTD), the thermocouple thermometer (T/C), and the optical pyrometer (OP). These common temperature measuring systems have been described in great detail in many professional society monographs, as well as in textbooks. The pertinent circuits and other operational details have been documented (see, e.g., [1]). In this section we propose to give only the most important correction factors for these various systems.

Liquid-in-Glass Thermometer. The LIG is rarely used for temperature measurements in pipes because of the breakage problem. However, if the LIG is used, the most important single correction that should be applied is the stem correction. Whenever the mercury column is improperly immersed, serious errors can be expected. Briefly, the thermometer can be of the *total immersion* type, meaning that just the portion of the thermometer containing the liquid should be exposed to the temperature being measured. Then, whenever a portion of the liquid in the stem is left

460 Temperature Measurement in Pipes

emergent from the temperature being measured, the reading of the thermometer will be either too high or too low, depending on the ambient temperature level with respect to that of the temperature being measured. On the other hand, the thermometer can be of the *partial immersion* type, meaning that the bulb and a specified portion of the stem should be exposed to the temperature being measured. Then, whenever such a thermometer is used at conditions other than specified, an error will be introduced.

Both of these LIG types can be corrected by the graph of Figure 12.13, where C_s is the stem correction (degrees), to be added algebraically to the indicated temperature; $(t_1 - t_2)$ is the *observed* temperature for total, or the *specified* temperature for partial, minus the emergent stem temperature; and N is the length of the emergent stem.

Example 12.6. A total immersion thermometer indicated 780°F when the mercury column was immersed to the 200° mark on the scale. The mean temperature of the emergent column was 170°F. What was the corrected temperature of the bulb?

Solution: By Figure 12.13: At $(t_1 - t_2) = 780 - 170 = 610°F$, at $N = 780 - 200 = 580°F$, read $C_s = +32°F$. Thus:

$$t_{corrected} = 780 + 32 = 812°F.$$

Example 12.7. A partial immersion thermometer indicated 250°F when used improperly as a total immersion thermometer. The equivalent length of the submerged column was 110°F, taken from the inscribed immersion mark to the top of the mercury column. The specified mean temperature of the emergent stem was 75°F, and the approximate mean temperature of the emergent stem was 250°F. What was the corrected temperature of the bulb?

Solution: By Figure 12.13: At $(t_1 - t_2) = 75 - 250 = -175°F$, at $N = 110°F$, read $C_s = -1.7°F$. Thus:

$$t_{corrected} = 250 - 1.7 = 248.3°F.$$

Resistance Temperature Detector. For the most common RTD, that is, the platinum resistance thermometer, temperature is derived in the range 0 to 630°C from the interpolating equation

$$t_{68} = t + \Delta t, \tag{12.24}$$

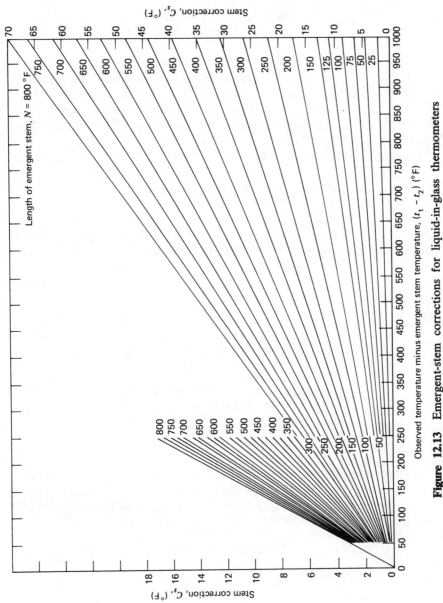

Figure 12.13 Emergent-stem corrections for liquid-in-glass thermometers (after ASME PTC 19.3, 1974, p. 51).

462 Temperature Measurement in Pipes

where t stands for the Callendar temperature, defined in turn by

$$t = \left(\frac{R_t - R_0}{R_{100} - R_0}\right) 100 + \delta\left(\frac{t}{100} - 1\right)\left(\frac{t}{100}\right) \, (°C), \qquad (12.25)$$

where the subscripts of R indicate the temperatures at which the resistance measurements are made. For example, R_0 is the resistance at the ice point (i.e., at 0°C). The characteristic constant (δ) of the thermometer is made available by calibration. The quantity Δt in (12.24) is the correction term to adjust t to t_{68}, the latter being temperature on the 1968 International Practical Temperature Scale.

The point here is that Δt is so small (i.e., $\pm 0.05°C$) as to be insignificant in most piping work. Thus there are no corrections for the RTD other than those arising from the usual heat transfer and recovery effects already discussed.

Example 12.8. A platinum RTD is characterized by the calibration constants, $R_{100}/R_0 = 1.3926$, $\delta = 1.495$, and $R_0 = 25.550 \, \Omega$. What is the temperature of the fluid in a pipe when the measured resistance of an adiabatic probe is 35.550 Ω?

Solution: By (12.25):

$$t^2 - \left(\frac{10^4}{\delta} + 10^2\right)t + \frac{10^6}{\delta}\left(\frac{R_t - R_0}{R_{100} - R_0}\right) = 0,$$

$$t^2 - 6788.9632t + \frac{10^6}{1.495}\left(\frac{35.550 - 25.550}{35.58093 - 25.550}\right) = 0,$$

$$t^2 - 6788.9632t + 666833.804 = 0,$$

$$t = 99.687°C.$$

Thermocouple Thermometer. Barring circuit errors, dealt with extensively in Reference 1, the thermocouple system usually yields temperature via a polynomial calibration equation in terms of measured electromotive force (emf, E). A typical calibration equation is of the form

$$t = A + BE + CE^2, \qquad (12.26)$$

the solution of which must be familiar to all once the constant coefficients, A, B, and C, are available, as from calibration.

Optical Pyrometer. This form of temperature measurement is based on the fact that the spectral radiance from an incandescent body is a function of its temperature. Planck's equation:

$$J_{b,\lambda} = \frac{c_1 \lambda^{-5}}{\exp(c_2/\lambda T) - 1} \tag{12.27}$$

describes the energy distribution for black body radiation as a function of temperature and wavelength. In general, to obtain the temperature of a test body, the intensity of its radiation at a *particular* wavelength is compared with that of a standard light source (usually internal to the optical pyrometer).

For nonblack bodies, the temperature so observed with an optical pyrometer is called the *brightness* temperature. This is always less than the actual temperature. The brightness temperature must be converted to the *actual* temperature through applying the emissivity of the test body. This correction is by far the most important one to be applied when measuring temperatures by an optical pyrometer.

Example 12.9. An optical pyrometer is sighted on a target inside a pipe and indicates a brightness temperature of 1327°C. The effective wavelength of the pyrometer is 0.655 μ, and the effective emittance of the target is 0.6. Estimate the actual temperature of the target.

Solution: At the brightness temperature of 1327°C, the target emissivity of 0.6, and the wavelength of 0.655 μ, standard correction tables (see e.g., [1]) indicate a temperature correction of 62°C. Hence

$$t_{actual} = t_{brightness} + \Delta t \tag{12.28}$$

or

$$t_{actual} = 1327 + 62 = 1389°C.$$

12.5 THE USUAL ENGINEERING APPROXIMATIONS

As a reminder of the relations between the various absolute and empirical temperature scales in common use, the following brief review is given:

$$T(K) = t(°C) + 273.15, \tag{12.29}$$

$$T(°R) = t(°F) + 459.67, \tag{12.30}$$

$$T(K) = \tfrac{5}{9}[t(°F) + 459.67] = \tfrac{5}{9} T(°R), \tag{12.31}$$

where $T(K)$ represents absolute temperature in kelvins, $T(°R)$ represents absolute temperature in degrees Rankine, and $t(°C)$ and $t(°F)$ represent empirical temperatures in degrees Celsius and in degrees Fahrenheit, respectively.

REFERENCES

1. R. P. Benedict, *Fundamentals of Temperature, Pressure, and Flow Measurements*, 2nd ed., Wiley-Interscience, 1977.
2. J. P. Joule and W. Thomson, "On the thermal effects of fluids in motion," Part 2, in *Mathematical and Physical Papers of W. Thomson*, Vol. 1, Cambridge, 1882, p. 357.
3. A. A. R. El Agib, "A general expression for the isentropic stagnation temperature of fluids," *Nature*, London, Vol. 204, 1964, p. 989.
4. H. W. Emmons and J. G. Brainerd, "Temperature effects in a laminar compressible-fluid boundary layer along a flat plate," *Trans. ASME*, Vol. 63, 1941, p. A-105.
5. E. Eckert and W. Weise, "The temperature of unheated bodies in high-speed gas stream," *NACA TM* 1000, December 1941.
6. A. A. R. El Agib, A. J. Binnie, and T. R. Foord, "Effects of recovery factor on measurement of temperature of moving fluids," *Proc. Inst. Mech. Eng.*, Vol. 180, Part 3F, 1965–1966, p. 174.

NOMENCLATURE

Roman

- a coefficient
- c constant
- c_p specific heat at constant pressure
- C constant
- C_s stem correction factor
- g_c gravitational constant
- G flow rate per unit area
- h heat transfer coefficient
- J Joule's mechanical equivalent of heat, spectral radiance
- k thermal conductivity
- K dynamic correction factor

N length of emergent stem
Pr Prandtl number
p absolute pressure
r flat plate recovery factor
R overall recovery factor, resistance
R_D pipe Reynolds number
S entropy, stagnation factor
T absolute temperature
v specific volume
V directed fluid velocity
X heat transfer term
Y heat transfer term

Greek

λ wavelength
δ thermometer constant
ε emissivity
μ fluid viscosity
Δ finite difference

Subscripts

c convection
k conductivity
r radiation
t total
v dynamic
w wall

13

Pressure Measurement in Pipes

> ...a mosquito has blood pressure, and if some insect physiologist has not yet gotten around to measuring it, he will... — *Gustav Eckstein*

13.1 GENERAL REMARKS

Here we give a brief introduction to the problem of measuring fluid pressure in a pipe or in piping components. We recall that we must distinguish between static, total, and dynamic conditions of the fluid. Now probe and wall tap geometries become extremely important, and various correction factors to account for—finite tap size, tap shape, probe blockage, flow angle variations, and so on—must be discussed. The static calibrations of various pressure transducers likewise must be considered as they affect the overall pressure measurement in a pipe.

13.2 PRESSURE MEASUREMENT IN MOVING FLUIDS [1]

Three definitions are required to cover the moving fluid situation concerning pressure measurement.

1. **Static pressure (p).** This is the actual pressure of the fluid at all times (whether in motion or at rest). In principle, it can be sensed by a very small hole drilled perpendicular to and flush with the flow boundaries so that it does not disturb the fluid in any way.

2. **Dynamic pressure** (p_v). This is the pressure equivalent of the directed kinetic energy of the fluid continuum and is defined by

$$(p_v)_{\text{inc}} = \frac{\rho V^2}{2g_c} \qquad (13.1)$$

for the incompressible case, in agreement with (2.4), and by

$$(p_v)_{\text{comp}} = \frac{\rho V^2}{2g_c}\left[1 + \frac{M^2}{4} + (2-\gamma)\frac{M^4}{24} + \cdots\right] \qquad (13.2)$$

for the compressible case, in agreement with (3.66).
3. **Total pressure** (p_t). This is the sum of the static and dynamic pressures. It can be sensed by a probe that is at rest with respect to the piping boundaries when it locally stagnates the fluid isentropically.

These three pressures are related by

$$p_t = p + p_v, \qquad (13.3)$$

as given by (2.5).

Sensing Static Pressure. The most common methods for sensing static pressures include *wall taps*, as first used extensively by Bernoulli; and *aerodynamic probes*, wherein small pressure taps are located at critical points on such surfaces as spheres, cylinders, wedges, and cones. These will be discussed briefly, primarily to point out the most important corrections that should be applied to such measurements.

Wall Taps. The effect of finite hole size has been discussed at great length in the literature; a recent summary is given in Reference 2. Briefly, tap size error is expressed as a function of the wall shearing stress (τ) since it is believed to arise because of a local disturbance of the boundary layer. Dimensional analysis (see, e.g., Section 5.4.2) leads to two acceptable pairs of nondimensional terms, depending on the grouping of variables chosen. For one grouping we have

$$\frac{e}{\tau} = f(R_d^*), \qquad (13.4)$$

where e is the error in static pressure caused by the tap size, and R_d^* is the tap Reynolds number, defined in turn by

$$R_d^* = \frac{V^* d}{\nu}, \qquad (13.5)$$

where d is the tap diameter, ν is the kinematic viscosity, and V^* is the friction velocity of (5.18), that is,

$$V^* = \sqrt{\frac{\tau g_c}{\rho}}. \qquad (13.6)$$

From another grouping of variables we obtain

$$p^* = g(\tau^*), \qquad (13.7)$$

where

$$p^* = \frac{ed^2}{\rho \nu^2} = \left(\frac{e}{\tau}\right) R_d^{*2}, \qquad (13.8)$$

and

$$\tau^* = \frac{\tau d^2}{\rho \nu^2} = R_d^{*2}. \qquad (13.9)$$

Both of these sets of dimensionless variables have been used to correlate the available data on static wall tap errors. Such graphs are shown in Figures 13.1 and 13.2. The straight lines of Figure 13.2 can be represented by the empirical equations

$$\frac{e}{\tau} = 0.000157 (R_d^*)^{1.604} \qquad (13.10)$$

to describe static tap error in the range $0 < R_d^* < 385$, whereas at tap Reynolds numbers greater than 385, we have

$$\frac{e}{\tau} = 0.269 (R_d^*)^{0.353}. \qquad (13.11)$$

Still another method for presenting static pressure tap error has been suggested [7]. Here e is nondimensionalized by the dynamic pressure (p_v) rather than the wall shear stress (τ), and the resulting quantity is given in terms of the pipe Reynolds number ($R_D = VD/\nu$), instead of the tap

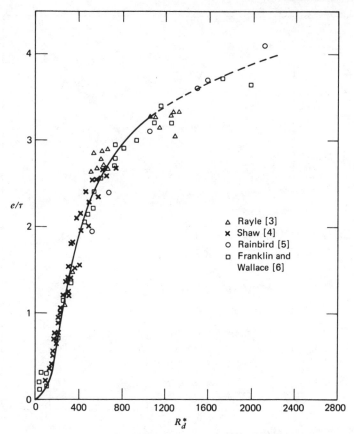

Figure 13.1 Static tap error as a function of wall shear stress and tap Reynolds number (after Benedict and Wyler [2]).

Reynolds number of (13.5), that is

$$\frac{e}{p_v} = f(R_D). \tag{13.12}$$

Such a plot is presented in Figure 13.3.

Example 13.1. Water at 70°F is flowing at a velocity of 10 ft/sec in a 4 in. smooth pipe. The density and kinematic viscosity of the fluid are $\rho = 62.3$ lbm/ft^3 and $\nu = 1.11 \times 10^{-5}$ ft^2/sec. Find the absolute error in the static pressure measurement sensed by a $\frac{1}{4}$ in. square-edged tap.

470 Pressure Measurement in Pipes

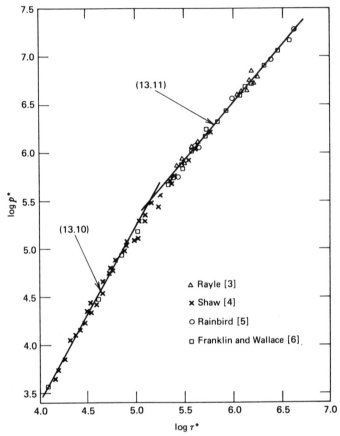

Figure 13.2 Static tap error in terms of p^* and τ^* (after Benedict and Wyler [2]).

Solution: The pipe Reynolds number is

$$R_D = \frac{VD}{\nu} = \frac{10 \times 4/12}{1.11 \times 10^{-5}} = 3 \times 10^5.$$

By (6.12), Figure 6.7, or Table 6.1: $f = 0.01447$.
By (13.5) and (5.40):

$$R_d^* = \sqrt{\frac{f}{8}} \left(\frac{d}{D}\right) R_D = 797.4.$$

Now, we have at least four ways to get e/p_v.

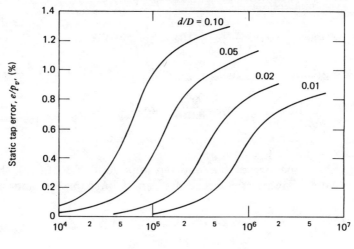

Figure 13.3 Static tap error as a function of pipe Reynolds number (after Wyler [7]).

1. By Figure 13.1:

$$\frac{e}{\tau} = 2.93.$$

2. By Figure 13.2: At $\log \tau^* = \log R_d^{*2} = 5.803$, we read $\log p^* = 6.2675$, which by (13.8) leads to

$$\frac{e}{\tau} = 2.91.$$

3. By (13.11):

$$\frac{e}{\tau} = 2.845.$$

Using an average value of $e/\tau = 2.9$ and (6.22), we have

$$\frac{e}{p_v} = \left(\frac{e}{\tau}\right)\frac{f}{4} = \frac{2.9 \times 0.01447}{4} = 0.0105 = 1.05\%.$$

4. By Figure 13.3:

$$\frac{e}{p_v} \simeq 1.00\%.$$

Then

$$e = \left(\frac{e}{p_v}\right)p_v = 0.01 \times \frac{\rho V^2}{2g_c} = 0.00672 \text{ psi.}$$

At $g = g_c$ and $T_{\text{mano}} = 76°\text{F}$, $w_{\text{H}_2\text{O}} = 0.036026$ lbf/in.^{3}, and we get

$$e = \frac{0.00672}{0.036026} = 0.186 \text{ in. H}_2\text{O,}$$

which is the error to be expected of this finite pressure tap.

Not only is the size of the wall tap important, but also its shape. In Figure 13.4 we indicate the effect of a variety of shapes on static pressure

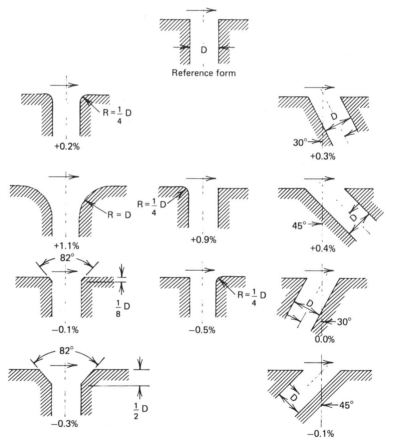

Figure 13.4 Effect of orifice shape on static pressure measurement (variation is in percentage of dynamic pressure) (after Rayle [8]).

measurement in terms of dynamic pressure [8]. Sometimes an 82° chamfer-edged tap to a depth of one-half the tap diameter is used in the hope that the −0.3% error of this shape will counteract the positive error of the finite tap size. However, the recommended practice in a piping installation is to drill square-edged wall taps and make suitable size corrections according to Example 13.1.

Aerodynamics Probes. The accuracy of static pressure measurement using pressure taps in aerodynamic bodies depends on the accuracy of location, the size of the holes, and the direction, and variation in the direction, of the flow. Cylinders inserted normal to the flow are common aerodynamic probes. Representative calibration curves for cylindrical probes are given in Figure 13.5.

Blockage Effect. Any probe inserted in a pipe *increases* the velocity near the probe by an amount ΔV. This effect has been characterized by a probe blockage factor, defined as

$$\varepsilon = \frac{\Delta V}{V}. \tag{13.13}$$

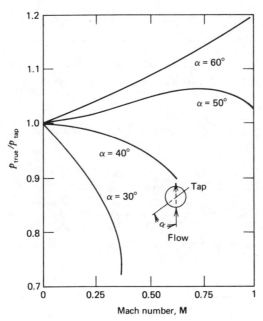

Figure 13.5 Calibration curves for a cylindrical pressure probe.

474 Pressure Measurement in Pipes

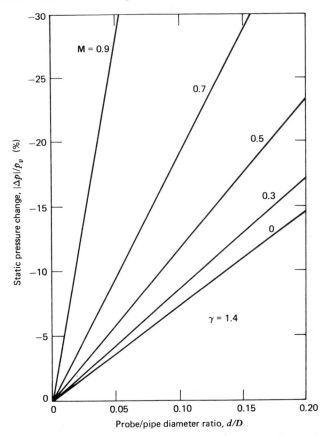

Figure 13.6 Effect of probe blockage on static pressure in a pipe (after Wyler [9]).

This parameter has been shown [9] to influence the static pressure as follows:

$$\frac{\Delta p}{p_v} = -2\varepsilon. \tag{13.14}$$

Thus, since ΔV is positive, ε is positive, and hence the static pressure *decreases*. This means that a probe will measure a static pressure lower than would have been in the pipe without the probe.

Wyler [9] has shown for subsonic pipe flow that

$$\frac{\Delta p}{p_v} = \frac{-2}{1-\mathrm{M}^2}\left[\frac{1.15+0.75(\mathrm{M}-0.2)}{2}\right]\left(\frac{2d}{\pi D}\right). \tag{13.15}$$

Equation 13.15 is plotted in Figure 13.6, and shows acceptable blockages or resultant pressure errors in terms of d/D, M, and $\Delta p/p_v$.

Example 13.2. A $\frac{1}{4}$ in. diameter probe is installed to the centerline of a 5 in. diameter pipe. Compare the static pressure error in water and air at a Mach number of 0.5.

Solution: In *water*, take Mach number $\simeq 0$. By (13.15):

$$\frac{\Delta p}{p_v} = -2\left(\frac{1.15-0.15}{2}\right)\left(\frac{2\times\frac{1}{4}}{\pi\times 5}\right) = -3.2\%.$$

In *air*, at $M=0.5$, get

$$\frac{\Delta p}{p_v} = \frac{-2}{1-0.5^2}\left[\frac{1.15+0.75(0.5-0.2)}{2}\right]\left(\frac{2\times\frac{1}{4}}{\pi\times 5}\right) = -5.8\%.$$

As a check, Figure 13.5 indicates that, at $d/D = \frac{1}{4}/5 = 0.05$,

for water at $M\simeq 0$: $\dfrac{\Delta p}{p_v} \simeq -3.2\%$,

for air at $M\simeq 0.5$: $\dfrac{\Delta p}{p_v} \simeq -5.8\%$.

Sensing Total Pressure. Total pressures are used to determine head loss data, to establish velocities, to establish state points, and even to determine flow rates. By definition, total pressure can be sensed by stagnating the fluid isentropically, as by a Pitot tube. This is simply a tube bent to face the flow direction. Of course, taps placed at aerodynamic stagnation points on probes also will sense the total pressure. Because flow direction is not always precisely known in a piping component, there is a definite advantage in using a total pressure tube that is relatively unaffected by alignment. The characteristics of several Pitot tubes in respect to flow alignment are shown in Figure 13.7.

Deducing Total Pressure. There are several good reasons for avoiding total pressure measurement in a pipe. For one thing, there is the probe blockage effect. For another, it is often the *effective* total pressure across the pipe that is required, rather than a *point* measurement. For these reasons we may want to deduce the total pressure from static pressure, total temperature, and flow rate measurements. These expressions are for

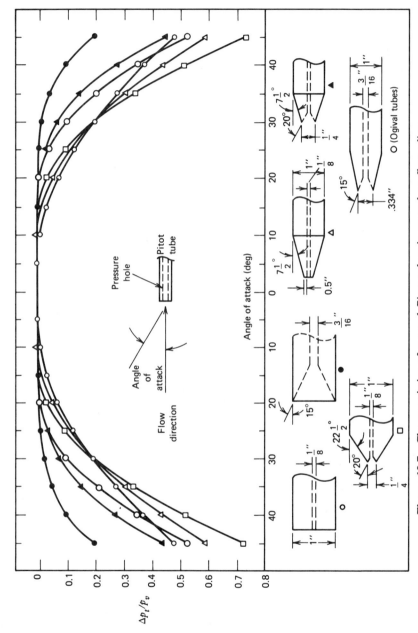

Figure 13.7 Characteristics of several Pitot tubes in regard to flow alignment (after Gracey et al. [10]).

constant density fluids:

$$(p_t)_{\text{inc}} = p + \frac{\dot{m}^2}{2g_c \rho A^2} \quad (13.16)$$

and for compressible fluids [from (3.69)]:

$$(p_t)_{\text{comp}} = p\left\{\frac{1}{2} + \left[\frac{1}{4} + \left(\frac{\gamma-1}{2\gamma}\right)\left(\frac{\dot{m}}{Ap}\right)^2\left(\frac{\overline{R}T_t}{g_c}\right)\right]^{1/2}\right\}^{\gamma/(\gamma-1)}. \quad (13.17)$$

A numerical solution of (13.17) was given in Example 3.3.

13.3 PRESSURE MEASURING SYSTEMS

Many instruments are used to measure pressures in pipes. Not the least are manometers, bourdon gages, and electrical pressure transducers. These common systems are so well described in various textbooks and in professional society monographs that only a cursory treatment need be given here.

Manometers. The manometer was used as early as 1662 by Boyle for the precise determination of steady fluid pressure. Bernoulli, in 1750, also used manometers extensively. The U-tube manometer serves as a pressure standard in the range from 0.1 in. of water to 100 psig, within a calibration uncertainty of 0.02 to 0.2% of the reading.

In the steady state the difference between the unknown pressure and the reference pressure (see Figure 13.8) is balanced by the weight per unit area of the equivalent displaced manometer liquid column. Thus

$$\Delta p_{\text{mano}} = w_M \Delta h_E, \quad (13.18)$$

where w_M is the corrected specific weight of the manometer fluid, and Δh_E is the equivalent manometer fluid height. The most common fluids used in manometers are mercury and water, because their specific weights are so well known (see Table 13.1). The values given are for standard gravity, and are corrected to local gravity according to

$$w_{\text{corrected}} = w_{\text{standard}}\left(\frac{g_{\text{local}}}{g_{\text{standard}}}\right). \quad (13.19)$$

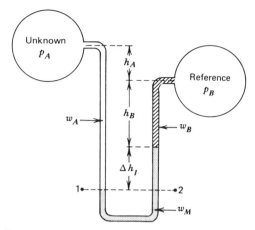

Figure 13.8 Generalized manometer notation.

Table 13.1 Specific Weights of Mercury and Water

Temperature (°F)	Specific Weight, $w_{s,t}$[a]	
	Mercury (lbf/in.^{3})	Water (lbf/in.^{3})
32	0.491154	0.036122
36	0.490956	0.036126
40	0.490757	0.036126
44	0.490559	0.036124
48	0.490362	0.036120
52	0.490164	0.036113
56	0.489966	0.036104
60	0.489769	0.036092
64	0.489572	0.036078
68	0.489375	0.036062
72	0.489178	0.036045
76	0.488981	0.036026
80	0.488784	0.036005
84	0.488588	0.035983
88	0.488392	0.035958
92	0.488196	0.035932
96	0.488000	0.035905
100	0.487804	0.035877

[a]At standard gravity value of 32.1740 ft/sec^2.

Example 13.3. Find the absolute static pressure in a water filled pipe located at $A = 5$ ft above the zero pressure position of a mercury U-tube manometer. The manometer shows a deflection of 10 in. of mercury with respect to the barometer, at a location where $g_l/g_c = 0.99$, at a manometer temperature of 72° F, and at a barometric pressure of 14.695 lbf/in.². Assume that the manometer tube bores are large enough to minimize capillary effects.

Solution: According to Figure 13.8, the reference pressure in this example is the barometric reading, w_B is the specific weight of air (which is negligible compared with that of water and mercury), w_A is the corrected specific weight of water at 72°F, w_M is the corrected specific weight of mercury at 72°F, Δh_I is 10 in. of mercury, and the sum of $h_A + h_B + \Delta h_I$ is $A + \frac{1}{2}(\Delta h_I)$.

By Figure 13.8: $p_1 = p_2$ (i.e., the same elevation in the same fluid, at rest, has the same pressure). But p_1 is made up of the pressures above it, that is,

$$p_1 = p_A + w_{H_2O}\left(A + \frac{\Delta h_I}{2}\right).$$

Similarly, p_2 is made up of the pressures above it, that is,

$$p_2 = p_B + w_{Hg}\Delta h_I.$$

Thus

$$p_A = p_{baro} + w_{H_2O}\left[\left(\frac{w_{Hg}}{w_{H_2O}} - \frac{1}{2}\right)\Delta h_I - A\right].$$

By Table 13.1 and (13.19):

$$w_{H_2O} = 0.036045 \times 0.99 = 0.035684 \text{ lbf/in.}^3,$$

$$w_{Hg} = 0.489178 \times 0.99 = 0.484286 \text{ lbf/in.}^3.$$

Thus

$$p_A = 14.695 + 0.035684\left[\left(\frac{0.484286}{0.035684} - 0.5\right)10 - 5 \times 12\right]$$

or

$$p_A = 14.695 + 2.523 = 17.218 \text{ psia}.$$

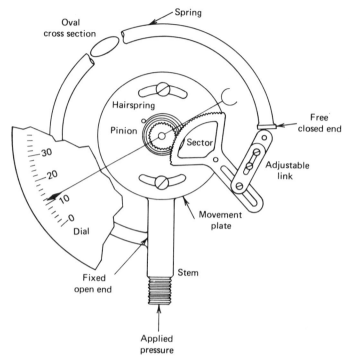

Figure 13.9 Common bourdon tube transducer (after ASME PTC Supplement 19.2, 1964).

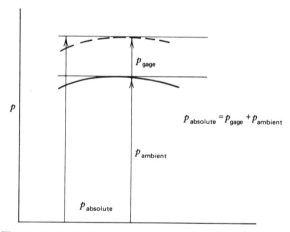

Figure 13.10 Relations between various pressure terms.

Bourdon Gages. A small volume tube is fixed and open at one end, and free and closed at the other end, as shown in Figure 13.9. The unknown pressure is applied to the open end, and the pressure difference across the tube walls deforms and displaces the free end. By an almost frictionless linkage the free end of the tube rotates a pointer over a calibrated scale to give a mechanical indication of pressure. The reference pressure in the case containing the bourdon tube is usually atmospheric, so that the pointer indicates gage pressures (see Figure 13.10).

Electrical Pressure Transducers. There is a wide variety of electrical pressure transducers. The variable resistance, variable capacitance, variable reluctance, and variable differential transformer types are the most common. The interested reader is referred to the many articles, textbooks, and professional documents on this specialized subject.

It is quite important to exercise the movable parts of the transducer before usage by applying a pressure to yield nearly full scale deflection and then releasing it. This procedure is followed several times. It is also necessary to then set the zero of the transducer when zero pressure is applied, and to adjust the span (the full scale output of the transducer) before the calibration can be used with confidence.

REFERENCES

1. R. P. Benedict, *Fundamentals of Temperature, Pressure, and Flow Measurements*, 2nd ed., Wiley-Interscience, 1977.
2. R. P. Benedict and J. S. Wyler, "Analytical and experimental studies of ASME flow nozzles," *Trans. ASME, J. Fluids Eng.*, September 1978, p. 265.
3. R. E. Rayle, "An investigation of the influence of orifice geometry on static pressure measurement," M. S. Thesis, M.I.T., 1949.
4. R. Shaw, "The influence of hole dimensions on static pressure measurements," *J. Fluid Mech.*, Vol. 7, Part 4, April 1960, p. 550.
5. W. J. Rainbird, "Errors in measurement of mean static pressure of a moving fluid due to pressure holes," *Natl. Res. Counc. Can. Rep.* DME/NAE, 1967 (3).
6. R. E. Franklin and J. M. Wallace, "Absolute measurement of static-hole error using flush transducers," *J. Fluid Mech.*, Vol. 42, Part 1, 1970, p. 33.
7. J. S. Wyler (Westinghouse Electric Corp.), Personal communication, March 1976.
8. R. E. Rayle, "Influence of orifice geometry on static pressure measurements," *ASME Pap.* 59-A-234, December 1959.

9. J. S. Wyler, "Probe blockage effects in free jets and closed tunnels," *Trans. ASME, J. Eng. Power*, October 1975, p. 509.
10. W. Gracey, W. Letko, and W. R. Russell, "Wind tunnel investigation of a number of total pressure tubes at high angles of attack," *NACA TN* 2331, April 1951.

NOMENCLATURE

Roman

- A pipe flow area
- d tap diameter
- D pipe diameter
- e static pressure error
- f function, friction factor
- g function
- g_l local acceleration of gravity
- g_c standard acceleration of gravity
- $\dot{m}$ mass flow rate
- $\mathbf{M}$ Mach number
- p static pressure
- p_v dynamic pressure
- p_t total pressure
- R_d^* tap Reynolds number
- R_D pipe Reynolds number
- T temperature
- V directed fluid velocity
- V^* friction velocity
- w specific weight

Greek

- α static tap angle
- Δ finite difference
- ε probe blockage factor
- γ ratio of specific heats
- ν kinematic viscosity
- ρ fluid density
- τ wall shear stress

14

Flow Measurement in Pipes

> ...each branch has less space than the artery it came from, but the summed space of the branches is always greater than the space in that artery. More space, but the same quantity of blood, so the stream must be moving slower and slower...
> — *Gustav Eckstein*

14.1 GENERAL REMARKS

Determination of the flow rate in a piping system is essential to an analysis of the performance of any of the piping components. Hence, in this chapter, we give definitions of flow rate for constant density and compressible fluids. Discharge coefficients, both empirical and theoretical, are discussed as they bear directly on the flow rate. Whenever the fluid is compressible, an expansion factor also is required, and this is discussed and tabulated. Flow rates where the fluid velocity equals or exceeds that of sound are also mentioned. We limit our discussion here to the basic differential pressure type fluid meters, including nozzles, orifices, and venturi meters (see Figure 14.1).

14.2 THE BASIC EQUATIONS [1]

Constant Density Fluids. The theoretical rate of flow in a pipe is given by

$$\dot{m}'_i = A_2 \left[\frac{2g_c \rho (p_1 - p_2)}{144(1 - \beta^4)} \right]^{1/2}, \qquad (14.1)$$

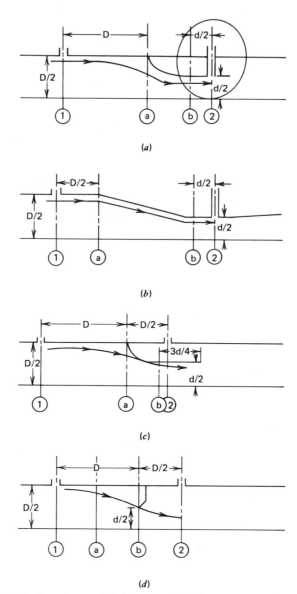

Figure 14.1 Various types of fluid meter. (*a*) Throat tap nozzle. (*b*) Venturi meter. (*c*) Pipe wall tap nozzle. (*d*) $1D$ and $\frac{1}{2}D$ tap orifices.

where $\dot{m}'_i$ is the ideal, incompressible flow rate (lbm/sec),
  A_2 is the throat area of the fluid meter (in.^{2}),
  ρ is the fluid density (lbm/ft^3),
  Δp is the pressure drop between inlet and throat (lbf/in.^{2}),
  β is the ratio of the throat diameter to the pipe diameter.

Compressible Fluids. The theoretical flow rate of an ideal gas in a pipe is given by

$$\dot{m}'_c = A_2 \left\{ \frac{2\gamma g_c p_1 \rho_1 [r^{2/\gamma} - r^{(\gamma+1)/\gamma}]}{(\gamma-1)(1-r^{2/\gamma}\beta^4)144} \right\}^{1/2} \quad (14.2)$$

with the same units as in (14.1), where r is the ratio of the static pressure at the throat to that at the inlet.

Note that in (14.1) and (14.2) the prime signifies the *ideal* condition, and the subscripts i and c signify the incompressible and compressible cases, respectively.

Critical Flow Relation. For a gas the critical static pressure ratio, that is, the one that yields the maximum isentropic flow rate for given fluid conditions at the inlet and for a given geometry, can be given as

$$r_*^{(1-\gamma)/\gamma} + \left(\frac{\gamma-1}{2}\right)\beta^4 r_*^{2/\gamma} = \frac{\gamma+1}{2}. \quad (14.3)$$

The flow rate of a gas through a convergent nozzle attains constancy and is maximized at the critical pressure ratio of (14.3). At this critical pressure ratio the fluid velocity equals the local velocity of sound, and the flow no longer responds to changes in the pressure downstream of the nozzle throat.

Whenever a nozzle or venturi is choked, that is, whenever the measured static pressure ratio is less than or equal to r^* of (14.3), the pressure ratio to use in (14.2) is r^*.

In orifices, on the other hand, the flow rate does not maximize, and the situation is too complex to discuss here. The pertinent literature must be consulted in this case (see, e.g., [2]).

14.3 THE DISCHARGE COEFFICIENT

Having now determined, via (14.1), (14.2), and (14.3), the *theoretical* flow rate in a pipe, we must realize that we still do not have the *actual* flow rate.

486 Flow Measurement in Pipes

For one thing, no real flow process is entirely reversible. Also, in fluid meters with abrupt area changes, the pipe may not always flow entirely full everywhere. The latter condition is described by the vena contracta effect, in which the fluid jet attains its minimum area (see Section 2.6.2). Finally, the static pressure downstream of the fluid meter can vary widely, depending on the location of the static taps, and this too affects the indicated flow rate. Thus it becomes evident that we need more information to establish the actual flow rate on a firm basis.

A discharge coefficient (C_D) is defined as the one factor into which are lumped all the imperfections of the fluid meter. Thus

$$C_D = \frac{\dot{m}_{actual}}{\dot{m}_{ideal}}. \tag{14.4}$$

The actual flow rate can be obtained by an experimental calibration process based on a "catching-weighing-timing" scheme. The conventional practice is to use liquids (like water) in the calibration test, and hence C_{Di} is usually obtained. Here, again, the subscript i signifies the incompressible case. The resultant C_{Di} is presented in terms of the fluid meter throat Reynolds number, defined as

$$R_d = \frac{\rho V d}{\mu} = \frac{48 \dot{m}_{actual}}{\pi d \mu}, \tag{14.5}$$

where $\dot{m}$ is in pounds mass per second, d is in inches, and μ is in pounds mass per foot per second. In applying (14.5), since the discharge coefficient is a function of the flow rate to be determined, an iterative procedure must be used. Fortunately, convergence is rapid, the correct C_D generally being obtained with one or two guesses.

Nozzles. Empirical equations have been determined based on thousands of tests of nozzles in air, water, and oil. One such, for ASME nozzles with pipe wall taps, has been given [3] as

$$C_D = 0.19436 + 0.152884 (\ln R_d)$$
$$- 0.0097785 (\ln R_d)^2 + 0.00020903 (\ln R_d)^3. \tag{14.6}$$

Solutions to this equation are given in Table 14.1.

Table 14.1 Most Probable Discharge Coefficients for ASME Nozzles Used with Pipe Wall Taps [based on (14.6)]

	1	2	3	4	5	6	7	8	9
$R_d \times 10^{-3}$:									
C_D:	—	—	0.89886	0.90898	0.91630	0.92195	0.92650	0.93028	0.93350
$R_d \times 10^{-4}$:	1	2	3	4	5	6	7	8	9
C_D:	0.93628	0.95242	0.96024	0.96512	0.96855	0.97113	0.97316	0.97482	0.97620
$R_d \times 10^{-5}$:	1	2	3	4	5	6	7	8	9
C_D:	0.97737	0.98373	0.98647	0.98804	0.98907	0.98979	0.99032	0.99074	0.99106
$R_d \times 10^{-6}$:	1	2	3	4	5	6	7	8	9
C_D:	0.99132	0.99251	0.99288	0.99305	0.99315	0.99322	—	—	—

Example 14.1. Find the flow rate of air ($\gamma = 1.4$, $\overline{R} = 53.35$ ft-lbf/lbm °R), assuming that the fluid is approximately incompressible, when $p_1 = 14.655$ psia, $p_2 = 11.347$ psia, $D = 10.136$ in., $d = 4.054$ in., $T_1 = 80°F$, and $\mu = 0.1237 \times 10^{-4}$ lbm/ft sec.

Solution:

$$\beta = \frac{d}{D} = 0.4, \quad \beta^4 = 0.0256,$$

$$A_2 = \frac{\pi}{4}(4.054)^2 = 12.9079 \text{ in.}^2,$$

$$r = \frac{p_2}{p_1} = \frac{11.347}{14.655} = 0.77427,$$

$$r^{2/\gamma} = 0.69386, \quad r^{(\gamma+1)/\gamma} = 0.64496,$$

$$\rho_1 = \frac{p_1}{RT_1} = \frac{14.655 \times 144}{53.35 \times 540} = 0.07325 \text{ lbm/ft}^3.$$

By (14.2): $\dot{m}'_c = 3.7318$ lbm/sec.
The discharge coefficient must be obtained by iteration. Assume that the throat Reynolds number is 10^6. Then, by Table 14.1: $C_D = 0.99132$.
By (14.4): $\dot{m}_{\text{actual}} = C_D \times \dot{m}'_c = 0.99132 \times 3.7318 = 3.6994$ lbm/sec.
By (14.5):

$$R_d = \frac{48 \times 3.6994}{\pi \times 4.054 \times 0.1237 \times 10^{-4}} = 1.13 \times 10^6.$$

Since this value of the Reynolds number differs slightly from the assumed value, a second trial is necessary. Using $R_d = 1.2 \times 10^6$, we get

$C_D = 0.99156$, $\dot{m} = 3.7003$, and $R_d = 1.13 \times 10^6$. Now, since R_d has not changed significantly, we can conclude that $\dot{m}_{\text{actual}} = 3.7$ lbm/sec.

Theoretical equations also have been suggested for use in predicting nozzle discharge coefficients. Although the same confidence cannot be placed in these equations as in the empirical ones, they have their place in fluid metering. One will be given here to relate this section to the sections on boundary layer theory.

Consider a general fluid meter such as the one shown in Figure 14.2. A discharge coefficient can be determined in terms of boundary layer parameters as follows [4]. By (14.4), we require $\dot{m}_{\text{actual}}$ and $\dot{m}_{\text{ideal}}$. Ideally, conservation of mass can be written as

$$\dot{m}_{\text{ideal}} = \rho A_2 U_2', \tag{14.7}$$

where U_2' signifies the ideal potential velocity at the throat in the absence of a boundary layer. The corresponding ideal energy equation between inlet and throat is

$$\frac{p_1}{\rho} + \frac{(U_1')^2}{2g_c} = \frac{p_2}{\rho} + \frac{(U_2')^2}{2g_c}. \tag{14.8}$$

Conservation of mass in the actual case, expressed in terms of the displacement thickness and the potential core velocity at the throat, is by (4.16)

$$\dot{m}_{\text{actual}} = \rho A_2^* U_2 = \rho \pi R_2^2 U_2 \left(1 - 2\frac{\delta_2^*}{R_2}\right). \tag{14.9}$$

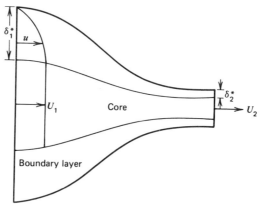

Figure 14.2 General fluid meter, showing boundary layer notation.

The energy flux equation between inlet and throat of the fluid meter is

$$\dot{E}_1 = \dot{E}_2 + (\dot{E}_{\text{loss}})_{1,2} \tag{14.10}$$

or, in expanded form,

$$\left(\dot{m}\frac{p_1}{\rho} + \dot{\text{K.E.}}_1\right) = \left(\dot{m}\frac{p_2}{\rho} + \dot{\text{K.E.}}_2\right) + (\dot{E}_{\text{loss}})_{1,2}. \tag{14.11}$$

The kinetic energy flux ($\dot{\text{K.E.}}$) has been expressed [4,5] by (4.28) as

$$\dot{\text{K.E.}} = \frac{\rho \pi R^2 U^3}{2g_c}\left(1 - 2\frac{\delta^*}{R} - 2\frac{\delta^{**}}{R}\right). \tag{14.12}$$

The energy flux loss ($\dot{E}_{\text{loss}}$) has been given [4] by (4.27) as

$$(\dot{E}_{\text{loss}})_{1,2} = \left(\frac{2\pi\rho}{2g_c}\right)(R_2 U_2^3 \delta_2^{**} - R_1 U_1^3 \delta_1^{**}). \tag{14.13}$$

Hence the energy flux balance of (14.10) becomes

$$\dot{m}\frac{p_1}{\rho} + \left(\frac{\rho\pi R_1^2 U_1^3}{2g_c}\right)\left(1 - 2\frac{\delta_1^*}{R_1} - 2\frac{\delta_1^{**}}{R_1}\right)$$
$$= \dot{m}\frac{p_2}{\rho} + \left(\frac{\rho\pi R_2^2 U_2^3}{2g_c}\right)\left(1 - 2\frac{\delta_2^*}{R_2} - 2\frac{\delta_2^{**}}{R_2}\right)$$
$$+ \left(\frac{2\pi\rho}{2g_c}\right)(R_2 U_2^3 \delta_2^{**} - R_1 U_1^3 \delta_1^{**}). \tag{14.14}$$

In reduced form, replacing $\dot{m}$ by (14.9), we have

$$\frac{p_1 - p_2}{\rho} = \left(\frac{U_2^2}{2g_c}\right)\left[1 - \left(\frac{R_1}{R_2}\right)^2\left(\frac{U_1}{U_2}\right)^3\left(\frac{1 - 2\delta_1^*/R_1}{1 - 2\delta_2^*/R_2}\right)\right]. \tag{14.15}$$

For the same pressure drop in both the ideal and actual cases, (14.8) and (14.15) can be equated to yield

$$\frac{U_2}{U_2'} = \left[\frac{1 - \beta^4}{1 - \beta^4\left(\frac{1 - 2\delta_2^*/R_2}{1 - 2\delta_1^*/R_1}\right)^2}\right]^{1/2}. \tag{14.16}$$

And, by (14.4), we have for the discharge coefficient

$$C_D = \frac{A_2^* U_2}{A_2 U_2'} \tag{14.17}$$

or

$$C_D = \left(1 - 2\frac{\delta_2^*}{R_2}\right)\left[\frac{1-\beta^4}{1-\beta^4\left(\frac{1-2\delta_2^*/R_2}{1-2\delta_1^*/R_1}\right)^2}\right]^{1/2}. \tag{14.18}$$

Thus we have arrived at a theoretical discharge coefficient based solely on the boundary layer parameters, δ_1^* and δ_2^*. In the plenum inlet case, where $\beta = 0$, we obtain from (14.18)

$$(C_D)_{\text{plenum}} = 1 - 2\frac{\delta_2^*}{R_2}, \tag{14.19}$$

in agreement with many other investigators [6–9].

A comparison of some of these theoretical discharge coefficients with actual experimental data is given in Figure 14.3 for the plenum inlet case.

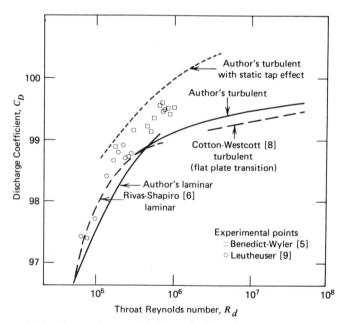

Figure 14.3 Comparison of theoretical and experimental discharge coefficients for a plenum inlet nozzle.

Orifices. Empirical equations also are available for square-edged orifices (see Figure 14.1). It is imperative that the static tap location be specified so that the correct set of equations or tables can be used. One of the most common choices of tap locations is the $1D$ and $\frac{1}{2}D$ taps. This means that the upstream tap is located just one pipe diameter from the orifice, while the downstream tap is just one-half pipe diameter from the orifice. For this choice the ASME [10] gives the following equations for the orifice:

$$b = 0.0002 + \frac{0.0011}{D} + \left(0.0038 + \frac{0.0004}{D}\right)\left[\beta^2 + (16.5 + 5D)\beta^{16}\right] \quad (14.20)$$

$$K_0 = 0.6014 - 0.01352 D^{-0.25} + (0.376 + 0.07257 D^{-0.25})$$
$$\times \left(\frac{0.00025}{D^2\beta^2 + 0.0025D} + \beta^4 + 1.5\beta^{16}\right) \quad (14.21)$$

and

$$K = K_0 + \frac{1000b}{R_D^{1/2}}, \quad (14.22)$$

where K is the flow coefficient, defined by

$$K = \frac{C_D}{(1-\beta^4)^{1/2}}. \quad (14.23)$$

These are complicated mathematical expressions, rarely used except in computer applications. Otherwise, ASME tables or graphs are used (see Table 14.2 and Figure 14.4.

In 1975, J. Stolz [12] presented a new equation for orifices, which has been shown by Miller and Cullen [13] to have a significantly better form than the ASME equations of 14.20–14.22. The Stolz equation is

$$C_{DS} = 0.5959 + 0.0312\beta^{2.1} - 0.184\beta^8$$
$$+ 0.0029\beta^{2.5}\left(\frac{10^6}{R_d\beta}\right)^{0.75} + \frac{0.09 L_1 \beta^4}{1-\beta^4}$$
$$- 0.0337 L_2 \beta^3 \quad (14.24)$$

Table 14.2 ASME Discharge Coefficients for Square-Edged Orifices versus Throat Reynolds Number (for $1D$ and $\frac{1}{2}D$ taps and 2 in. pipe)[a]

β	R_d						
	30,000	40,000	50,000	75,000	100,000	500,000	1,000,000
0.1500	0.6035	0.6019	0.6008	0.5990	0.5980	0.5943	0.5934
0.2000	0.6025	0.6009	0.5999	0.5982	0.5973	0.5938	0.5930
0.2500	0.6024	0.6009	0.5998	0.5982	0.5973	0.5938	0.5930
0.3000	0.6030	0.6015	0.6004	0.5988	0.5978	0.5943	0.5935
0.3500	0.6042	0.6026	0.6015	0.5999	0.5989	0.5953	0.5944
0.4000	0.6061	0.6044	0.6033	0.6015	0.6005	0.5967	0.5958
0.4500	0.6085	0.6068	0.6056	0.6038	0.6027	0.5987	0.5978
0.5000	0.6116	0.6098	0.6085	0.6066	0.6054	0.6013	0.6003
0.5500	0.6151	0.6132	0.6119	0.6098	0.6086	0.6042	0.6032
0.5750	0.6170	0.6150	0.6136	0.6115	0.6103	0.6058	0.6047
0.6000	0.6189	0.6168	0.6154	0.6133	0.6120	0.6073	0.6062
0.6250	0.6207	0.6186	0.6172	0.6149	0.6136	0.6088	0.6077
0.6500	0.6226	0.6204	0.6189	0.6165	0.6151	0.6102	0.6090
0.6750	0.6243	0.6220	0.6204	0.6180	0.6165	0.6113	0.6101
0.7000	0.6260	0.6236	0.6219	0.6193	0.6178	0.6122	0.6109
0.7250	0.6278	0.6252	0.6234	0.6205	0.6188	0.6128	0.6114
0.7500	0.6299	0.6269	0.6249	0.6217	0.6198	0.6130	0.6114

[a]From Bean [10].

where C_{DS} is the Stolz discharge coefficient,

L_1 is the dimensionless location of the upstream pressure tap with respect to the upstream face of the orifice, and takes on values of 0 for corner taps, $1/D$ for flange taps, and 1 for $1D$ and $\frac{1}{2}D$ taps. Note: when $L_1 \geqslant 0.4333$, the coefficient of the $\beta^4/(1-\beta^4)$ term becomes 0.039,

L_2 is the dimensionless location of the downstream pressure tap with respect to the downstream face of the orifice, and takes on values of 0 for corner taps, $1/D$ for flange taps, and $(0.5 - E/D)$ for $1D$ and $\frac{1}{2}D$ taps. Note: E represents the thickness of the orifice plate, and is nominally 0.25 inches.

The Stolz equation (14.24) is 'universal' in the sense that it applies equally well to flange tap, $1D$ and $\frac{1}{2}D$ tap, and corner tap orifice installations.

Example 14.2. Using ASME tables, find the flow rate of water through an orifice with $1D$ and $\frac{1}{2}D$ taps when $p_1 = 17.742$ psia, $p_2 = 14.634$ psia, $D = 2$ in., $d = 1.209$ in., $\mu = 0.4584 \times 10^{-3}$ lbm/ft sec, and $\rho = 62$ lbm/ft^3.

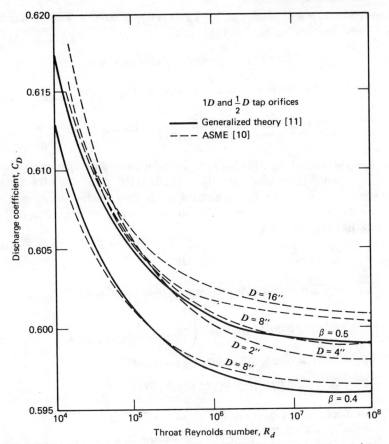

Figure 14.4 Comparison of theory with ASME practice for $1D$ and $\tfrac{1}{2}D$ tap orifices.

Solution:

$$\beta = \frac{d}{D} = 0.6045, \qquad (1-\beta^4)^{1/2} = 0.93084,$$

$$A_2 = \frac{\pi d^2}{4} = 1.148 \text{ in.}^2,$$

$$p_1 - p_2 = 3.108 \text{ lbf/in.}^2$$

By (14.1):

$$\dot{m}' = \frac{1.148}{0.93084}\left(\frac{64.34 \times 62 \times 3.108}{144}\right)^{1/2} = 11.4436 \text{ lbm/sec}.$$

494 Flow Measurement in Pipes

Assume that $R_d = 10^5$; then C_D, by Table 14.2, is 0.6123. The actual flow rate, for the first try, is, by (14.4),

$$\dot{m} = 0.6123 \times 11.4436 = 7.0069 \text{ lbm/sec.}$$

A check of the Reynolds number indicates, by (14.5), that

$$R_d = \frac{48 \times 7.0069}{1.209\pi \times 0.4584 \times 10^{-3}} = 1.932 \times 10^5.$$

Since this calculated R_d differs from the assumed value, a second trial is necessary. Using the calculated R_d of 1.932×10^5 gives $C_D = 0.6105$ and $\dot{m} = 6.9863$ lbm/sec. The new value of R_d is now 1.926×10^5. Since this does not differ significantly from the previous value, we conclude that the flow rate has been determined.

When the final Reynolds number of 1.926×10^5 of Example 14.2 is used in the Stolz equation of (14.24) we obtain

$$C_{DS} = 0.5959 + 0.0312(0.6045)^{2.1} - 0.184(0.6045)^8$$

$$+ 0.0029(0.6045)^{2.5}\left(\frac{10^6}{1.926 \times 10^5 \times 0.6045}\right)^{0.75}$$

$$+ 0.039(1)(0.6045)^4/(1 - 0.6045^4)$$

$$- 0.0337(0.5 - 0.25/2)(0.6045)^3$$

or $C_{DS} = 0.6108$, which compares favorably with $C_{ASME} = 0.6105$.

14.4 THE EXPANSION FACTOR

Whenever we must meter the flow of a compressible fluid through a piping system, the simple expression of (14.4) will not suffice because the C_D values available, as from Tables 14.1 and 14.2, are really those determined from *liquid* tests. When we settle on the use of these single-liquid C_D-R_d relationships for all fluids, we must introduce an expansion factor (Y) to account for deviations between compressible fluids and the constant density fluid of (14.4). Thus

$$Y = \frac{\dot{m}_{\text{actual}}}{\dot{m}_{\text{actual, incompressible}}}. \tag{14.25}$$

It follows for *any* fluid through *any* fluid meter that the actual flow rate can be determined from

$$\dot{m}_{\text{actual}} = Y(C_{Di}\dot{m}'_i), \tag{14.26}$$

where Y is given by (14.25), C_{Di} by (14.4), and $\dot{m}'_i$ by (14.1). It further follows that $Y_{\text{liquid}} = 1$.

Experimental Determination of Y. When the compressible fluid is condensible, as is steam, we can still determine the actual flow rate by catching and weighing the flowing fluid. When the compressible fluid is a permanent gas, as is air, we usually install a standard fluid meter in series with the meter whose expansion factor is required. In either case the factor so determined experimentally is $C_{Di}Y$, as shown in Figure 14.5, that is

$$Y_{\text{experimental}} = \frac{\dot{m}_{\text{actual}}/\dot{m}'_i}{C_{Di}} = \frac{C_{Di}Y}{C_{Di}}. \tag{14.27}$$

Analytical Determination of Y. For nozzles and venturis, the so called adiabatic expansion factor can be derived [14] from fluid flow relations already given as

$$Y' = \left\{ \left(\frac{\gamma}{\gamma - 1} \right) \frac{r^{2/\gamma}[1 - r^{(\gamma-1)/\gamma}](1 - \beta^4)}{(1-r)(1 - r^{2/\gamma}\beta^4)} \right\}^{1/2}, \tag{14.28}$$

where the prime signifies the ideal, zero contraction case. This expansion factor is given in Table 14.3.

For orifices, where contractions are the rule, an empirical relation has been given [15] for use with $1D$ and $\frac{1}{2}D$ taps as

$$Y_{\text{orifice}} = 1 - \frac{(0.41 + 0.35\beta^4)(1-r)}{\gamma}. \tag{14.29}$$

Example 14.3. Find the actual flow rate of steam through a nozzle of $C_{Di} = 0.98$, when $\dot{m}'_i = 4$ lbm/sec and $Y' = 0.89286$.

Solution: By (14.26):

$$\dot{m}_{\text{actual}} = 0.89286 \times 0.98 \times 4 = 3.5 \text{ lbm/sec.}$$

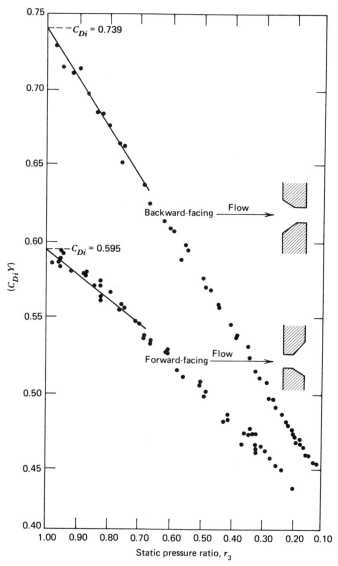

Figure 14.5 An experimental determination of C_{Di} and Y for several orifices.

Table 14.3 Adiabatic Expansion Factors for Nozzles and Venturis

r	\β 0 to 0.2	0.4	0.6
0.98	0.989228	0.988858	0.987139
0.96	0.978337	0.977610	0.974237
0.94	0.967325	0.966252	0.961289
0.92	0.956187	0.954780	0.948292
0.90	0.944920	0.943190	0.935240
0.88	0.933518	0.931478	0.922130
0.86	0.921979	0.919640	0.908955
0.84	0.910296	0.907671	0.895713
0.82	0.898466	0.895567	0.882396
0.80	0.886483	0.883321	0.868999
0.78	0.874341	0.870929	0.855518
0.76	0.862035	0.858385	0.841945
0.74	0.849558	0.845683	0.828274
0.72	0.836903	0.832815	0.814499
0.70	0.824064	0.819776	0.800612
0.68	0.811033	0.806557	0.786606
0.66	0.797801	0.793151	0.772472
0.64	0.784301	0.779548	0.758202
0.62	0.770702	0.765740	0.743787
0.60	0.756814	0.751717	0.729216
0.58	0.742687	0.737467	0.714480
0.56	0.729789	0.724217	0.699827
0.54	0.713882	0.708353	0.684497
0.52	0.698851	0.693439	0.670078
0.50	0.684731	0.679426	0.656436
0.48	0.671434	0.666227	0.643495
0.46	0.658883	0.653765	0.631193
0.44	0.647010	0.641974	0.619476
0.42	0.635757	0.630796	0.608294
0.40	0.625071	0.620180	0.597608
0.38	0.614907	0.610081	0.587360
0.36	0.605223	0.600456	0.577580
0.34	0.595982	0.591272	0.568179
0.32	0.587152	0.582495	0.559154
0.30	0.578704	0.574095	0.550483
0.28	0.570610	0.566048	0.542147
0.26	0.562846	0.558328	0.534130
0.24	0.555391	0.550915	0.526417
0.22	0.548224	0.543789	0.518994
0.20	0.541328	0.536932	0.511852
0.18	0.534685	0.530328	0.504978

Table 14.3 Continued

r	β 0 to 0.2	0.4	0.6
0.16	0.528282	0.523961	0.498366
0.14	0.522103	0.517819	0.492007
0.12	0.516136	0.511867	0.485894
0.10	0.510369	0.506156	0.480022
0.08	0.504791	0.500614	0.474386
0.06	0.499392	0.495251	0.466983
0.04	0.494162	0.490058	0.463812
0.02	0.489094	0.485027	0.458874
0	0.484178	0.480150	0.454183

REFERENCES

1. R. P. Benedict, *Fundamentals of Temperature, Pressure, and Flow Measurements*, 2nd ed., Wiley-Interscience, 1977.
2. R. P. Benedict, "Generalized contraction coefficient of an orifice for subsonic and supercritical flows," *Trans. ASME, J. Basic Eng.*, June 1971, p. 99.
3. R. P. Benedict, "Most probable discharge coefficients for ASME flow nozzles," *Trans. ASME, J. Basic Eng.*, December 1966, p. 734.
4. R. P. Benedict, "Generalized fluid meter discharge coefficient based solely on boundary layer parameters," *Trans. ASME, J. Eng. Power*, October, 1979, p. 572.
5. R. P. Benedict and J. S. Wyler, "Analytical and experimental studies of ASME flow nozzles," *Trans. ASME, J. Fluids Eng.*, September 1978, p. 265.
6. M. A. Rivas and A. H. Shapiro, "On the theory of discharge coefficents for rounded-entrance flow meters and venturis,"*Trans.ASME*, April 1956, p. 489.
7. G. W. Hall, "Application of boundary layer theory to explain some nozzle and venturi flow peculiarities,"*Proc.Inst. Mech. Eng.*, Vol. 173, No. 36, 1959, p. 837.
8. K. C. Cotton and J. C. Westcott, "Throat tap nozzles used for accurate flow measurements," *Trans. ASME, J. Eng. Power*, October 1960, p. 247.
9. H. J. Leutheuser, "Flow nozzles with zero beta ratio," *Trans. ASME, J. Basic Eng.*, September 1964, p. 538.
10. H. S. Bean, Ed., *Fluid Meters—Their Theory and Application*, Report of ASME Research Committee on Fluid Meters, 6th ed., 1971.
11. R. P. Benedict and J. S. Wyler, "A generalized discharge coefficient for differential pressure type fluid meters,"*Trans. ASME, J. Eng. Power*, October 1974, p. 440.

12. J. Stolz, "Refitting of the universal equation for discharge coefficients of orifice plate flowmeters," IS0/TC30/SC2 (France 12), 90E, October 1977.
13. R. W. Miller and J. T. Cullen, "The Stolz and ASME-AGA orifice equations compared to laboratory data," ASME Paper 78-WA/FM-2, December 1978.
14. R. P. Benedict, "Generalized expansion factor of an orifice for subsonic and supercritical flows," *Trans. ASME, J. Basic Eng.*, June 1971, p. 121.
15. E. Buckingham, "Notes on the orifice meter; the expansion factor for gases," *J. Res. Natl. Bur. Stand.*, Vol. 9, 1932, p. 61.

NOMENCLATURE

Roman

A area
C_D discharge coefficient
d throat diameter
g_c gravitational constant
$\dot{m}$ mass flow rate
p static pressure
r static pressure ratio
R_d throat Reynolds number
$\bar{R}$ specific gas constant
T absolute static temperature
U ideal velocity
V directed fluid velocity
Y expansion factor

Greek

β diameter ratio
γ ratio of specific heats
ρ fluid density
μ fluid viscosity
δ^* displacement

Subscripts

1,2 axial stations
c compressible
i incompressible

15

Special Measurements in Pipes

> ...of course, the circulation adds no unnecessary resistance, as the rich throw nothing away...
> —*Gustav Eckstein*

15.1 GENERAL REMARKS

In this chapter we discuss briefly some of the special measurements that are made in pipes. These include skin friction measurements by three methods, namely, by Preston tubes, by balance methods, and by inference from velocity measurements; boundary layer measurements by hot wire anemometers and by Pitot tubes; velocity distribution measurements by the laser-Doppler velocimeter; and nonintrusive flow measurements by a cross-correlation technique using acoustic sensors. Of course, we can only skim the surface of this major field of endeavor. Our interest here is simply to give some indication of the existing work in this area of measurements in pipes and in piping components.

15.2 SKIN FRICTION

15.2.1 The Preston Tube

In determining friction factors, velocity distributions, boundary layer parameters, or even static pressure tap corrections, we need some knowledge of the shear stress at the wall of the piping component.

Skin Friction

J. H. Preston, in 1954, described a simple method for obtaining the turbulent skin friction by means of small round Pitot tubes resting on the wall surface [1]. Such tubes are now called Preston tubes after the originator. He defined a universal nondimensional relation for the difference between the total pressure recorded by such a tube and the static pressure at the wall in terms of the skin friction as

$$\frac{(p_t - p_0)d^2}{4\rho\nu^2} = f\left(\frac{\tau_0 d^2}{4\rho\nu^2}\right), \tag{15.1}$$

where p_t is the total pressure indicated by a Preston tube,
p_0 is the static pressure at the wall,
τ_0 is the wall shear stress,
d is the Preston tube outside diameter.

Note that the form of (15.1) is analogous to that of (13.7) through (13.9), as was determined by a dimensional analysis.

The raw results of Preston's initial investigation are shown in Figure 15.1, based on the general arrangement of the pressure sensors given in Figure 15.2. The small scatter of the data and the fact that the curves are well separated speak well for the care taken in the experiment.

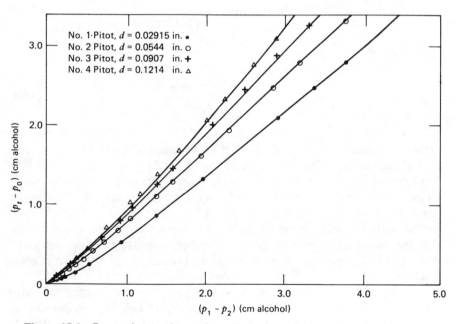

Figure 15.1 Preston's experimental results for round Pitot tubes resting on a pipe wall (after Preston [1]).

502 Special Measurements in Pipes

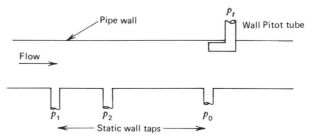

Figure 15.2 General arrangement of pressure sensors.

The same results are replotted in Figure 15.3, where Preston used the logarithmic scales to cover the wide range of variables, and to effectively collapse all the data to a single curve. The equation given to represent this single curve was

$$\tau^* = -1.396 + 0.875 p^* \tag{15.2}$$

for p^* greater than 5, where τ^* is defined by

$$\tau^* \equiv \log\left(\frac{\tau_0 d^2 g_c}{4\rho v^2}\right) \tag{15.3}$$

and p^* is defined by

$$p^* \equiv \log\left[\frac{(p_t - p_0) d^2 g_c}{4\rho v^2}\right]. \tag{15.4}$$

Thus the general function of (15.1) is made specific by (15.2).

Preston's claim is that the curve defined by (15.2) can be used to convert readings of $(p_t - p_0)$ to values of τ_0 on other walls under falling or rising pressure gradients, independently of whether rough or smooth surfaces are involved, provided only that the wall protrusion heights are small compared with the least Pitot diameter. In other words, the Preston tube reading, taken together with the wall static pressure tap reading, can yield, through (15.2), the wall shear stress and, through (6.21), the skin friction coefficient.

Hsu [2] confirms Preston's work and suggests that the Preston tube also can yield reliable results in adverse pressure gradients and in developing boundary layers. He further states that the wall thickness of the Preston tube is not important in the measurement of the local skin friction coefficient.

Skin Friction 503

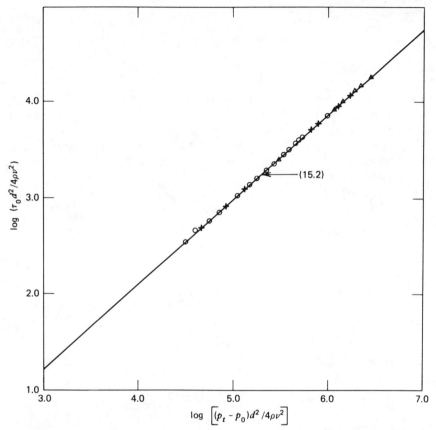

Figure 15.3 Nondimensional plotting of results of Figure 15.1 (after Preston [1]).

Head and Rechenberg [3], in 1962, tested Preston tubes in fully developed turbulent pipe flow and in developing turbulent boundary layers. By ingenious experiments they showed that for given skin friction the Preston tube readings were the same for both boundary layer and pipe flows. They did suggest, however, that a slight modification of the Preston calibration [of (15.2)] was in order.

In 1965, V. C. Patel provided the definitive calibration for the Preston tube. His results are given in terms of τ^* of (15.3) and p^* of (15.4) as

$$\tau^* = 0.5 p^* + 0.037 \tag{15.5}$$

for $0 < p^* < 2.9$;

$$\tau^* = 0.8287 - 0.1381 p^* + 0.1437 p^{*2} - 0.006 p^{*3} \tag{15.6}$$

504 Special Measurements in Pipes

for $2.9 < p^* < 5.6$; and

$$p^* = \tau^* + 2\log(1.95\tau^* + 4.1) \tag{15.7}$$

for $5.6 < p^* < 7.6$.

Example 15.1. Find the wall shear stress, the skin friction coefficients, and the friction factor in a pipe if a Preston tube of 0.12 in. diameter is installed on the inner surface of a pipe carrying air at a mean velocity of 100 ft/sec and a centerline velocity of 130 ft/sec, at a temperature of 80 °F, when the Preston tube manometer deflection is 1.111 in. of water.

Solution: By Table 13.1: At $T = 80$ °F, $w_{H_2O} = 0.036$ lbf/in.^{3}
By (13.8): $p_t - p_0 = w_{H_2O} \Delta h = 0.036 \times 1.111 = 0.04$ psi.
By Table (8.6):

$$\rho_{air} = 0.0735 \text{ lbm/ft}^3,$$

$$\nu_{air} = 16.9 \times 10^{-5} \text{ ft}^2/\text{sec}.$$

By (15.4):

$$p^* = \log\left[\frac{(p_t - p_0)d^2 g_c}{4\rho\nu^2}\right]$$

$$= \log\left(\frac{0.04 \times 0.12^2 \times 32.174}{4 \times 0.0735 \times 16.9^2 \times 10^{-10}}\right) = 6.3438.$$

By (15.2): $\tau^* = -1.396 + 0.875(6.3438) = 4.1548$.
By (15.3):

$$\frac{\tau_0 d^2 g_c}{4\rho\nu^2} = 10^{4.1548} = 14{,}283.38,$$

and hence

$$\tau_0 = 2.589 \times 10^{-4} \text{ psi},$$

which is the wall shear stress by the Preston calibration.
For the *same* p^*, Patel's calibration of (15.7) yields, by a short iteration of $6.3438 = \tau^* + 2\log(1.95\tau^* + 4.1)$,

$$\tau^* \simeq 4.17, \quad \text{and} \quad \tau_0 = 2.681 \times 10^{-4} \text{ psi}.$$

Continuing with Preston's original calibration, we have by (6.25):

$$C_f = \frac{\tau_0}{\rho V_c^2/2g_c} = 1.93 \times 10^{-3},$$

which is the *local* skin friction coefficient.

By (6.21):

$$C_{fM} = \frac{\tau_0}{\rho V^2/2g_c} = 3.26 \times 10^{-3},$$

which is the *mean* skin friction coefficient.

By (6.23):

$$f = 4 C_{fM} = 4 \times 3.26 \times 10^{-3} = 0.013,$$

which is the friction factor.

15.2.2 Balance Method for Obtaining τ_0

Smith and Walker [5] measured local surface shear stress by a floating element skin friction balance, employing a device similar to that used by Dhawan [6] and Schultz-Grunow [7] (see Figure 15.4). Specifically, the floating element is returned to its no-load neutral position by an electromagnet whose strength is variable. The position of the floating element is indicated by a differential transformer having a sensitivity to a position change of a few millionths of an inch. The electromagnetic force is equal and opposite to the drag force exerted on the element, so the average surface shear stress on the floating element can be deduced from the measured electromagnetic force.

Allen [8] investigated the effects of gap size, protrusion effects, lip forces, normal forces, and friction forces on the floating balance (see Figure 15.4). He found no advantage to maintaining a small gap size.

All in all, the more complex floating balance method has proved to confirm the simpler Preston tube method for determining wall shear stress.

15.2.3 Inference from Velocity Measurements

Having already discussed the Preston tube method for determining wall shear stress, as well as the direct force method via a floating element, we next consider briefly several methods for *inferring* local wall shear stress by using near-wall velocity profile measurements.

506 Special Measurements in Pipes

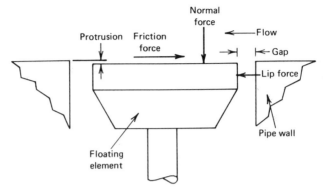

Figure 15.4 Balance method for determining skin friction, showing aerodynamic forces acting on floating element.

In Section 4.5 we applied the continuity and momentum principles to velocity profiles in the boundary layer to obtain by inference the frictional force, the wall shearing stress, and the skin friction coefficient for both laminar and turbulent boundary layers. Several examples were given, and results were compared with accepted experimental results.

In Section 6.10 we considered the Clauser plot, in which local shear stresses were inferred from a plot of u/V_c versus $V_c y/\nu$ (see Figure 6.10).

Neither the mathematical integration method of Chapter 4 nor the graphical method of Chapter 6 will be discussed further here. However, it is well known that the "law of the wall," on which the Clauser plot is based, does not correlate the velocity data right down to the wall, so that some adjustment must be made to account for the velocity profile *very* near the wall. Pierce and Gold [9] consider that the latter effect accounts for the inconsistencies noted when inferring skin friction coefficients from velocity profiles. Some of their results can be summarized as follows:

1. Mild pressure gradients do not significantly affect the near-wall data.
2. For a given velocity profile the specific choice of the similarity formulas and constants (see Table 5.5) has a significant effect on the prediction of the shear velocity (variations up to 3.7% were observed).
3. Changes in very-near-wall velocity data can be identified with different measuring probes and different sizes and geometries of these probes.

15.3 BOUNDARY LAYER MEASUREMENTS

Of the variety of measurements that can be made in the boundary layer, we consider only two instruments here, namely, the hot wire anemometer and the Pitot tube. Since whole books and monographs have been written to describe these methods, we can only hope to touch the surface in this section.

15.3.1 Hot Wire

Since 1914 the hot wire anemometer has proved to be the most important instrument for measuring the time dependent velocity components of fluid flow because of the remarkable sensitivity of the hot wire to the fluctuating velocities of turbulent flow, as compared with the prohibitively slow response of pressure probes. The hot wire's rapid response results from the use of a very thin, electrically heated wire in an electrical bridge circuit. Two basic modes of the hot wire circuit are in use (see Figure 15.5).

Briefly, the *constant current* hot wire circuit is relatively simple, the wire temperature and resistance changing as the rate of convective cooling changes; whereas the *constant resistance* hot wire circuit requires electronic linearization techniques, the wire response consisting of a change in current which keeps the temperature and resistance of the wire constant. Of the two circuits the constant temperature hot wire anemometer shows higher frequency response and greater accuracy for large velocity fluctuations.

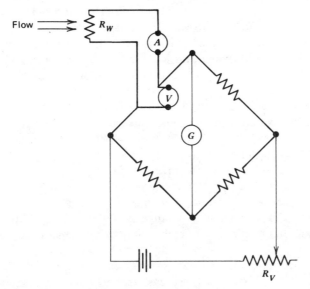

Figure 15.5 Typical hot wire anemometer Wheatstone bridge circuit. *Note* (a) In the constant current node, anemometer deflection is kept constant by adjusting R_V. *Note* (b) In the constant resistance mode, galvanometer deflection is kept zero by adjusting R_V. A = ammeter; V = voltmeter; G = galvanometer; R_W = hot wire; R_V = variable resistor.

508 Special Measurements in Pipes

In respect to heat transfer, it is well known that a heated wire placed in a moving fluid stream will be cooled, depending primarily on the fluid flow rate past the wire. King [10], in 1914, first gave the theoretical equation for the heat loss from a small cylindrical wire placed in a moving fluid stream as

$$q = \left(k + \sqrt{2\pi c_p D} \ \sqrt{\rho V}\right)(T_w - T_f), \tag{15.8}$$

where q is the rate of heat loss per unit length of wire,
k is the thermal conductivity of the fluid,
c_p is the specific heat capacity of wire,
D is the wire diameter,
ρV is the mass flow rate per unit area,
T_W is the wire temperature,
T_f is the fluid temperature.

In (15.8) the wire temperature is assumed constant because of the high thermal conductivity of the wire, and the fluid is assumed to be incompressible.

Problem areas to be faced when using hot wires include the basic calibration, calibration drift, wire sensitivity, and noise. Also, the wires are fragile and difficult to work with. Shock, vibration, particles in the fluid, and fluid drag all contribute to wire breakage.

Some of the basic references on hot wire anemometry, in addition to King's article, include Dryden and Kuethe [11], Weske [12], and Laurence and Landes [13]. More recent works on hot wires are those of Morrow and Kline [14] and Wyler [15].

Flow direction can be measured in several ways by the hot wire anemometer. If a single wire is placed perpendicular to the flow direction, the reading will correspond to the maximum velocity. When the same wire is rotated 90° to be parallel to the flow direction, a reading corresponding to the minimum velocity is obtained. It has been found that the heat transfer from a heated wire in the constant temperature circuit varies nearly as the sine of the angle of incidence between the wire and the fluid flow direction. Of course, a V-arrangement can be used to determine velocity components directly. Typical wire arrangements are shown in Figure 15.6.

Common wire materials in use are tungsten and platinum. Tungsten is the stronger, but platinum is more easily attached to the necessary wire supports.

For more details on the application of hot wires for determining velocities and turbulence, the extensive literature on this subject should be consulted.

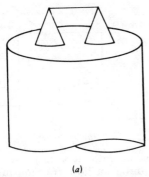 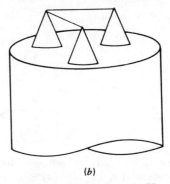

Figure 15.6 Typical hot wire arrangements. (*a*) Single wire. (*b*) V-array.

15.3.2 Pitot Tubes

Pitot tubes with very small openings have long been used in boundary layer studies. For incompressible flow, by (2.5) we have

$$p_v = \frac{\rho u^2}{2g_c}, \qquad (15.9)$$

which is the Pitot-static tube response when there is complete conversion of kinetic energy to pressure. In practice, a coefficient of from 0.98 to 0.99 is applied to (15.9) to account for incomplete conversion.

We next consider the effect of a variation in the point velocity across the finite opening of the Pitot tube.

According to (5.29), the velocity variation in the laminar sublayer is given by

$$u = C_L y, \qquad (15.10)$$

where the subscript L signifies the laminar case. In the turbulent portion of the boundary layer, the velocity variation is given by (5.13) as

$$u = C_T y^{1/7}, \qquad (15.11)$$

where the subscript T signifies the turbulent case. Figure 15.7 illustrates this point for these two cases.

Combining (15.9) and (15.10), we obtain for the response of a Pitot-static tube placed with its centerline at a distance y_0 from the pipe wall

$$(p_v)_{L,\mathcal{C}} = \left(\frac{\rho C_L^2}{2g_c}\right) y_0^2 \qquad (15.12)$$

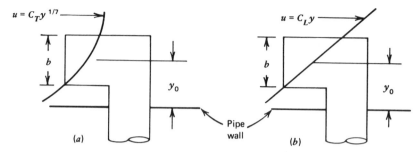

Figure 15.7 Effect of velocity distribution on Pitot tube response (after Knudsen and Katz [16]). (a) Pitot tube in turbulent flow. (b) Pitot tube in laminar flow.

and

$$(p_v)_{T,\mathcal{C}} = \left(\frac{\rho C_T^2}{2g_c}\right) y_0^{2/7}, \qquad (15.13)$$

both based on the centerline velocities.

For the Pitot response based on the integrated (i.e., average) velocities and on a Pitot tube opening of b, we have for these two boundary layer situations:

$$(\bar{p}_v)_L = \left(\frac{\rho C_L^2}{2g_c}\right) \int_{y_0-(b/2)}^{y_0+(b/2)} \left(\frac{by^2 \, dy}{b^2}\right) \qquad (15.14)$$

and

$$(\bar{p}_v)_T = \left(\frac{\rho C_T^2}{2g_c}\right) \int_{y_0-(b/2)}^{y_0+(b/2)} \left(\frac{by^{2/7} \, dy}{b^2}\right). \qquad (15.15)$$

For the laminar sublayer case we set y_0, the location of the Pitot tube centerline, realistically at $b/2$, which represents the tube resting on the pipe walls. Then (15.14) becomes

$$(\bar{p}_v)_L = \left(\frac{\rho C_L^2}{2g_c}\right) \frac{b^2}{3}, \qquad (15.16)$$

and (15.12) becomes

$$(p_v)_{L,\mathcal{C}} = \left(\frac{\rho C_L^2}{2g_c}\right) \frac{b^2}{4}. \qquad (15.17)$$

Boundary Layer Measurements 511

Following Reference 16, we see from the ratio of (15.16) and (15.17) that the *average* response of a Pitot-static tube is 1.33 times as great as the *centerline* response in the laminar sublayer case. This means that a correction must be applied to the Pitot tube position when used in the laminar sublayer.

For the turbulent boundary layer case we set y_0 realistically at $2b$, which represents the tube close to the pipe wall. Then (15.15) becomes

$$(\bar{p}_v)_T = \left(\frac{\rho C_T^2}{2g_c}\right)\left(\frac{7b^{2/7}}{9}\right)(1.56392), \qquad (15.18)$$

and (15.13) becomes

$$(p_v)_{T,\,\mathfrak{C}} = \left(\frac{\rho C_T^2}{2g_c}\right)\left(\frac{b}{2}\right)^{2/7}. \qquad (15.19)$$

From a ratio of (15.18) and (15.19) we see that the *average* response of a Pitot-static tube is now within 0.2% of the *centerline* response in the turbulent boundary layer case. Thus a small Pitot tube closely yields the point velocity in a turbulent portion of the boundary layer.

Example 15.2 (after [16]). Find the difference between the *indicated* velocity and the *point* velocity at the Pitot tube centerline, if a tube of opening $b = 0.005$ in. is resting on the pipe wall in the laminar sublayer, and $C_L = 86.3$ ft/sec in. and $\rho = 0.0314$ lbm/in.^{3}

Solution: By (15.16):

$$(\bar{p}_v)_L = \left(\frac{0.0314 \times 86.3^2}{2 \times 32.174}\right)\left(\frac{0.005^2}{3}\right) \times 1728 = 0.052 \text{ psf.}$$

By (15.17):

$$(p_v)_{L,\,\mathfrak{C}} = \left(\frac{0.0314 \times 86.3^2}{2 \times 32.174}\right)\left(\frac{0.005^2}{4}\right) \times 1728 = 0.039 \text{ psf.}$$

The indicated velocity corresponding to $(p_v)_L$ is, from (15.9),

$$V_{\text{ind}} = \sqrt{\frac{2g_c p_v}{\rho}} = 0.249 \text{ ft/sec.}$$

The centerline velocity corresponding to $(p_v)_{L,\,\mathfrak{C}}$ is

$$V_\mathfrak{C} = 0.216 \text{ ft/sec.}$$

512 Special Measurements in Pipes

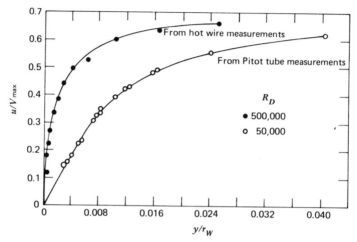

Figure 15.8 Mean velocity distribution in boundary layer adjacent to a pipe wall of I.D. = 9.72 in. (after Knudsen and Katz [16] and Laufer [17]).

Thus we see that the measured velocity is about 15% higher than the desired point velocity.

A comparison of data obtained by Laufer [17] with a hot wire anemometer and with a Pitot tube in air flowing in a circular pipe is shown in Figure 15.8. This is given to indicate the deviations possible between these two methods of measurement in the boundary layer immediately adjacent to the pipe wall.

15.4 LASER DOPPLER VELOCIMETER (LDV)

The LDV provides a powerful technique for obtaining fluid velocity measurements at a point, and serves as an alternative to the hot wire and Pitot tube methods.

Advantages of the LDV over the more conventional methods include its nonintrusive nature (meaning that it does not disturb the flow field), its inherently high response to fluctuating velocities, its provision of essentially a point measurement, and the attractive feature that it does not require calibration.

The beginnings of the LDV can be traced to Christian Doppler, who, in 1843, analyzed the pitch of a train whistle with respect to its motion. Doppler's results indicate that, as a train approaches a station, the pitch of its whistle increases, whereas the pitch decreases as the train recedes. In a practical sense the Doppler shift of scattered electromagnetic radiation has long been applied to velocity determination in radar systems. With the advent of the laser light source, the Doppler technique is now used to measure localized velocities in moving fluids.

15.4.1 The Laser

Lasers are light sources that produce highly monochromatic light of high frequency which is well collimated and has a high degree of temporal and spatial coherence. A. L. Schawlow [18] has described lasers somewhat as follows. Lasers produce coherent light by stimulating excited atoms or molecules to emit radiation. They are based on C. H. Townes's [19] idea of combining excited stimulable atomic systems with a resonator to produce *coherent* emission. The word "coherent" here means that the phase difference between two points in a beam of light is independent of time. Laser light is coherent because most of the atoms are forced to contribute to the wave that is temporarily stored between the mirrors (see Figure 15.9). The output of the laser is directional because only waves traveling along the axis of the laser remain long enough to build up in intensity. Finally, laser light is *monochromatic* because of the resonant nature of the process of stimulated amplification. "Monochromatic" here signifies that the light is composed of one wavelength only.

15.4.2 The LDV

In 1964, Yeh and Cummins [20] first reported on the use of a LDV as an anemometer for providing fluid velocity measurements in pipes and ducts. A basic LDV system includes a laser light source, focusing optics, light collecting optics, a photomultiplier to detect and convert the light signal to an electrical signal, and a signal processor to convert the initial electrical signal to a more useful form.

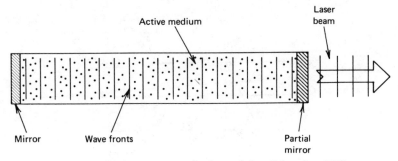

Figure 15.9 Basic structure of a laser (after Schawlow [18]).

514 Special Measurements in Pipes

The LDV system makes use of two beams of laser light, focused to form a very small measurement volume (wherein the "point" velocity is determined) at their intersection (see Figure 15.10). A standing wave pattern is formed where the beams cross and interfere with each other. This interference causes fringes to form wherein light cancels in some regions and complements in others.

The separation distance (d) of these fringes depends only on the angle of intersection (2α) of the two beams of light and on the wavelength (λ) of the light. Thus

$$d = \frac{\lambda}{2 \sin \alpha}. \tag{15.20}$$

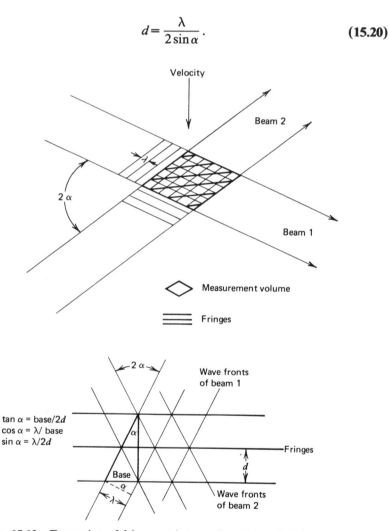

Figure 15.10 Formation of fringes at intersection of two light beams.

When very small particles carried by the moving fluid, either as impurities or as seeded matter, cross the fringes in the measurement volume, the coherent light is scattered from the particles. A stationary observer will see a Doppler shift in the frequency of the scattered light, that is, a particle passing through the measurement volume reflects an alternating intensity of illumination whose frequency is proportional to the particle velocity across the fringes.

The Doppler frequency ($f = 1/T$), generated by the particles as they scatter the incident wave fronts in the measurement volume, is detected by a photomultiplier tube. A typical output, characteristic of scattered light from two beams, is shown schematically in Figure 15.11. This signal is further processed by the electronics of the LDV system to yield the component of the particle velocity (V_p) perpendicular to the fringes, that is, since

$$V_p = \frac{d}{T} = df, \qquad (15.21)$$

(15.20) and (15.21) combine to yield

$$V_p = \frac{\lambda f}{2 \sin \alpha}. \qquad (15.22)$$

Fluid velocity is inferred from the particle velocity of (15.21)—hence the requirement for very small particles that move essentially at the fluid velocity.

Example 15.3 (after [21]). A LDV using a helium-neon laser of wavelength 6328 Å provides a beam intersection angle of 11.42°. What is the indicated velocity when the mean output frequency is 35 MHz?

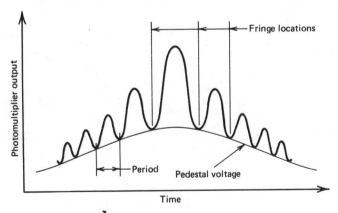

Figure 15.11 Scattered light output from both beams for a single particle.

516 Special Measurements in Pipes

Solution: The fringe spacing is given by (15.20) as

$$d = \frac{6328 \text{ Å} \times 10^{-10} \text{ m/Å}}{2\sin(11.42/2)°} = 3.18 \times 10^{-6} \text{ m}.$$

The particle velocity is given by (15.21) as

$$V_p = 3.18 \times 10^{-6} \text{ m} \times 35 \times 10^6 \text{ Hz} = 111.3 \text{ m/sec}.$$

15.4.3 LDV Arrangements

Two basic arrangements of LDV's are in use. The older is the reference beam system, wherein one beam reaches the photomultiplier detector directly, while the second beam reaches the detector after scattering from the particles moving with the fluid. These two beams are mixed so that only the difference frequency is measured. Otherwise, the actual light frequency would be too high to be measurable directly (see Figure 15.12). The more popular arrangement is the dual-scatter system. Here the laser beam is divided into two parallel beams that are incident on a lens. The lens focuses the two beams to a common point, and scattered light from both beams is collected (see Figure 15.13).

The dual-scatter system can be further divided into two arrangements: the forward scatter mode, which requires a smaller laser (typically 15 mW), with the photomultiplier detector located on the side of the measuring volume opposite to that of the laser; and the back scatter mode, which requires a more powerful laser (typically 4 W) but offers the advantage of being able to measure inside a flow passage with only one access hole, and with the photomultiplier detector located on the same side of the measuring volume as the laser.

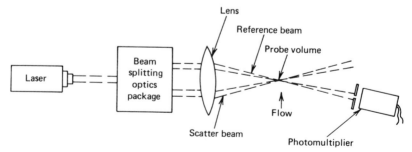

Figure 15.12 LDV reference beam system.

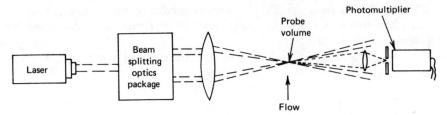

Figure 15.13 LDV dual-scatter system—forward mode.

15.4.4 LDV Results

The literature on laser Doppler velocimetry is growing rapidly. Only a few typical results are given here. Blake, of the National Engineering Laboratories in East Kilbride, Scotland, has reported [22] on comparisons made between water tests via weigh tanks and LDV traverses across pipes. For laminar flow he reports agreement within ±0.5%, while for turbulent flow the agreement is within ±1%. Bates has reported [23] on typical results obtained from laser anemometer measurements of velocity profiles in fully developed pipe flow and has compared these results with theory. He further compares flow rates derived from velocity profiles with weigh tank flow rates, obtaining about ±2% agreement. He also determines the displacement thickness (δ^*), the shape factor (H), and the skin friction coefficient (C_f) to relate the LDV work to boundary layer theory. Durgin and Neale [24], of the Alden Research Laboratories in Holden, Mass., have reported on similar tests, where they show from 1 to 1.8% agreement between LDV and weigh tank results. Their setup is shown schematically

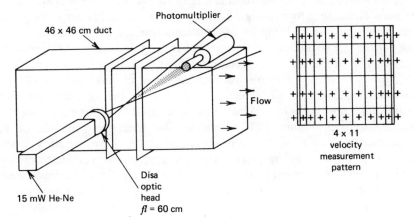

Figure 15.14 Alden Research Laboratories LDV setup (after Durgin and Neale [24]).

in Figure 15.14, along with their velocity measurement pattern. Incidentally, they follow a Gaussian integration scheme such as is described in References 25 and 26.

Recent review papers on LDV systems are those of Wyler [27], Keyser [28], Wang [29], and Durst et al. [30].

15.5 NONINTRUSIVE FLOW MEASUREMENT

Ultrasonics has been applied to flow measurement for some time. In the usual form, intrusion into the piping system must be made in the form of wells, into which are placed the ultrasonic transmitters and receivers [31,32]. In another form, the ultrasonic devices are strapped on the outside of the pipe, in the so-called sing-around mode, wherein much of the signal is bypassed through the pipe walls rather than passing through the fluid [33]. Both of these types depend on accurately measuring the transit time of an acoustic signal through the moving fluid.

A new type of acoustic flow meter is considered here, as an example of a nonintrusive flow measurement method. It is based on the use of two diametral ultrasonic beams, which interact with the fluid to detect the passage of such inherent tracers as turbulence eddies. The method used for determining the transit time of the eddy pattern (and hence by inference the fluid velocity) is cross-correlation analysis.

Early work on this cross-correlation acoustic flow meter was done by M. S. Beck et al. [34] at the University of Bradford, England. Later developments are described by R. S. Flemons of Canadian General Electric Co. [35] somewhat as follows. Ultrasonic transducers are clamped to the outside of existing pipes ranging from 4 to 30 in. in diameter; pipe wall thickness and material impose no problems because the ultrasonic energy propagates across the pipe diameter and is not refracted at the pipe walls. Loose paint and scale are first removed from the outer pipe surface. Long, straight pipe runs upstream are not required in this method of flow measurement. In fact, the preferred location for the acoustic sensors is within five pipe diameters of an elbow so that eddies inherent in the flow can be utilized. Figure 15.15 illustrates the operating principle of the nonintrusive acoustic flow meter. It is based on the probability that a turbulence pattern that intercepts the upstream ultrasonic beam will also intercept the downstream beam. The voltage pattern from the downstream channel will lag that of the upstream channel by an amount closely corresponding to the fluid transit time. This time, which ultimately yields the flow velocity, is determined by means of a cross-correlation analyzer.

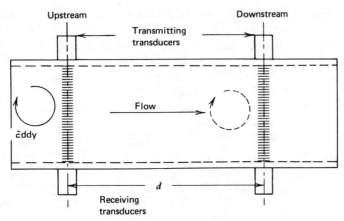

Figure 15.15 Operating principle of a nonintrusive ultrasonic flow meter (after Flemons [35]).

Some of the factors that affect the measurement accuracy of this ultrasonic flow meter are uncertainty as to the transducer separation distance (d in Figure 15.15), uncertainty as to the pipe flow area, and uncertainty as to the velocity profile in the pipe. For this type of flow rate detector, uncertainties of $\pm 2\%$ have been reported.

REFERENCES

1. J. H. Preston, "The determination of turbulent skin friction by means of pitot tubes," *J. R. Aeronaut. Soc.*, Vol. 58, February 1954, p. 109.
2. E. Y. Hsu, "The measurement of local turbulent skin friction by means of surface Pitot tubes," *Davis Taylor Model Basin* Rep. 957, August 1955.
3. M. R. Head and I. Rechenberg, "The Preston tube as a means of measuring skin friction," *J. Fluid Mech.*, Vol. 14, 1962, p. 1.
4. V. C. Patel, "Calibration of the Preston tube and limitations on its use in pressure gradients," *J. Fluid Mech.*, Vol. 23, Part 1, 1965, p. 185.
5. D. W. Smith and J. H. Walker, "Skin friction measurements in incompressible flow," *NASA TR* R-26, 1959.
6. S. Dhawan, "Direct measurement of skin friction," *NACA Rep.* 1121, 1953.
7. F. Schultz-Grunow, "New frictional resistance law for smooth plates," *NACA TM* 986, 1941.
8. J. M. Allen, "Experimental study of error sources in skin friction balance measurements," *ASME Trans., J. Fluids Eng.*, March 1977, p. 197.

520 Special Measurements in Pipes

9. F. J. Pierce and D. S. Gold, "Near-wall velocity measurements for wall shear inference in turbulent flows," *Natl. Bur. Stand. Spec. Publ.* 484, October 1977.
10. L. V. King, "On the convection of heat from small cylinders in a stream of fluid," *Phil. Trans. R. Soc. London*, Vol. 214, 1914, p. 373.
11. H. Dryden and A. Kuethe, "The measurement of fluctuations of air speed by the hot wire anemeter," *NACA TN* 320, 1929.
12. J. R. Weske, "Methods of measurement of high air velocities by the hot wire method," *NACA TN* 880, 1943.
13. J. Laurence and C. Landes, "Auxiliary equipment and techniques for adapting the constant temperature hot wire anemometer to specific problems in air flow measurements," *NACA TN* 2843, 1952.
14. T. B. Morrow and S. J. Kline, "The evaluation and use of hot wire and hot film anemometers in liquids," *Rep.* MD-25, Thermosciences Division, Stanford University, August 1971.
15. J. S. Wyler, "Practical aspects of hot wire anemometry," *Instrum. Control Syst.*, January 1974, p. 57.
16. J. C. Knudsen and D. L. Katz, *Fluid Dynamics and Heat Transfer*, McGraw-Hill, 1958, Chapter 8.
17. J. Laufer, *NACA Rep.* 1174, 1954.
18. A. L. Schawlow, "Lasers and coherent light," *Am. Sci.*, Vol. 55, No. 3, Autumn-September 1967, p. 197.
19. J. P. Gordon, H. J. Zeiger, and C. H. Townes, *Phys. Rev*, Vol. 95, 1954, p. 282.
20. Y. Yeh and H. Z. Cummins, "Localized fluid flow measurements with a He-Ne laser spectrometer," *Appl. Phys. Lett.*, Vol. 4, 1964, p. 176.
21. *TSI Tech. Bull.* 25, "An introduction to the TSI laser anemometer," Thermo Systems Inc., St. Paul Minn., 1975.
22. A. K. Blake, "Development of the NEL laser velocimeter," International Conference on Modern Developments in Flow Measurement, Atomic Energy Research Establishment, Harwell, U. K., September 1971.
23. C. J. Bates, "Experimental pipe flow analysis using a laser Doppler anemometer," *DISA Inf. Bull.* 16, Disa Electronics, Franklin Lakes, N. J., July 1974.
24. W. W. Durgin and L. C. Neale, "Laser Doppler anemometer for flow measurement," *Natl Bur. Stand. Spec. Publ.* 484, October 1977, p. 471.
25. R. P. Benedict and J. S. Wyler, "Determining flow rate from velocity measurements," *Instrum. Control Syst.*, February 1974, p. 47.
26. J. S. Wyler and R. P. Benedict, "Gaussian integration for real time data processing," *Instrum. Control Syst.*, May 1973, p. 67.
27. J. S. Wyler, "State of art survey of optical measurement techniques," *Westinghouse STD Rep.* EM-1374, December 1973.
28. D. R. Keyser, "Laser flow measurement," *ASME Pap.* 77-WA/FM-2, December 1977.

29. C. P. Wang, "Laser anemometry," *Am. Sci.*, Vol. 65, May–June 1977, p. 289.
30. F. Durst, A. Melling, and J. H. Whitelaw, "Laser anemometry," *J. Fluid Mech.*, Vol. 56, Part 1, 1972, p. 143.
31. C. R. Hastings, "LE flowmeter—a new device for measuring liquid flow rates," *Westinghouse Eng.*, November 1968.
32. J. L. McShane, "Ultrasonic flowmeters," Paper 2-10-214, in *Flow, Its Measurement and Control in Science and Industry*, Symposium, Pittsburgh, Pa., May 10–14, 1971.
33. D. J. Dunn, "The sing-around velocimeter and its use in measuring the size of turbulent eddies in the sea," *Electron. Eng.*, Vol. 37, No. 449, July 1965, p. 432.
34. M. S. Beck et al., "Total volume and component flow measurement in industrial slurries and suspensions using correlation techniques," *Meas. Control*, Vol. 4, No. 8, August 1971, p. T-133.
35. R. S. Flemons, "A new non-intrusive flowmeter," *Natl. Bur. Stand. Spec. Publ.* 484, October 1977, p. 319.

NOMENCLATURE

Roman

b Pitot tube opening
c_p constant pressure specific heat
C_f skin friction coefficient
C proportionality constant
d diameter,
distance
D diameter
f friction factor,
frequency
g_c gravitational constant
h height
k thermal conductivity
p pressure
q rate of heat transfer
T temperature
u x component of velocity
w specific weight
y distance

Greek

- α angle
- Δ finite difference
- λ wavelength
- ν kinematic viscosity
- ρ density
- τ shear stress

Name Index

Allen, D. N. deG., 63
Allen, J. M., 505, 519
Allen, W. F., Jr., 361
Alvi, S. H., 418
Au, S. B., 162, 174, 196, 209, 210, 224, 258, 418

Bates, C. J., 517, 520
Baumann, H., 44, 63
Baumann, H. D., 419
Bean, H. S., 64, 128, 417, 492, 498
Beck, M. S., 518, 521
Beij, K. H., 416
Benedict, R. P., 18, 63, 64, 128, 174, 175, 224, 259, 290, 309, 310, 340, 341, 369, 388, 416-418, 438, 464, 469, 470, 481, 490, 498, 499, 520
Benson, R. S., 373, 416
Bernardi, R. T., 414, 415, 419
Bernoulli, D., 1, 5, 17, 18, 20, 22, 24-29, 46, 54, 62, 92, 114, 133-136, 173, 281, 427, 428, 430
Binder, R. C., 214, 225, 417
Binnie, A. J., 464
Bizon, R. R., 438
Blake, A. K., 517, 520
Blasius, P. R. H., 139, 140, 155, 158, 159, 174, 189, 232, 233, 235, 247, 251, 257
Boger, H. W., 361, 418
Borda, J. C., 60, 61, 267, 391-393, 396, 405
Boyle, R., 81-83, 477
Bragg, S. L., 120, 122, 123, 129
Brainerd, J. G., 464
Brandt, G. B., 418
Brodgesell, A., 419
Bronsted, J. N., 128
Brower, W. B., 63
Bryant, R. A. A., 128

Buckingham, E., 128, 191, 224, 229, 499
Buresh, J. F., 419
Burkhart, T. H., 258

Caratheodory, C., 69, 78, 128
Carlucci, N. A., 64, 290, 340, 341, 416
Carmody, T., 18
Carnot, S., 78, 83, 128, 267, 396, 405
Charles, J. A. C., 81-83
Clapeyron, E., 82, 83
Clauser, F. H., 196, 225, 228, 249, 258, 506
Clausius, R. J. E., 79
Colebrook, C. F., 239-242, 247, 250-252, 258, 271, 354, 363, 364, 367, 370
Coles, D., 196, 225
Cornell, W. G., 415, 419
Cotton, K. C., 418, 490, 498
Coxe, E. B., 417
Crocco, L., 341
Crockett, K. A., 418
Cross, H., 430, 438
Cullen, J. T., 491, 499
Cummins, H. Z., 513, 520
Cunningham, R. G., 128

Daniel, P. T., 438
Darcy, H. P. G., 133, 136, 216, 225, 228, 231, 236, 257, 271, 284, 293, 301-304, 307, 308, 425, 426, 428, 429
da Vinci, L., 4
Dean, R. B., 224
Deissler, R. G., 206, 208, 224, 256, 259
de Pitot, H., 28
Dhawan, S., 505, 519
Dill, W. P., 416
Dodge, L., 411, 417, 418
Doppler, C., 512, 513, 515

523

Drew, T. B., 250, 251, 258
Dryden, H., 508, 520
Dudek, J. A., 290, 340, 417
Dunn, D. J., 521
Durgin, W. W., 517, 520
Durst, F., 518, 521
Dussinberre, G. M., 64

Eagle, A., 198, 224
Eckert, E., 464
El Agib, A. A. R., 448, 464
Emmons, H. W., 63, 175, 464
Euler, L., 1, 4, 11, 16, 113, 114, 133-136

Ferguson, R. M., 198, 224
Flemons, R. S., 518, 519, 521
Foord, T. R., 464
Franklin, R. E., 469, 470, 481
Freeman, J. R., 385-387, 417
Frossel, W., 257
Froude, W., 114
Fukaya, H., 438

Gallagher, R. H., 64
Gay-Lussac, J. L., 81-83
Giles, R. V., 426, 438
Gleed, A. R., 290, 340, 416-418
Goff, J. A., 128
Gold, D. S., 506, 520
Gordon, J. P., 520
Gracey, W., 476, 482
Gruschwitz, E., 161, 174

Hagen, G. H. L., 133, 136, 141, 230, 231, 256, 257
Hall, G. W., 401, 418, 498
Hall, W. B., 340
Hamilton, L. H., 419
Hansen, M., 142, 174
Hastings, C. R., 521
Head, M. R., 201, 224, 503, 519
Herning, F. R., 418
Hickox, G. H., 417
Hilbrath, H., 378, 416
Hoshino, H., 438
Howarth, L., 139, 140, 174
Hoyt, J. W., 128, 254, 258
Hsu, E. Y., 502, 519
Hughes, H. J., 384, 417
Hunsaker, J. C., 417
Hutton, S. P., 417

Ince, S., 17, 173
Ippen, A. T., 245, 258
Ito, H., 365-368, 370, 371, 416
Ito, T., 438

Jain, A. K., 250-252, 258
James, R. E., 258
Jobson, D. A., 64, 120, 129
Johnston, J. P., 341
Joule, J. P., 75, 83, 86, 89, 128, 446, 464

Karplus, W. J., 63
Katz, D. L., 212, 214, 225, 250, 251, 258, 290, 510, 512, 520
Kelvin, see Thomson, W.
Keyser, D. R., 518, 520
King, L. V., 508, 520
Kirchhoff, G., 36, 63, 417
Kline, S. J., 255, 256, 259, 508, 520
Klomp, E. D., 378, 405, 416
Kobus, H., 18
Koo, E. C., 258
Kneisel, O., 412, 419
Knudsen, J. G., 212, 214, 225, 250, 251, 258, 290, 510, 512, 520
Kuethe, A., 508, 520
Kusic, G., 438

Lakshmana Rad, N. S., 418
Lam, S., 63, 174
Landes, C., 508, 520
Langhaar, H. L., 256, 257
Lansberg, P. T., 128
Laplace, P. S., 29, 34, 43, 48, 49, 51, 52, 56, 88, 136, 167
Larson, J. K., 361
Laufer, J., 512, 520
Lavi, A., 438
Laurence, J., 508, 520
Letko, W., 482
Leutheuser, H. J., 169-172, 175, 490, 498
Levy, S., 361
Li, W., 63, 174
Liao, G. S., 361
Lindley, D., 196, 224, 418
Linehan, J., H., 419
Loh, W. H. T., 128
Ludwig, H., 196, 225, 253, 258

McAdams, W. H., 258
Mach, E., 91, 92, 114

Name Index 525

McShane, J. L., 521
Mariotte, E., 81, 82
Martin, J. D., 258
Massey, J. L., Jr., 340
Maxwell, J. C., 89
Melling, A., 521
Merrill, E. W., 258
Meyer, C. A., 63, 175
Mickley, H. S., 258
Miller, B., 197, 202, 224
Miller, R. W., 491, 499
Mollo-Christensen, E. L., 258
Moody, L. F., 243-245, 250-252, 258, 271, 272
Morrow, T. B., 508, 520

Navier, L. M. H., 133, 135-137, 173, 230, 257
Neale, L. C., 517, 520
Newton, I., 4, 88, 96, 156, 312
Nikuradse, J., 195, 196, 201, 212-214, 216, 224, 228, 235-245, 250, 251, 257, 258

O'Brien, M. P., 417
Oden, J. T., 64
Ohm, G. S., 35
Okamoto, Z., 438
Orme, E. M., 340

Pannell, J. R., 215, 225, 232, 257
Pao, R. H. F., 196, 225
Pardoe, W. S., 417
Patel, V. C., 196, 210, 224, 503, 504, 519
Patterson, G. N., 378, 405, 416
Pierce, F. J., 506, 520
Pigott, R. J. S., 416
Planck, M., 463
Poiseuille, J. L. M., 133, 136, 230, 231, 256, 257
Preiswirk, E., 128
Preston, J. H., 196, 225, 500-505, 519

Rainbird, W. J., 196, 225, 469, 470, 481
Rayle, R. E., 469-472, 481
Rechenberg, I., 201, 224, 503, 519
Regnault, V., 83, 128
Reichardt, H., 206, 207, 224
Reynolds, A. J., 250, 251, 253, 258
Reynolds, O., 134, 141, 174, 224, 230, 232, 257, 272
Richtmire, B. G., 417
Rivas, M. A., 170, 171, 175, 418, 490, 498
Roberts, D. W., 341
Robertson, J. M., 243, 258

Rothfus, R. R., 225
Rotty, R. M., 340
Rouse, H., 17, 173, 241-243, 246, 247, 258, 417
Russell, W. R., 482

Safford, A. T., 384, 417
Sajben, M., 361
Saph, V., 232, 233, 257
Schawlow, A. L., 513, 520
Schinzinger, R., 438
Schlichting, H., 142, 170, 174, 213, 225, 235, 243, 256, 259
Schoder, E. W., 232, 233, 257
Schuder, C. B., 419
Schulte, R. D., 340, 416, 418
Schultz-Grunow, F., 505, 519
Senecal, V. E., 225
Shapiro, A. H., 128, 170, 171, 175, 225-257, 259, 340, 361, 417, 418, 490, 498
Shaw, R., 469, 470, 481
Showmann, A. R., 340
Siegel, R., 259
Sisson, W., 416
Smith, D. W., 225, 505, 519
Smith, K. A., 258
Smith, R. D., 196, 257, 417
Soroka, W. W., 63
Southwell, R. V., 63
Sovran, G., 378, 405, 416
Squire, H. B., 253, 258
Sridharan, K., 418
Stanton, T. E., 215, 225, 232, 247, 257
Steltz, W. G., 128, 290, 341
Stokes, G. G., 133, 136, 137, 173, 230, 257
Stolz, J., 491, 492, 494, 499
Street, R. L., 174, 290, 340, 437
Streeter, V. L., 63, 437
Swamee, P. K., 250-252, 258
Swetz, S. D., 64, 290, 416

Taylor, C., 64
Techo, R., 250, 251, 258
Thompson, M. J., 417
Thomson, W., 83, 84, 128, 446, 464
Ticker, R. R., 258
Tietjens, O. G., 174
Tillman, W., 196, 225, 253, 258
Tollmien, W., 141, 145, 170, 174
Townes, C. H., 513, 520
Truesdell, C., 17, 173

Upp, E. L., 418

Vazsonyi, A., 372-375, 416
Vennard, J. K., 174, 290, 340, 437
Virk, P. S., 254, 258, 259
von Karman, T., 160, 162, 174, 199, 201, 202, 224, 236-241, 247, 253, 258
von Mises, R., 386, 417

Wacker, W. A., 416
Walker, J. H., 196, 225, 505, 519
Wallace, J. M., 469, 470, 481
Walz, A., 163-167, 170, 174
Wang, C. P., 518, 521
Weisbach, J., 133, 136, 228, 231, 257, 271, 284, 293, 301, 303, 385-387, 417, 425-429
Weise, W., 464
Weske, J. R., 508, 520

Westcott, J. C., 418, 490, 498
Wheatstone, C., 37, 38, 507
White, C. M., 239
Whitelaw, J. H., 521
Wieghardt, K., 163, 175
Woollott, D., 373, 416
Wood, C. F., 438
Wyler, J. S., 64, 174, 224, 290, 340, 417, 418, 469-471, 474, 481, 482, 490, 498, 508, 518, 520
Wylie, E. B., 63, 437

Yeh, Y., 513, 520
Young, A. D., 253, 258

Zeiger, H. J., 520
Zienkiewicz, O. C., 64
Zworykin, V. K., 44, 63

Subject Index

Abrupt contractions, 369, 383-390
Abrupt enlargements, 21, 62, 267-270, 308, 369, 380-383
Absolute pressure, 74
Absolute roughness, 180, 236
Absolute temperature, 78
Acoustic velocity, 91, 92, 114
Adiabatic expansion factor, 495
Adiabatic loss coefficient, 294-298
Adiabatic process, 283, 311-314
Analog:
    conducting sheet, 35-43
    electrolytic tank, 42, 43
    hydraulic-gas, 107-116
ASME coefficients:
    nozzle, 53, 397-399
    orifice, 399-401
ASME loss parameter, 397
Average velocity, 179, 184, 234

Basic identity, 80, 81
Bernoulli energy equation, 17-24, 54, 92, 134, 428, 430
Bernoulli pressure equation, 28, 29
Bernoulli work equation, 24-28
Blasius solution, 139, 140, 232
Blockage, 473-475
Borda tube, 60, 61, 390-393
Boundary layer:
    flat plate, 139, 140, 153, 154, 156-160
    laminar, 140, 158, 159
    pipes, 162, 501-512
    turbulent, 159, 160
Bourdon tube, 480, 481
Boyle's law, 81
Branching network, 431-434
Bridge circuits, 37, 507

Brightness temperature, 463
Buckingham π theorem, 191, 229
Buffer zone, 198, 199, 200

Callendar equation, 462
Carnot-Borda equation, 268, 396
Carnot cycle, 78
Cauchy-Riemann equation, 32-34
Celsius scale, 463
Charles' law, 81
Circulation, 33
Clapeyron equation, 82
Clauser plot, 228, 249
Colebrook friction factor, 239-243, 363, 364
Compressible fluid flow, 1, 4, 5, 66, 96, 292, 485
Compressible loss coefficient, 293-301
Conservation of energy, 13-16, 114, 308
Conservation of mass, 4, 6-9, 112, 308
Conservation of momentum, 4, 9-13, 113, 308
Constant density loss coefficient, 265
Continuity equation, 4, 6-9, 30, 48, 134, 155, 425, 430
Contraction coefficient, 56, 58, 120-127, 384-386
Control valve, 353-359, 411
Critical pressure ratio, 84, 347
Critical Reynolds number, 186, 231
Critical zone, 231
Cylinders, 40, 41, 473

D and ½D taps, 491-494
Darcy-Weisbach equation, 136, 228, 284, 303, 425, 428, 429
Deissler equation, 206, 208
Density, 4, 7, 178
Developing boundary layers:

528 Subject Index

laminar, 193, 255, 256
turbulent, 193, 256
Diabatic flow, 5, 283
Diaphragm pressure transducer, 481
Diffusers, efficiency, 377-379
Dimensional analysis, 191, 196, 229
Discharge coefficient:
  flow nozzle, 397-399, 486-490
  square-edged orifice, 399-401, 491-494
  venturi, 401, 486-490
Displacement thickness, 146-148, 153, 161
Dynamic correction factor, 449-451
Dynamic pressure, 97, 467
Dynamic temperature, 90, 444

Elbows, 21, 364-372
EMF, thermocouple, 462
Emissivity, 453-458, 463
Energy:
  equation, 5, 13-16, 49, 263
  integral equation, 163
  kinetic, 75
  potential, 10, 76
  thickness, 150-153
Enthalpy, 86
Entropy, 79
Equation of state, 82-86
Euler equation, 11, 16
Expansion factor, 99-111, 494-498

Fahrenheit temperature scale, 463
Fanno flow, 319, 330, 335
Film coefficient, 455-458
Finite element methods, 52, 55, 57
Flat plate, 156-160
Flow:
  coefficient, 56
  flashing, 343-361
  networks, 45, 47
  numbers, 99-102
  rate, 359
  work, 75
Fluid meters:
  nozzle, 396-399, 484
  orifice, 396-401, 484
  venturi, 397-401 484
Fluid properties:
  compressible, 337
  incompressible, 274, 478
Force balance, 4, 121
Friction:

factor, 231, 235, 238, 250-253
Reynolds number, 193
velocity, 159, 192, 196
Froude number, 114, 115

Gas constant, specific, 82-86
Gas properties, 337
Gas, ideal, 83
Gay-Lussac law, 81, 82
General energy equation, 15, 16, 75, 76
Generalized flow function:
  compressible, 314-323
  constant density, 280-288
Graphical flow net, 43-47
Graphical solutions:
  heat transfer, 326, 327, 454-458
  two phase flow, 349-352
Gravity:
  local, 4, 477
  standard, 4, 477
  wave, 114
Green's theorem, 31

Hazen-Williams coefficient, 426-429
Heat, transfer effects, 5, 451-459
Hot wire anemometer, 507-509
Hydraulic-gas analogy, 107-116
Hydraulic radius, 271
Hydrodynamics, 1, 134

Ideal fluid, 134
Ideal gas, 81-86
Immersion correction, 460, 461
Incompressible flow, 1, 66, 100
Incompressible loss coefficient, 265
Inlet, 390-393
Internal energy, 69
Inviscid flow, 3, 56
Isentropic:
  exponent, 88, 350-352
  process, 88, 89, 281, 282, 319, 343, 344
Isobaric process, 88
Isothermal:
  loss coefficient, 298-301
  process, 88, 284, 320, 334, 339, 340

Joule-Thomson effect, 445, 446

Kelvin:
  ideal gas, 83, 84
  temperature, 83

Kinematic viscosity, 178, 274, 337
Kinetic energy coefficient, 217-221, 223, 264, 293

Laminar:
  boundary layer, 140, 158
  flow, 230
  sublayer, 196, 198
Laplace:
  equation, 29-35, 46-52, 136
  notation, 49, 51, 136
Laser Doppler velocimeter, 512-518
Law of the wall, 190, 193
Laws:
  Boyle-Mariotte, 81
  Charles-Gay Lussac, 81
  ideal gas, 83
  perfect gas, 85
Liquid-in-glass thermometer, 459-461
Liquids, 274, 478
Liquid/vapor mixtures, 345
Logarithmic layer, 193, 201
Logarithmic law of the wall, 193, 199, 201, 203, 206, 208
Loss coefficient:
  abrupt contraction, 383-390
  abrupt enlargement, 267-270, 308-311, 380-383
  diffuser, 377-379
  elbow, 366-372
  fluid meter, 401-407
  general, 264-266, 293-301
  inlets and exits, 390-393
  reducers, 376
  screen, 413-416
  tee, 372-376
  valve, 407-412
  viscous pipe, 363, 364

Mach number, 91, 92, 96, 232
Macroscopic, 66
Manometer, 477-479
Mariotte law, 81
Mass flow rate, 8
Momentum:
  correction factor, 221-223
  equation, 3, 4, 10, 13, 56, 58, 134, 156
  integral equation, 160-163
  thickness, 148-150, 153, 161
Moody:
  equation, 250
  plot, 243-247
Moving fluid effect, 444-451

Navier-Stokes equation, 135
Networks:
  complex, 434-437
  parallel, 424-431
  series, 422-424
Newton-Raphson method, 95, 312
Newton's second law, 3, 4
Nikuradse:
  equations, 195, 212
  shift, 201, 202
Noncircular pipes, 271
Nozzles:
  ASME, 53-58, 61
  discharge coefficient, 397-399, 486-490
  expansion factor, 495, 497, 498
  loss, 401-407
Numerical solutions, 47-53
Nusselt number, 455, 456

Ohm's law, 35, 38
One D and ½D taps, 491-494
One dimensional:
  equations, 8, 20 ,263, 292
  flow, 12, 15
Optical pyrometer, 462, 463
Orifice meter:
  discharge coefficient, 399-401, 491-494
  expansion factor, 495, 496
  loss, 401-407

Parobola, 185, 218, 222
Paraboloid, 184
Pi theorem, 191, 196, 229
Pipe factor, 213-215
Pipes:
  noncircular, 271
  roughness, 180
Pitot tubes, 28, 509-512
Planck's law, 463
Platinum resistance thermometer, 460, 462
Poiseuille's law, 136, 230
Polytropic process, 87
Potential:
  energy, 10
  function, 33, 34, 36
  solutions, 35, 53-56
Power law, 187, 189, 203
Prandtl:

530 Subject Index

boundary layer equations, 138, 194
friction factor, 235, 240
number, 445
Pressure:
  absolute, 74
  critical, 347-350
  static, 91, 96, 97, 466
  total, 91, 96, 97, 467, 475
Preston tube, 500-505
Profile, velocity, 184-223
Properties of fluids:
  compressible, 337
  constant density, 274

Radiation pyrometer, 462, 463
Rankine temperature, 463
Ratio:
  critical pressure, 84, 347
  specific heats, 86, 89
Rayleigh process, 320, 333, 336-338
Recovery factor, 446-449
Regnault's ideal gas, 83
Reichardt's expression, 206-208
Relaxation methods, 47, 53
Resistance, thermometer, 460, 462
Reversible process, 71-73, 79
Reynolds number, 134, 141, 468
Rough pipe, 210, 211, 236, 250, 253
Roughness values:
  absolute, 180, 236
  relative, 214, 236, 245, 246, 426
Roughness Reynolds number, 212
Rouse:
  limit line, 241-247
  plot, 246, 247

Screens, 413-416
Separation, 143, 144
Series network, 422-424
Shaft work, 76, 77
Shape factors, 154, 155, 161
Sharp-edged orifice, *see* Meter
Shear stress:
  Newtonian, 137
  Wall, 157, 192, 248, 500-506
Shock process, 321-323
Skin friction coefficient, 247-249, 253, 500-505
Smooth pipe, 186, 205, 232, 250, 251
Solidity, 413-416
Sound velocity, 91, 92
Stagnation:

concept, 89, 90
factor, 447-449
Static:
  pressure, 466
  temperature, 90, 444
Steady flow, 8
Stream function, 30-33, 36
Streamlines, 21, 36, 43
Stolz equation, 491-493
Stress:
  compressive, 3
  shear, 137
  tensile, 3
Sublayer, laminar, 196
Subsonic flow, 92, 124, 355
Supercritical flow, 92, 124, 435

Tees, 372-376
Temperature in moving fluids, 444-451
Temperature scales:
  absolute, 78
  dynamic, 444
  empirical, 69
Thermocouple, 462
Thermodynamic laws:
  first law, 69-74, 264, 292
  second law, 77-81
  zeroth law, 68, 69
Thermodynamic temperature scale, 463
Thermometer:
  liquid-in-glass, 459-461
  platinum resistance, 460, 462
Total immersion, 459, 460
Total pressure, 467, 475
Total temperature, 90, 91, 444
Transducer, 481
Transition, 141, 185, 237, 242, 250, 252
Turbulent boundary layer, 142, 159
Turbulent velocity profile, 186, 200, 205
Two dimensional flow, 29

Ultrasonic flow meter, 518, 519
U-tube manometer, 477-479

Valve:
  critical pressure, 353, 354
  flow coefficient, 408
  loss coefficient, 409
Vapor pressure, 345
Velocity:
  acoustic, 91, 92, 114

Subject Index 531

average, 179, 184, 234
coefficient, 385
defect, 146
distribution, 40, 184-223
potential, 34
Velocity profile:
  laminar, 180, 183
  turbulent, 180, 181, 187-189, 200, 201, 212
Vena contracta, 56, 58
Venturi:
  discharge coefficient, 401, 486-490
  expansion factor, 495, 497, 498
  loss coefficient, 401-407
Viscosity:
  dynamic, 178
  kinematic, 178

of selected fluids, 274, 337, 478
Viscous pipe, 363, 364
Volumetric average velocity, 9, 179, 184
von Karman:
  equation, 199, 240
  friction factor, 237

Wall:
  roughness, 211
  tap, 467
  tap correction factor, 468, 471
Walz stepping equations, 164-166
Wheatstone bridge, 37, 507
Work:
  flow, 75
  shaft, 76, 77